Communications in Computer and Information Science 2032

Rationale

The CCIS series is devoted to the publication of proceedings of computer science conferences. Its aim is to efficiently disseminate original research results in informatics in printed and electronic form. While the focus is on publication of peer-reviewed full papers presenting mature work, inclusion of reviewed short papers reporting on work in progress is welcome, too. Besides globally relevant meetings with internationally representative program committees guaranteeing a strict peer-reviewing and paper selection process, conferences run by societies or of high regional or national relevance are also considered for publication.

Topics

The topical scope of CCIS spans the entire spectrum of informatics ranging from foundational topics in the theory of computing to information and communications science and technology and a broad variety of interdisciplinary application fields.

Information for Volume Editors and Authors

Publication in CCIS is free of charge. No royalties are paid, however, we offer registered conference participants temporary free access to the online version of the conference proceedings on SpringerLink (http://link.springer.com) by means of an http referrer from the conference website and/or a number of complimentary printed copies, as specified in the official acceptance email of the event.

CCIS proceedings can be published in time for distribution at conferences or as post-proceedings, and delivered in the form of printed books and/or electronically as USBs and/or e-content licenses for accessing proceedings at SpringerLink. Furthermore, CCIS proceedings are included in the CCIS electronic book series hosted in the SpringerLink digital library at http://link.springer.com/bookseries/7899. Conferences publishing in CCIS are allowed to use Online Conference Service (OCS) for managing the whole proceedings lifecycle (from submission and reviewing to preparing for publication) free of charge.

Publication process

The language of publication is exclusively English. Authors publishing in CCIS have to sign the Springer CCIS copyright transfer form, however, they are free to use their material published in CCIS for substantially changed, more elaborate subsequent publications elsewhere. For the preparation of the camera-ready papers/files, authors have to strictly adhere to the Springer CCIS Authors' Instructions and are strongly encouraged to use the CCIS LaTeX style files or templates.

Abstracting/Indexing

CCIS is abstracted/indexed in DBLP, Google Scholar, EI-Compendex, Mathematical Reviews, SCImago, Scopus. CCIS volumes are also submitted for the inclusion in ISI Proceedings.

How to start

To start the evaluation of your proposal for inclusion in the CCIS series, please send an e-mail to ccis@springer.com.

Miguel Mujica Mota · Paolo Scala
Editors

Simulation for a Sustainable Future

11th Congress, EUROSIM 2023
Amsterdam, The Netherlands, July 3–5, 2023
Proceedings, Part I

Springer

Editors
Miguel Mujica Mota
Amsterdam University of Applied Sciences
Amsterdam, Noord-Holland, The Netherlands

Paolo Scala
KLM
Hoofddorp, The Netherlands

ISSN 1865-0929 ISSN 1865-0937 (electronic)
Communications in Computer and Information Science
ISBN 978-3-031-68434-0 ISBN 978-3-031-68435-7 (eBook)
https://doi.org/10.1007/978-3-031-68435-7

This Springer imprint is published by the registered company Springer Nature Switzerland AG
The registered company address is: Gewerbestrasse 11, 6330 Cham, Switzerland

Preface

The EUROSIM Congress 2023, the 11th EUROSIM Congress, was organized by the Dutch Benelux Simulation Society, in Amsterdam (DBSS) on July 3–5, 2023. EUROSIM (www.eurosim2023.eu) is the Federation of European Simulation Societies; it provides a European forum for regional and national simulation societies to promote the advancement of modeling and simulation in industry, research, and development. Under the EUROSIM umbrella, EUROSIM Member Societies and cooperating societies and groups organize conferences, produce publications on modeling and simulation, and work in standardizing or technical committees, among other activities.

This volume contains a selected number of full articles of papers presented at the congress. Each submission was reviewed by at least by 3 scientific committee members of Ph.D. level. Each paper included in this volume is at the forefront of development in its scientific area, and each was presented during the conference where there were discussions for improvement. The topics are diverse, but in all of them simulation is a key component of the solution presented.

During the conference we could witness the evolution of simulation as a tool that can be complemented with novel data science techniques like ML and optimization to develop future tools that will influence our world. Some of the applications aim at solving problems that will make the future more sustainable. The readers will find in this volume very interesting applications, solutions, and developments where simulation, or different techniques together with simulation, are applied for diverse topics ranging from medicine to transportation.

We thank all the authors for participating in the conference and the effort to shape their work to fit the requirements of this volume. Special thanks go to the International Scientific Committee which allowed us to put together all the material on time and provided valuable feedback to the authors. We also thank our sponsors Sentient Hubs, Wolfram Research, IGAMT, Incontrol, and UNAM. Finally, we thank Amsterdam University of Applied Sciences for providing the location for the congress.

July 2023

Miguel Mujica Mota
Paolo Scala

Organization

General Chair

Miguel Mujica Mota	Amsterdam University of Applied Sciences, The Netherlands

Program Committee Chairs

Arnold Heemink	Delft University of Technology, The Netherlands
Miguel Mujica Mota	Amsterdam University of Applied Sciences, The Netherlands
Paolo Scala	KLM, The Netherlands

Steering Committee

Margarita Bagamanova	Amsterdam University of Applied Sciences, The Netherlands
Arnold Heemink	Delft University of Technology, The Netherlands
Hai Xiang Lin	Delft University of Technology, The Netherlands
Miguel Mujica Mota	Amsterdam University of Applied Sciences, The Netherlands
Paolo Scala	KLM, The Netherlands

Program Committee

Margarita Bagamanova	Amsterdam University of Applied Sciences, The Netherlands
Abdel El Makhloufi	Amsterdam University of Applied Sciences, The Netherlands
Idalia Flores de la Mota	National Autonomous University of Mexico, Mexico
Arnold Heemink	Delft University of Technology, The Netherlands
Gieta Inderdjiet	Amsterdam University of Applied Sciences, The Netherlands
Hai Xiang Lin	Delft University of Technology, The Netherlands

Miguel Mujica Mota	Amsterdam University of Applied Sciences, The Netherlands
Alejandro Murrieta Mendoza	Amsterdam University of Applied Sciences, The Netherlands
Paolo Scala	KLM, The Netherlands

Additional Reviewers

Sameer Alam
David Al-Dabass
Asteris Apostolidis
Margarita Bagamanova
Sahil Belsare
Agostino Bruzzone
Roman Buil
Erik Dahlquist
Mohammad Dehghani
Bruno Desart
Abdel El Makhloufi
Javier Faulin
Luis Manuel Fernández-Ahumada
Idalia Flores de la Mota
Jorge Luis Garcia-Alcaraz
Egils Ginters
Edmond Hajrizi
Arnold Heemink
Glenn Jenkins
Emilio Jimenez
Esko Juuso
Gorazd Karer
Juan Ignacio Latorre
Bernt Lie
Vito Logar
Edgar Lopez-Rojas
José Antonio Marmolejo
Marina Massei
Peter Meincke
Miguel Mujica Mota
Gašper Mušič
Allan Nõmmik
Jorge Ochoa
Stephan Onggo
Jose Maria Ortiz-Gomez
Antonio Padovano
Mercedes Perez
Miquel Angel Piera
Edgar Possani
David Post
Blaž Rodič
Rodrigo Romero Silva
Henrik Rothe
Christina Rott
Cristina Ruiz
Yuri Senichenkov
Michael Schultz
Konstantinos Stamoulis
Gary Tan
Susana C. Tellez
Ricardo Torres
Fausto Vieira
Edward Williams
Richard Wu
Borut Zupančič
Raid Al-Aomar
Goran Andonovski
Benedikt Badanik
Markus Bans
Felix Breitenecker
Branko Bubalo
Jordi Casas
Fabio De Felice
Daniel Delahaye
Alejandro Di Bernardi
Rosario Enriquez
Roberto Felix Patron
Jaume Figueras i Jové
Luiz Augusto Franzese
Amir Ghasemi
Antoni Guasch Petit
Cathal Heavey

David Martin Herold
Edgar Jimenez
Angel Juan
Sofia Kalakou
Andreas Korner
Sanja Lazarova-Molnar
Hai Xiang Lin
Francesco Longo
David Lucido
Luis Martin-Domingo
Rina Mary Mazza
Yuri Merkurev
Alejandro Murrieta Mendoza
Letizia Nicoletti
Jenaro Nosedal
Elias Olivares Benitez
Angel Orozco
Taha Osman
Maria Papanikou
Antonella Petrillo
Jorge Pinho de Sousa
Nikki Popper
Juan Jose Ramos
Alicia Roman-Alonso
Kristjan Roosipold
Olivier Rose
Mario Ruz
Raquel Salamanca
Paolo Scala
Joh Shortle
Adriano Solis
Artis Teilans
Ali Torabi
Ashish Verma
Gabriel Wainer
Bogusz Wisnicki
Alejandro Zarate Perez

Sponsors

Sentient Hubs

WOLFRAM

IGAMT

INCONTROL

Amsterdam University of Applied Sciences

Contents – Part I

Environmental Sustainability

Healthcare

Contents – Part II

Monitor, Control, and Theoretical Systems

Environmental Sustainability

Irrigation Efficiency Using a Systems Dynamics Model: A Case Study of a Pepper Crop in La Merced - Peru

Seleni Lara Jauregui[1], Oroselfia Sánchez[2(✉)], Idalia Flores-De la Mota[1], and Francisca Irene Soler Anguiano[1]

[1] Universidad Nacional Autónoma de México, 04510 Coyoacán, CDMX, México
[2] Department of Chemical, Industrial and Food Engineering, Universidad Iberoamericana, 01219 Mexico City, Mexico
oroselfia.sanchez@ibero.mx

Abstract. This paper presents a system dynamics simulation model for an irrigation system applied to bell pepper crops, integrating various components of a cropping system, including water usage and plant growth. The study focuses on Latin American regions with limited irrigation facilities, highlighting the importance of simulation techniques to analyze different levels of analysis for irrigation components, including water wastage. The increasing global population, food security, and water management concerns require innovative approaches to overcome difficulties and paradigms in crop irrigation. The simulation model proposed in this study provides insights into the irrigation system's behavior and explores various scenarios to optimize crop yield while minimizing water wastage. Results obtained through simulation demonstrate the dynamics between irrigation time and water discarded in a crop, considering the complex behavior of living systems. The paper's contribution highlights the value of simulation models as a powerful tool for analyzing and improving crop irrigation systems, ultimately contributing to sustainable agriculture and water management practices.

Keywords: Irrigation Model · Pepper Crop · System Dynamic · Waste Management

1 Introduction

1.1 The Role of Agriculture

Agriculture plays a crucial role in feeding the world's growing population and contributing to economic development [1]. However, water scarcity and inefficient use in agriculture have become significant challenges globally, particularly in arid and semi-arid regions [2]. Irrigation is an essential component of agricultural production and has contributed to improving productivity and profitability [3]. Nevertheless, it also faces challenges such as unequal water distribution and inadequate irrigation methods [4]. According to the International Atomic Energy Agency [5], improving water use efficiency in agriculture is critical to meeting growing food demand while also preserving

M. Mujica Mota and P. Scala (Eds.): EUROSIM 2023, CCIS 2032, pp. 3–18, 2024.
https://doi.org/10.1007/978-3-031-68435-7_1

water resources. System dynamics modeling has been used to analyze the interactions and feedbacks between water, agriculture, and other factors, providing insights into the system's behavior and identifying potential solutions [6].

1.2 Irrigated Agriculture

Irrigated agriculture has played a critical role in enhancing agricultural productivity in Latin America. It accounts for more than 60% of the region's agricultural production [7]. However, it also faces various challenges, including inefficient water use, inadequate infrastructure, and poor management practices [8]. Efforts to improve irrigation efficiency in Latin America have focused on adopting best management practices, including modernizing irrigation infrastructure, using precision irrigation systems, and promoting water-saving technologies. Such practices have been shown to increase water use efficiency and reduce water waste [9].

The use of system dynamics modeling can help identify the potential impacts of different management scenarios on the system and can inform decision-making processes. For example, modeling can be used to simulate the impact of climate change on water availability and crop production, and to assess the effectiveness of different adaptation strategies [10]. Efficient irrigation practices and the adoption of best management practices can also contribute to more sustainable and resilient agriculture systems, which can enhance their productivity, economic viability, and environmental sustainability. In turn, this can contribute to the development of more vibrant rural economies, including increased job opportunities and income generation.

Types of Irrigation. Irrigation can be classified into several types based on the method of water delivery to the crops [11]. According to the FAO report [12], irrigation efficiency is defined as the ratio of water beneficially used in crops and the total water used in irrigation practice. The following Table 1 gives a summary of some of the most common types, and the percentage of efficiency in each type of irrigation.

Table 1. Type of irrigation and efficiency.

Irrigation Type	Cost	Efficiency in water use	Use of technology	Crop yield	Percentage of efficiency
Flooding	Very Low	Very Low	Null	Low / Medium	40 – 50%
Furrow	Low	Low	Low	Medium	50 – 70%
Sprinkler	Medium	Medium	Medium	Medium / High	70 – 80%
Drip or underground	Very High	Very High	Very High	High	90 – 95%

Irrigation Efficiency. Irrigation efficiency refers to the amount of water applied to crops that is used by the plants for growth and development. A high irrigation efficiency means that more water is being used for crop growth and less is being lost to non-productive sources, leading to higher crop yields and less water waste [13]. Improving irrigation efficiency is crucial for sustainable water management and food security, especially in regions with limited water resources. The irrigation system is complex. On the one hand, it is necessary to use less water for the crop in a shorter time but, on the other hand, the farmer seeks to maximize crop productivity with the available water and losses from the irrigation system. To solve this dilemma, the use of a system dynamics approach and determination of the optimal amount of water for the crop is proposed.

Consequently, the main problem is the failure to use water resources and the amount of water needed for agricultural use. The purpose of this study is to apply a dynamic water management model to a bell pepper crop to determine the amount of water available for the crop in La Merced and the future trend under conditions of climate change and a growing demand for water. The contribution of this article is to highlight the value of simulation models as a powerful tool to analyze and improve crop irrigation systems, ultimately contributing to farmer decision making and efficient water use with sustainable management.

2 Literature Review

System dynamics is a methodology that uses computer simulation models to analyze and understand complex systems over time [14]. It can be applied to irrigation systems to identify key variables and feedback loops, and to test different scenarios and interventions for improving system performance. Table 2 shows the literature reviewed, detailing the approach evaluated in each case.

The classical causal model is a type of system dynamics model that uses causal loops to represent the feedback relationships between variables in a system [18]. In the context of irrigation, the classical causal model can be used to identify the factors that affect irrigation efficiency, such as irrigation scheduling, crop water requirements, soil type and moisture, and irrigation system design. Based on these reviews and the models used, we propose the use of the system dynamics approach with different scenarios, varying the discard fraction and irrigation time.

The objective of this article is to value simulation models as a powerful tool to analyze and improve crop irrigation systems, ultimately contributing to farmer decision making and sustainable water management in agriculture.

Table 2. Reviewed literature on irrigation efficiency with a system dynamics approach.

Reviewed papers	Summary
Dynamic Simulation of Irrigation Efficiency in Surface Irrigation Systems [15]	This article presents a dynamic simulation model for irrigation efficiency in surface irrigation systems. The model uses a system dynamics approach and considers factors such as soil infiltration, crop water uptake, and irrigation system design. The model is tested on a hypothetical field and results show that it can accurately predict irrigation efficiency under various conditions
Modeling Irrigation Efficiency with a System Dynamics Approach [16]	This article presents a system dynamics model for analyzing irrigation efficiency in agricultural systems. The model incorporates factors such as weather conditions, soil characteristics, crop water requirements, and irrigation system design. The authors apply the model to a case study in the Texas High Plains and show that it can accurately predict irrigation efficiency under different conditions
Using System Dynamics to Improve Water Use Efficiency in Irrigated Agriculture [17]	This article presents a system dynamics model for improving water use efficiency in irrigated agriculture. The model considers factors such as soil characteristics, crop water requirements, and irrigation system design. The authors apply the model to a case study in Iran and show that it can be used to identify opportunities for improving water use efficiency in the region

3 Methodology

The case study will be based on the systems thinking methodology [18]; there are four stages involved with the execution of the system dynamics model. It begins with the identification and definition of the problem; followed by system conceptualization, model formulation, model testing, and evaluation, model use, implementation and dissemination, and learning/strategy/infrastructure design. The steps required to model and simulate the complex system based on systems thinking in the case study are detailed in Fig. 1.

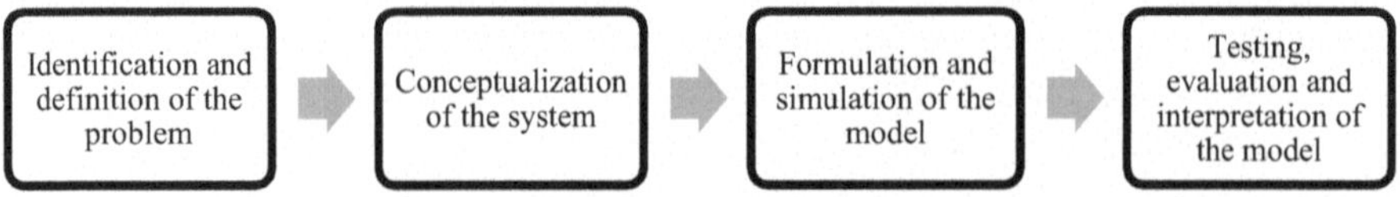

Fig. 1. Systems thinking methodology.

4 Case Study

For the identification and definition of the problem, a Randomized Complete Block Design (RBLD) was used to observe and analyze plant behavior and water consumption in localized irrigation for a small plant (less than 30 days, before flowering) and a large plant (more than 30 days, after flowering until fruit development) versus flood irrigation in a bell pepper crop in La Merced-Peru (Fig. 2).

Fig. 2. Bell pepper plant growth.

The system dynamics simulation model was applied to the data obtained in the experiment. The significant variables for the simulation were: irrigation time, amount of water used, irrigation area, and plant behavior.

The bell pepper (Capsicum annum L.) is a fruit vegetable. It is very important from a commercial and nutritional point of view, as it is used both fresh and processed and is rich in vitamin C [19]. It represents an essential source of income for family farming in Peru. Furthermore, as its cultivation, production and trade rises, the demand for it also increases.

The simulation model applied to the problem considering different amounts of water and the available time for irrigation is analyzed using system dynamics methodology. During the experiment, we observed that the irrigation time for small pepper plants can be divided into specific intervals: quarter, half, and no more than an hour. However, for big pepper plants, we estimated that the irrigation time should vary between 30 min, one hour, and one and a half hours based on our observations during the experiment. It is important to note that these observations were made during the real experiment. Sections 4.1 to 4.3 present the general description of the irrigation system. Then, Sects. from 4.4 to 4.7 present the verification and validation of the applied model simulation. Finally, the results are shown in Sect. 5. The following sections present the models for analyzing an irrigation system and the simulation results.

4.1 The Classical Causal Loop Diagram for Irrigation Systems

In system dynamics, a Causal Loop Diagram is used to represent a conceptual model for a particular system. Modelling an irrigation system using this approach allows us to start with basic variables and simulate behaviors to study dynamics. The Causal Loop Diagram (CLD) for an irrigation system is based on six interrelated variables.

The variable **Irrigated Area** is considered a stock variable that accumulates all behaviors. The **Increase Rate** and **Discard Rate** variables impact how the irrigated area behaves. The irrigation system is controlled by limiting elements, including the **Planned Irrigated Area**, **Irrigated Area Adjustment Time**, and **Discard Fraction**, which can be established or determined by the analyst. In Sect. 4.3, we provide more context on the related variables. For the irrigation system studied in this work, the variables in the Classic Irrigation System Causal Loop represent the following elements.

- **Stock variable:** The irrigated area accumulates the dynamics for the increase rate and discard rate showing the main goal of the conceptual model [18].
- **Secondary variables:** The increase rate is a function of the planned irrigated area and irrigated area adjustment time that could be deterministic or not. The behavior of the discard rate is limited by the discard fraction that represents how much water can be discarded in the system and depends on the area being irrigated at the time.
- **Limiting variables:** Planned irrigated area is the specific value that is going to be irrigated. Irrigated area adjustment time is related to the time that we expect to have for irrigating the planned irrigated area. Discard fraction represents the fraction of water that is allowed to spend during the irrigation.

Both feedbacks in the conceptual model correspond to balancing loops because the model is seeking a specific goal which is to irrigate the planned area within a specific time (Fig. 3).

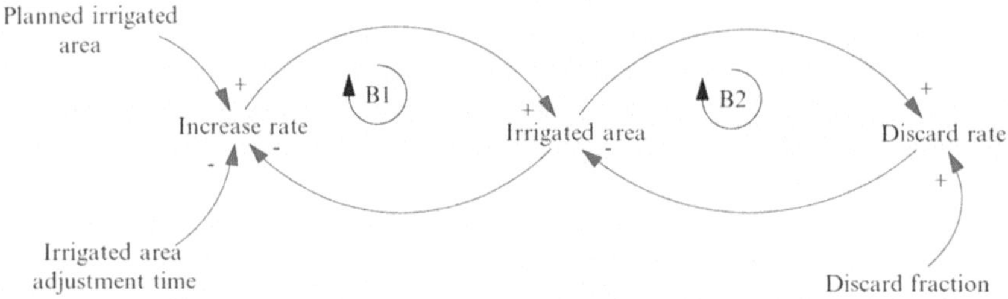

Fig. 3. Causal Loop Diagram from a classic irrigation system.

The Irrigation CLD explains how an irrigated area depends on different variables and can stress other behaviors. Nowadays it is important to grow different crops but considering that not too much water should be discarded. This literature model can be transformed and applied to a real crop, as we propose in this paper.

4.2 The Mathematical Model for the Irrigation System

The classic CLD for irrigation systems is defined by a mathematical model that represents the irrigation system dynamics using the main variables. Equations from 1 to 5 explain the mathematical relations for the classical irrigation model.

$$Irrigated\ area = A = \int \frac{A}{t_0} dt' = \frac{A}{t_0} t' \Big|_0^t = A \frac{t}{t_0} \tag{1}$$

$$A = \int_0^t increase\ rate(dt) - \int_0^t discard\ rate(dt) \tag{2}$$

$$t_0 = irrigated\ area\ adjustment\ time(time\ units) \tag{3}$$

$$Increase\ rate = \frac{(Planned\ area - Irrigated\ area)}{t_0} = \frac{A}{t_0} \tag{4}$$

$$Discard\ rate = Irrigated\ area * Discard\ fraction \tag{5}$$

4.3 Stock and Flow Diagram for Irrigation System

Figure 4 shows the Stock and Flow Diagram (SFD) for the irrigation system that we analyzed. This model is the interpretation of the theoretical CLD for an irrigation system and is programmed into the Vensim simulator. The SFD or the system of the differential equations represents the feedback structure about the CLD for irrigation systems [18]. For the verification and validation conceptual model, we propose general data just to obtain the behaviors and use them with the real parameters of the crop. There are three main variables: irrigated area, irrigated area increases rate and irrigation area discard rate. The stock variables area limited by the condition of irrigation, the inflow-increasing rate and decreased-by-one outflow discard rate [18]. Also, irrigated area adjustment time and discard fraction are variables that limit the irrigation system. In this section, we present the results of the verification process. We programmed two irrigation models: the first is a General Model based on SFD (Fig. 4), which uses general data for verification and validation.

The time units used for the model were years and hours (h), and areas were measured in square meters (m^2). Table 3 presents the parameters for each variable of the Irrigation Models.

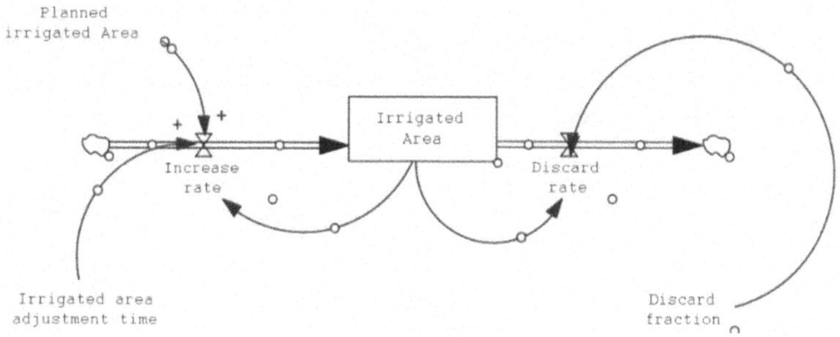

Fig. 4. Vensim Lay-Out of Stock and Flow Diagram for Irrigation Model.

An ongoing simulation has been developed using the six classical variables defined by experimental parameters. A study on the cultivation of real pepper crops is being conducted in Peru. Simulation is an optional way of obtaining valuable data in a shorter time. It also allows us to observe the different behaviors when testing various water percentages, depending on the irrigation technique being studied. In this paper, a General Irrigation Model was established to verify and validate the simulation. With appropriate data from the real crop and literature, we programmed the Applied Irrigation Model. Following simulation methodologies from the literature [20], we propose to verify and validate a general conceptual model and its application. Sections 4.4 and 4.5 present the verification and validation tests of the irrigation system simulation. To verify the simulation model, we conducted structural and behavioral verification tests, parameter verifications, and simulations under extreme conditions based on Kleijnen [21] y Sargent [22]. Additionally, for model validation, we planned the irrigation area and set the simulation time to run the experiments. To validate the simulation results of the applied simulation model, we presented the results to a pepper crop expert.

4.4 Verification of Irrigation Models

In this section, we present the results of the verification process. We programmed two irrigation models: the first is a General Model based on SFD (Fig. 4), which uses general data for verification and validation.

The second model, named the Applied Irrigation model, uses data from the real crop and literature. To verify the models, we applied four different tests based on the Verification and Validation process (V&V process) mentioned by Sargent [22] and Kleijnen [21]. Firstly, we verified the structure of the conceptual model in Vensim Layout, followed by reviewing the equations and parameters proposed in the programmed model. Then, we observed the behavior of partial and final outputs, such as the increase rate and irrigated area. We started with a simplified model using general data, and the General Irrigation Model was simulated using the same variables and relations programmed in Vensim software from the Stock and Flow Diagram. However, the set parameters were general data for testing the final simulation time for the real parameters (see Table 3). The following table shows the data programmed for this simulation model. Finally, we used the animation of the behavior to observe the operation of the irrigation simulation model [22]. To determine if the model is sensitive to the simulation time, we found that this parameter cannot be excluded from verification because the results need a particular simulation time to be consistent. Thus, the simulation time will be validated for the Applied Irrigation Model in Sect. 4.5.

Table 3. Parameters and units for the General and Applied Irrigation Models.

Variable	Units	General Irrigation Model	Applied Irrigation Model (Results)	Units
Planned Irrigated Area	Hectare (ha)	1000	205	(m^2)
Irrigated Area	(ha)	0.25	–	(m^2)
Discard fraction	(1/year)	0.1	0.35	(1/year)
Discard rate	(ha/year)	Max (Irrigated Area * discard fraction, 0)		(m^2/h)
Increase rate	(ha/year)	Max (Planned Irrigated Area - Irrigated Area / Irrigated area adjustment time, 0)		(m^2/h)
Irrigated Area adjustment time	(year)	0.20	(h)	(h)
Final Time	(year)	1	3	(year)

4.5 Validation of Irrigation Models

Based on the V&V process, we conducted several tests to validate the Irrigation Model, including graphical comparisons of data, animation, and expert opinion, following the guidelines suggested by Sargent [22] and Kleijnen [21]. In Peru, the irrigated area for pepper crops is typically around 180 m^2. To ensure the validity of our simulation results, we validated the parameter for this variable. Table 3 presents the values programmed in Vensim, and we compared the obtained irrigated area with the objective of the simulation (180 m^2), as shown in Fig. 5a. We used 205 m^2 as the parameter for the simulation model and observed how the discard rate changed during different simulation times (Fig. 5b; Table 4). We found that a simulation time of three years was acceptable for the experiments. Furthermore, the general irrigation model also resulted in a simulation time of three years (Fig. 5c).

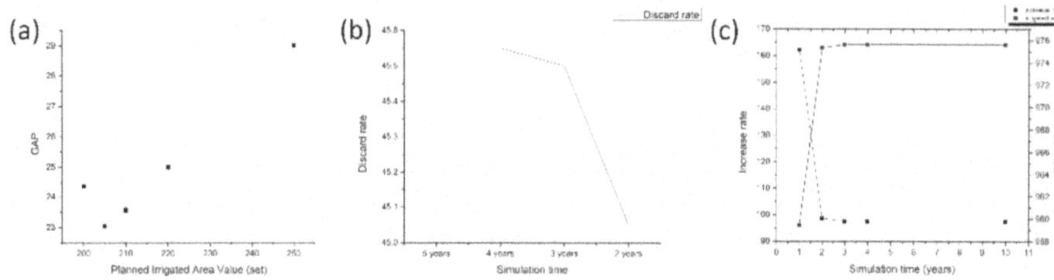

Fig. 5. Validation results from Irrigation Models. (a) represents the gap between the set planned irrigation area and the simulation value. (b) shows the consistency of discard rate value obtained in different simulations. (c) compares the simulation time programmed into Vensim.

Table 4. Validation results of planned irrigation and for simulation time.

	Validation results of planned irrigation area		Validation results for simulation time	
Set planned irrigation area (m^2)	Simulation value (m^2)	Discard rate	Simulation time (years)	Discard rate
200	175.63	43.91	5	45.55
250	221.00	55.49	4	45.55
220	195.00	48.00	3	45.50
210	186.43	46.60	2	45.05
205	181.96	45.49	–	–

The behavior presented in the CLD is validated by the SFD simulation, as shown in Fig. 6. This indicates that the structure of the model is consistent with the behavior of the system. To verify the parameter for an irrigated area of 205 m^2, we conducted a simulation of the model and obtained the irrigation time required for this area. The verification process included comparing the simulated results with real-world data, as well as analyzing the sensitivity of the model to different input parameters. To ensure the accuracy of the model, a warm-up simulation time of three years was required, which is consistent with the simulation results.

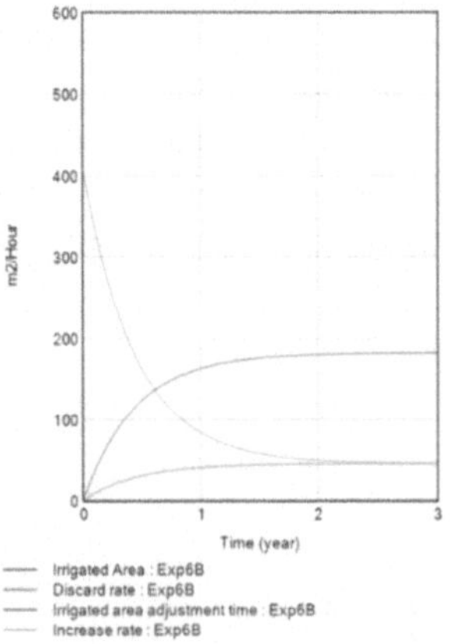

Fig. 6. Behaviors obtained from validation.

4.6 Application of an Irrigation Simulation Model for a Crop

The following behaviors correspond to the results of a simulation for a pepper crop cultivated in Peru. For about four months, the simulation used parameters obtained by means of real experimental design under ambient conditions. To find the answers to the above questions, we proposed different scenarios that estimated the possible behavior of one plant in a crop. We decided to run the model with parameters that belong to one plant. We designed two main scenarios for the simulation:

- **Scenario 1**. What are the system dynamics for a small pepper plant from crop A?
- **Scenario 2**. What are the system dynamics for a big pepper plant from crop A?

With both scenarios, we propose 30 different simulation experiments, considering the combination of the different possible values of the time setting for an irrigation area and the discard fraction depending on irrigation technique by zones. The simulation was performed using the following model assumptions:

- A small plant has a shorter irrigation time than a bigger one.
- The discard fraction is higher when the plant is smaller because they have smaller roots.
- The crops can be irrigated by different techniques, such as flooding and localization. The value of the water discarded by crop facilities is reflected by the percentage of water wasted. And more, these values are different between a global parameter than Latin America or Peru areas.
- The irrigation time for a pepper plant is from 10 to 60 min depending on the size.

4.7 Data for the Model

To ensure the simulation model achieves its objectives, two types of data were used and are listed in Table 3. The use of both data sets ensures that the water volumes discarded from irrigating the crop at various times are accurately determined (see Table 3, Table 4 and Table 5). The specific data set for the pepper crop simulation was used to simulate and study the behavior of different irrigation scenarios. These parameters were based on the literature recommendations for irrigation efficiency in pepper crops (see Table 5). For global behavior, 0.10 is the value presented in the literature for the localization technique and 0.25 for the flood irrigation technique. In the case of Peru, records show that 45% of the water is wasted by flood irrigation and 35% of the water is wasted in the case of the rest of Latin America. The simulation model allows us to explore different discard fractions to waste less water in the cultivation of a crop. Furthermore, we use this data to propose different experiments to study different discard fractions and irrigation times for the crop. In summary, using both the generic and specific data sets allowed for accurate determination of water volumes discarded from irrigating the crop at various times, and helped achieve the objectives of the simulation model. As outlined in Sect. 4, we derived the variation of irrigation time from our observations during the real experiment. The values tested in the Irrigation Model were obtained through a combination of various parameters, and the details of each experiment are available in a publicly accessible repository called "Irrigation Efficiency_EUROSIM2023" [23]. We represented the discrete values for irrigation time and discard fraction based on the relevant literature and the real experiment as follows:

- Irrigation time: discrete values from 0.25 to 1.5 h
- Discard fraction: from 0.10 to 0.45 depending on the crop irrigation technique and the geography (see Table 5).

Table 5. Parameters of different discard fractions used in the model.

Discard fractions used for the simulation model	Parameter
Discard fraction for small plants	0.25
Discard fraction for big plants	0.18 0.15
Discard fraction in Peru	0.45
Discard fraction in Latin America	0.35
Sustainable Discard fraction (by localization)	0.10

The values of 0.10 and 0.25 for the localization and flood irrigation techniques, respectively, are commonly reported in the literature for the global behavior. However, in Peru, empirical records demonstrate that as much as 45% of water is wasted through flood irrigation, while the rest of Latin America experiences a 35% waste rate. To address this challenge, the simulation model is used to explore different discard fractions that may help to reduce water wastage during crop cultivation. Based on this data, several experiments are designed to study the impact of different discard fractions and irrigation times on the crop. Table 5 provides five distinct values for the discard fraction that are used in the simulation experiments. To determine the parameters for larger pepper plants, we rely on real-world experiments conducted over four months. To evaluate the efficiency of the proposed irrigation system, we simulated the above parameters using the developed model. The obtained results are presented and analyzed in detail in the following sections.

5 Results

5.1 Analysis of Discard Rate Results

The results of the simulation experiments are classified under these different criteria: not obtain adequate parameters for an effective irrigation system, the time setting for an irrigation the discard rate, and sustainable irrigation options. Moreover, the experiment results are given for small and bigger pepper plants. In this first step of the analysis of the experiment results, only the behavior of the discard rate is considered.

Figure 7 shows that not every experiment is adequate for an effective irrigation system. The results classified as sustainable show that a long irrigation time and low water discard fraction mean less water wastage. The details about the simulation parameters for each experiment can be found in repository presents the details about the simulation parameter for each experiment [23] (Irrigation Efficiency_EUROSIM2023). The simulation results indicate that the irrigation system is more sensitive for small pepper plants than for bigger ones. Moreover, if the goal is to determine better conditions for less water waste, the irrigation time should not be rushed. Thus, the efficiency of the irrigation system and the water wasted in the zones can radically decrease. Based on the simulation results, we propose to analyze the results from its dynamics rather than just focusing on how much water can be discarded. The next section presents the second step of the experiment analysis.

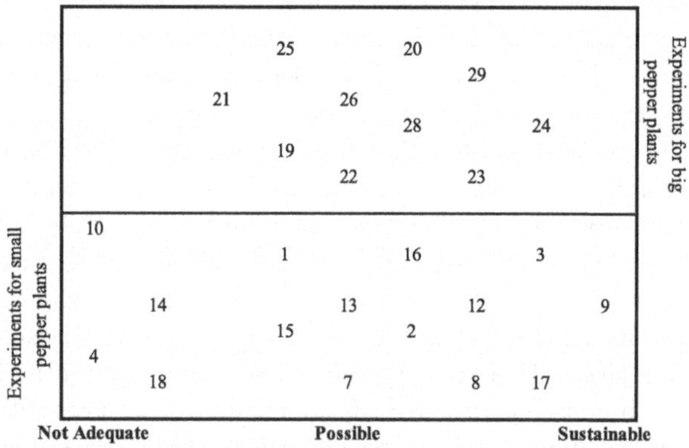

Fig. 7. Classification of simulation experiments results for small and big pepper plants.

5.2 Analysis of Irrigation System Results

The second step of the analysis of the experimental design is to study the dynamics that happen when different amounts of water can be used during many irrigation times. The simulation results allowed us to investigate the effects of shortening or lengthening irrigation time on small and big pepper plants. Table 6 presents the feasible experiments for the Irrigation Model experiment design based on parameter combinations.

Table 6. Classification of the experiments for an efficient irrigation system.

Discard fractions parameter	Experiments for small pepper plants			Experiments for big pepper plants		
Real experimental value	1	2	3	19	20	21
Sustainable discard fraction	7	8	9	22	23	24
Peru discard fraction	11	12	13	25	26	27
LA discard fraction	15	16	17	28	29	30

The results were divided into feasible and non-feasible categories based on irrigation time, discard rate behavior, and irrigation area achievement. The feasible experiments are shown in white cells, indicating that the parameters used in those experiments can produce an efficient irrigation system. For small pepper plants, experiments 1, 2, 7, 8, 11, and 15 provided information about water discard during crop irrigation. For big pepper plants, experiments 19, 22, 23, 25, and 28 yielded results about water volumes and irrigation time. Experiments 4, 10, 14, and 18 were found to be non-feasible for an effective irrigation system as they did not reach the planned irrigation area in the expected time. More details about the results can be found in repository [23] (Irrigation Efficiency_EUROSIM2023). Experiments 5 and 6 served as a starting point for combining certain parameters to validate the simulation models.

The simulation results indicate that increasing irrigation time does not always result in complete coverage of the irrigation area. Besides, not all the irrigated adjustment time parameters are feasible for every discard fraction. For example, if the system allows more water (0.45 or 0.35) to be discarded, then the irrigation area could not be completed, even if the irrigation time is increased. In other words, the increment of time is just for the wasted water and does not mean that the irrigation system is more efficient.

Contrary to expectations, the simulation results indicate that increasing irrigation time may not always lead to greater efficiency in the irrigation system. Experiments 25 and 26 demonstrate that insufficient increments in irrigation time can result in inadequate water usage for the crops. That strategy only allows for wasting a higher amount of water. This happens because the irrigated adjustment time is not a variable that enhances the efficiency of an irrigation system, using a high discard fraction. Rather, according to experiment 23, a lower value of the discard fraction can improve efficiency. These results propose changes in irrigation techniques to achieve an effective irrigation system and sustainability. Every modification implies costs, infrastructure, and agricultural labor. The crop used for this simulation is cultivated in Latin America, where financial resources are limited. It is worth noting that the irrigation system can be sustained without relying on crop technification, as demonstrated by Experiment 19 which achieved a lower discard fraction than the current value used in Peru. About the possible amount of water that can be used in the real crop, the following Table 7 present the water quantities that can be discarded depending on the irrigation time and the discard fraction.

Table 7. Feasible values for the crop.

	Experiments for small pep-per plants (For an irrigation time of 15 and 30 min)		Experiments for big pepper plants (For an irrigation time of 30 and 60 min)	
	Location Irrigation	Flood Irrigation	Location Irrigation	Flood Irrigation
Real experimental value	2.50 L	5.50 L	2.20 L	3.80 L
Sustainable discard fraction	1.00 L	2.20 L	1.15 L	2.16 L
Peru discard fraction	4.50 L	9.90 L	5.18 L	10.8 L
LA discard fraction	3.50 L	7.70 L	4.00 L	7.56 L

Through simulation, we verified that the only result of increasing irrigation time is more water being wasted. It also allowed us to test different discard fractions to evaluate irrigation systems. This would be very complicated to carry out in real life as it would require the farmers to have the necessary time, funds, know-how and re-sources.

6 Discussion and Conclusion

In conclusion, irrigated agriculture in Latin America requires careful management and optimization for its sustainability and productivity. System dynamics modeling and best management practices can enhance the efficiency of water use, improve crop yields, and increase the economic and social benefits for farmers and rural communities. Our results demonstrate that the implementation of a state-of-the-art irrigation system, managed by a skilled technician, is necessary to achieve a sustainable discard rate and an appropriate discard fraction. This will ultimately help to minimize water waste in crops, promoting sustainable agriculture. However, an irrigation system's efficiency cannot be based solely on the discard fraction, as the time required to irrigate the entire crop area also plays a crucial role. Our simulation results show that increasing irrigation time results in more water wastage, making it challenging to evaluate different irrigation systems in real life. Non-conventional techniques such as system dynamics pro-vide information about water efficiency and pave the way for future studies to help Latin American crop production, including finding the proper irrigation times for different zones and adapting irrigation systems to crop conditions, climate, and economic possibilities. Future work can also focus on analyzing water availability, amounts, and budget using different techniques. Ultimately, our study highlights the importance of investing in infrastructure and adopting innovative approaches to improve the sustainability and productivity of irrigated agriculture in Latin America.

References

1. The World Bank: Globally, 70% of Freshwater is Used for Agriculture (2022). [Online] Available: https://www.worldbank.org/en/topic/water-in-agriculture
2. FAO: The future of food and agriculture – Alternative pathways to 2050 (Summary version). Food and Agriculture Organization of the United Nations, Rome (2018)
3. FAO: Irrigation in Latin America and the Caribbean in figures. Food and Agriculture Organization of the United Nations, pp. 239–281 (2000)
4. The Economic Commission for Latin America and the Caribbean: Water Resources Management in Latin America and the Caribbean, ECLAC, Subregional Headquarters for the Caribbean (2015)
5. International Atomic Energy Agency: Agricultural water management. IAEA (2022)
6. Turner, B.L., Menendez, H.M., Gates, R., Tedeschi, L.O., Atzori, A.S.: System dynamics modeling for agricultural and natural resource management issues: review of some past cases and forecasting future roles. In: Resources, pp. 3–4 (2016)
7. Gonzalez-Torres, R., Barrera, J., Munoz-Carpena, R.: Water, agriculture, and climate change in Latin America: challenges and opportunities for sustainable development. In: Sustainability, p. 13 (2021)
8. FAO: Irrigated agriculture in Latin America and the Caribbean: current status, challenges and prospects. AQUASTAT – FAO's Global Information System on Water and Agriculture (2020)
9. Birichi, G., Abebe, H., Sereke, F.: Effect of irrigation best management practices on water use efficiency and crop yield: a review. In: Advances in Agriculture, pp. 2–4 (2017)
10. Fernández-García, D., Rey, D., Garrido, A., Rodríguez-Díaz, J.A.: System dynamics for irrigation water management: a review of current knowledge and practice. In: Water Resources Management, pp. 1876–1884 (2016)

11. International Irrigation Management Institute: Irrigation management in Latin America: current status, problem areas and potential areas for improvement. In: IIMI, pp. 106–110 (1990)
12. FAO: Irrigation in Southern and Eastern Asia in Figures: AQUASTAT Survey- 2011. FAO, Rome (2012)
13. Fang, Q., Ma, L., Green, T.R.: Impacts of irrigation methods on groundwater table and crop yield. In: Gricultural Water Management, pp. 53–62 (2013)
14. Kibret, K.S., McCartney, M.: Improving irrigation water use efficiency through system dynamics modeling. In: Agricultural Water Management, pp. 272–284 (2017)
15. Razavi, S., Haghani, A., Mirchi, A.: Dynamic simulation of irrigation efficiency in surface irrigation systems. In: Journal of Irrigation and Drainage Engineering, pp. 1–3 (2014)
16. Sowell, R.S., Hossain, M.M.: Modeling irrigation efficiency with a system dynamics approach. In: Journal of Irrigation and Drainage Engineering, pp. 3–6 (2013)
17. Seyed Morteza, M.L., Abdolmajid, L., Ataie-Ashtiani, B.: Using system dynamics to improve water use efficiency in irrigated agriculture. In: Water Resources Management, pp. 1–4 (2016)
18. Bala, B.K., Arshad, F.M., Noh, K.M.: System Dynamics: Modelling and Simulation, pp. 21–24. Springer, Malaysia (2018)
19. Fornaris, G.J.: Conjunto Tecnológico para la Producción de Pimiento. Colegio de Ciencia Agrícolas (2005)
20. Banks, J.: Handbook of Simulation: Principles, Methodology, Advances, Applications, and Practice. John Wiley & Sons, Inc. (1998)
21. Kleijnen, J.P.C.: Verification and validation of simulation models. Eur. J. Oper. Res. **82**, 145–162 (1995)
22. Sargent, R.: Verification and validation of simulation models: an advanced tutorial. Winter Simulation Conference (WSC) (2020)
23. Sánchez, O.: Irrigation Efficiency_EUROSIM2023 [GitHub repository] (2023). [Online] Available: https://github.com/Oroselfia/Irrigation-Efficiency_EUROSIM2023#irrigation-efficiency_eurosim2023

Simulation and Optimization in Cross-Sectoral Water: A Review

Adrielly Nahomee Ramos Alvarez, Idalia Flores-De la Mota(✉), and Francisca Irene Soler Anguiano

National Autonomous University of Mexico, Coyoacan Mexico City, Mexico
idalia@unam.mx, francisca.soler@ingenieria.unam.edu

Abstract. Simulation and optimization models have been of pivotal importance in bringing solutions to major problems. However, their impact is determined by the accuracy and level of detail of the conceptual models describing the systems. In water management these tools have provided valuable solutions, nonetheless, throughout the years models have simplified critical aspects that curb efforts to deliver more effective results. That is the case of cross-sectoral relationships that are often misrepresented in the models despite their relevance with the aim to reduce complexity. Cross-sectoral perspectives have taken place in recent years due to the valuable insights they deliver by addressing those commonly unseen relationships. As a result, the objective of the present article is to introduce a brief review of the cross-sectoral studies that using simulation and optimization techniques have contributed to mitigate the effects of the current and future water crisis through a change of paradigm in models' constructions.

Keywords: Cross-sectoral optimization · water simulation · water optimization · sustainability

1 Introduction

Water is a fundamental resource in every economical sector, and its adequate management soon will constitute the difference between success and failure. Nowadays countries experience a hydric crisis exacerbated by poor regulation, lack of initiatives, insufficient strategies, inadequate technology, and the absence of accurate models and data.

This crisis is not particular of a certain sector or region, the scarcity and contamination of the resource in one affect, both directly and indirectly, the others due to the existing interactions among sectors. These interactions are not always taken into consideration while modelling the system or implementing solutions, being rather simplified in most simulations. Throughout the years several studies have focused in describing, understanding, and facing the problem in parts, by looking at the relationships inside the sectors but not necessarily among them, leaving important aspects of the system unattended. According to Lara-Rosano "One of the difficulties in focusing a problem is to think that the problem lays in the parts, when generally lays in the relationships between the parts" (Lara-Rosano, 2017). Considering these relationships has provided

M. Mujica Mota and P. Scala (Eds.): EUROSIM 2023, CCIS 2032, pp. 19–30, 2024.
https://doi.org/10.1007/978-3-031-68435-7_2

important findings leading to the development of more effective solutions. This is visible in water systems where cross-sectoral approaches have directed efforts towards more sustainable practices improving the hydric situation significantly.

Modelling systems from a cross-sectoral perspective has contributed to fill gaps with vital information. United Nations affirms that "water-related data and information are central to understanding and valuing the resource" and that "Further efforts and investments are required to sustain the supply chain of data … across sectors and scales." (UN Water, 2021) to have a wider vision of the problem and the situation providing more elements to complete the hydric scenario. Although "A degree of information/knowledge deficiency is generally present in any model development activity" (Birta and Arbez, 2019) in complex issues such as water management, lacking the awareness of the relationships between sectors, can lead to an important "…misrepresentation of impacts" (Harrison et al., 2016) building considerable obstacles in developing efficient simulation models, optimization strategies and effective solutions. Consequently, water simulation models that work under the assumption and simplification that interactions with other sectors are despicable, fall short to properly describe the system.

Cross-sectoral studies are still scarce, but their results are of crucial importance. Specially in environmental matters where correct representation of relationships offers insights that otherwise would remain invisible by estimating more precisely the magnitude and behavior of the interactions and influences across sectors. Therefore, the aim of the present paper is to introduce an overview of cross-sectoral simulation and optimization investigations highlighting their contributions and opportunity areas in order to set a starting point for developing simulations that effectively describe cross-sectoral interactions to bring more impactful optimizations. Helping confront long lasting complex problems such as the ones the world faces in environmental and sustainability matters.

2 Water Context

2.1 Global Water Context

Although water is a fundamental resource for every human activity it has not been valued nor represented properly as it can be seen in the current hydric situation. Water levels have been decreasing around the world, in 1967 renewable water per capita was 13,400 cubic meters, while in 2017 it declined to 5,732 cubic meters (Oxford, 2018).

It is important to highlight that hydric challenges vary according to the region (UN-Water, 2021):

In Sub-Saharian Africa water resources are unequally distributed, despite having 9% of global fresh water, access to water is scarce, only 24% of the region's population has access to drinking water (UN-Water, 2019). Coupled with rapid population growth, inappropriate water governance, environmental degradation, deforestation, the socioeconomic situation, and unsustainable financial investments, there is a lack of infrastructure and plans adapted to the needs of the continent.

Transboundary water management in the Pan-European Region is a major challenge, countries use different approaches for its administration and the benefits of cooperation between countries are substantial; however, determining the value of water represents a substantial challenge, especially in collecting quantitative data.

The growing water stress in the Latin America and the Caribbean region has triggered conflicts between sectors that compete for scarce resources. Poor regulation, lack of initiatives and investment are factors that show the value attributed to water in the region despite its strategic value in multiple production processes or in the context of climate change.

Due to population growth, urbanization and rapid industrialization in Asia and the Pacific competition for water between sectors has become more severe, threatening agricultural production and food security as well as water quality. Unsustainable water withdrawals have in some cases exceeded half of supply availability. Wastewater remains under used and there is an urgent need to reduce aquifer pollution and promote water efficiency.

It is important to mention that a crucial part of water management is in the adequate valuation of the resource, that comes from a systemic approach and a proper representation of the relationships among the elements of the system which is obtained through the accurate collection of data and information with its correct analysis. Despite the usefulness of hydrological data for the generation of solutions, the levels of reported public data have decreased significantly with respect to the 20th century, for 2010 the number of stations reporting data decreased by 31% with respect to the reports of the decade of 1980 (UN-Water, 2021). Proper data collection and analysis specially among sectors, basins and frontiers is crucial to find and understand the existing relationships that are affecting the system in different levels. This is crucial for developing strategies and solutions.

Simulation and optimization models potential.

One of the hardest things to accomplish is to determine the interactions among the elements of the system. It is through simulation that those cross-sectoral interactions can be addressed, quantified, and explored. While optimization techniques provide significant value for a sustainable use of the resources.

2.2 Sectorial Water Context

Global water use has increased substantially in recent years, and it is projected to grow more while aquifer levels decrease. Even though the exact amount of required water is not certain, additional volumes of water will be necessary since food demand is expected to grow in 50% by 2050 relative to 2012 levels (UN Water, 2022), industry and domestic demands are also expected to raise. Just in 2016 an estimated of four billion people lived in areas that suffered critical water scarcity (Mekonnen and Hoekstra, 2016) and due to demographic growth, this scarcity will become more critical.

The use of groundwater in agriculture accounts for 59% of the areas equipped for irrigation in North America and 57% in South Asia. Water is rarely quantified in food production despite of its high value and the amount of water that is needed to guarantee its security. It accounts for 70% of water use globally (UN Water, 2021–2022).

Industry and energy use 19% of global freshwater withdrawals, with major differences between developed and underdeveloped countries, while Europe industrial withdrawals is of 57% in Africa industrial extractions account for 5% (UN Water, 2022).

Groundwater dependence for domestic use is intensifying, almost 50% of global urban population demand is supplied by groundwater. Major challenges in developing countries are addressed from unplanned settlements that lack proper hydric infrastructure, low sewer coverage to lack of adequate data and management (UN Water, 2022).

Water panorama in every sector is still far from the ideal and the actions in one affect the others in several ways. Although these impacts are not always visible or measured their effects are significant. As a result, cross-sectoral perspectives are of great importance for bringing wider solutions.

3 Cross-Sectoral Water Simulation and Optimization

Cross-sectoral analysis in environmental problematics have brought important insights for bridging knowledge gaps and creating more sustainable solutions. Cross-sectoral perspectives in water usage are not the exception, by establishing awareness of the interlinkages across sectors and giving visibility, models have been built with vital information that have boosted optimization and simulation outcomes.

Relationships across sectors are impactful, adaptation and mitigation actions by one sector can directly influence water demand, increase, or reduce water availability, change water quality for other sectors, and affect energy use. For example, pollution originated by agriculture "…has overtaken contamination from settlements and industries as the major factor in the degradation of inland and coastal waters." (UN Water, 2020). Chemicals, from fertilizers, are the most prevalent anthropogenic contaminant in groundwater globally. "Insecticides, herbicides, and fungicides, when improperly applied or disposed of, can pollute groundwater with carcinogens and other toxic substances." Persistent contamination of rural groundwater supplies with pathogens is estimated to affect 30% of the total domestic installations. (UN Water, 2022).

In small urban and rural settlements, use of water for agriculture and in some cases industrial applications results in reduced availability for domestic uses. (UN Water, 2020). These are some of the most noticeable relationships across sectors but many of them have not even been described. For this reason, it is essential that more studies may be performed from a cross-sectoral perspective.

3.1 Simulation and Optimization

Some cross-sectoral investigations are summarized by author showing the optimization tools and the simulation models used in every study considering the principal variables and highlighting the contribution made in every research.

Wang et al. (2022) used the Leontief input-output model in an extended multiscale framework for identifying the characteristic energy-water-land links and transboundary relationships. This to reflect the connections across sectors and regions from a multisectoral perspective.

Yue et al. (2022) developed a stochastic multi-objective optimization model for crop-livestock-biogas-crop recycling through the coordination of resources allocation considering resources input reduction, agricultural wastes recycle, global warming mitigation, socio-economic security in order to promote sustainable circular agricultural development.

Deng et al. (2022) implemented a machine learning model using a multilayer perceptron in an Elman neural network for predicting the water level in the Dongting lake from a cross-sectoral perspective. Their contribution is to provide a reliable water level prediction for better hydric resources management.

The objective of Tian et al. (2022) was to understand the operational issues that can be found in complex water systems. Their contribution lays in partitioning large water distribution networks by creating partitions for the allocation of multiple sources, and for a multi-objective network decomposition. This optimization being achieved by a depth-first-search metaheuristic algorithm.

Dombrowsky et al. (2022) made an analysis of the effects of cross-sectoral policies by addressing challenges among different sectors. They deliver a process analysis of river conversion and rehabilitation, agricultural water consumption reduction, mitigation of contamination due to mining, catchment management strategy to manage water conflicts, competition coordination for scarce water resources.

Xiao et al. (2021) proposed a differential game model with a two-player finite-horizon representing both agricultural and industrial sectors. The objective was to minimize toxic gas emissions, pollutants, and damage while maximizing utility and production through cross-sectoral game theory.

Pronk et al. (2021) constructed different scenarios for water reuse optimization. Interlinkage description was a pivotal aspect for quantifying water use among sectors and propose a framework for cross-sectoral water reuse. Reusing industrial effluent for irrigation, industry, and aquifer replenishment.

Langergraber et al. (2021) had the objective of analyze the multi-functionality potential of nature-based solutions for water management. Maximizing the use and reuse of resources while reducing the need of external resource inputs. Creating an optimization framework for Nature Based solutions.

Krapez et al. (2020) detected agricultural water leakages from an airplane and drone using the Triangle/Trapezoid method. With the objective of reducing losses in water transmission through a feasible surveillance method for water transportation mains outside urban areas. This monitoring method provides important data for the system modelling and understanding.

Quesada et al. (2019) studied existent chemicals in water that potentially affect basins. The investigation highlights the need to monitor residual pharmaceuticals in water and its implications in cross-sectoral uses. Through a review of pharmaceuticals' current situation as emerging contaminants in surface water. It also provides detail in adsorption processes as an alternative to remove these compounds from aquifers.

Seibel and Kuhlmann (2018) developed the model of a pilot chemical plant using multi-objective optimization to minimize electricity purchase, maximize grid balancing services, maximize product sale, minimize cost of poor-quality water and minimize gas emissions using cross-sectoral resource optimization.

Chuenchum et al. (2017) performed rainfall projections under climate change scenarios, building a rainfall-runoff simulation within an Integrated Flood Assessment System.

The main contribution with this simulation was the Regional Input-Output Model for determining the economic value of water.

Cook and Webber (2016) seek to improve irrigation efficiency in the agriculture sector to make significant volumes of water available for other purposes through the optimization of the resource conservation. The investigation contributed to the determination of water savings in the implementation of best management practices for irrigation efficiency across 2 counties in Texas.

Koutroulis et al. (2016) developed a hydrological simulation to assess the impact of climatic and socioeconomic futures on the water resources of a Mediterranean island, and constructed a cross-sectoral framework.

Wimmer et al. (2015) assembled an exploratory tool to investigate future adaptation needs in Europe's water-related sectors through the simulation of the effect of water management on agricultural productivity, hydric stress, demand.

The investigations listed above have intensely used simulation and optimization models for a more sustainable use of hydric resources contributing significantly to solve the complex water problem from a cross-sectoral water approach. Table 1 condenses some of the most relevant articles, however there are many more investigations that use cross-sectoral simulation and optimization models for water management and other resources.

The following Table 1 gives an overview of the simulation and optimization models for cross-sectoral water.

Table 1. Overview of models and authors.

Authors	Model	Simulation/Optimization
Wang et al., (2022)	Leontief input-output model. Extended multiscale input-output model (EE-MSIO)	Optimization
Yue et al., (2022)	Stochastic multi-objective optimization. Stewart model, analytic hierarchy process, carbon footprint	Optimization
Deng et al. (2022)	Machine learning model: multilayer perceptron, Elman neural network and optimization	Optimization
Tian et al., (2022)	Multilevel partitioning. Multi-objective network decomposition. Depth-first-search algorithm	Optimization
Dombrowsky et al., (2022)	Analysis of the effects of cross-sectoral water policies	Simulation basis

(*continued*)

Table 1. (*continued*)

Authors	Model	Simulation/Optimization
Xiao et al., (2021)	Differential game model, two-player finite-horizon, game theory	Optimization
Pronk et al. (2021)	Scenario construction for water reuse optimization using Sankey diagrams	Optimization
Langergraber et al., (2021)	Optimization framework for Nature Based Solutions	Optimization
Krapez et al., (2020)	Triangle/Trapezoid method (T-VI) for detection of water leaks from airplane	Simulation
Quesada et al. (2019)	Drug concentration determination, highlighting minimal, maximal, and mean values	Simulation basis
Seibel and Kuhlmann, (2018)	Simulation-based optimization, multi-objective optimization	Simulation & optimization
Chuenchum et al. (2017)	Simulation of runoff under climate change scenarios. Integrated Flood Assessment System. Regional Input-Output Model for economic value of water	Simulation
Cook and Webber, (2016)	Water conservation optimization for more availability. Analysis and implementation	Optimization
Koutroulis et al. (2016)	Hydrological simulation to determine impacts of climatic and socioeconomic futures	Simulation
Wimmer et al. (2015)	Simulation of the effect of WAS on agricultural productivity, stresses on aquatic ecosystems, unsatisfied demand, domestic	Simulation

Cross-sectoral perspectives have contributed significantly to provide more accurate simulations and more impactful optimizations by properly addressing pivotal interactions among sectors which has also potentiated the use of intelligent tools such as Machine Learning, Big Data, among others.

3.2 Variables

To determine the interrelations across sectors it is vital to properly identify the most relevant variables and parameters within the system. The presented investigations consider these variables.

- Economic: service and goods utility, savings, energy, treatment, damage, infrastructure, pumping costs, GDP.
- Pollutants: CO2 NO2 emissions, chemical, fertilizers, drugs present in water and soil, water quality.
- Volume: treated water, leakages, aquifer levels (superficial and underground water), water reuse, domestic, agricultural, and industrial water use, water demand.
- Physical: soil, atmospheric, water temperature, humidity, pressure.
- Energy: type of energy, use, cost, demand, generation.
- Social: policies, management, socioeconomic growth, innovations, strategies, stakeholders, organizational objectives.

Other variables are also of pivotal importance and therefore taken into consideration according to the needs and circumstances of every problem. Some variables will take precedence in determined cases. It is important to notice that as the number of interactions increase, the variables to be considered will augment as well. This resulting into more complex models, that will require to incorporate additional tools for properly and effectively manage information and insights.

3.3 Opportunities in Cross-Sectoral Models

Cross-sectoral approaches are still scarce as compared to other perspectives that have been widely implemented. For there are many gaps yet to fill. Among the opportunity areas that can be found in the present review are: cross-sectoral coordination policies, sustainable intensification schemes that minimize hydric stress, identification of nonobvious connections across sectors and boundaries, solutions for megacities' size and complexity, optimization, and decision support tools for different scenarios and weight objectives, considering more connections, activities, procedures, and elements along with the condition variability among sectors.

Other opportunity areas are also important to highlight, one of them is the lack of generalization and portability of models and methods to apply techniques in different aquifers and maximize coordination in cross-sectoral environmental policies (Deng et al., 2022). Control and monitoring of cross-sectoral water pollution including setting objectives in stages at all levels with adaptation options are still missing and a significant challenge to face. And it is also required to implement further optimization techniques to empower strategies.

Dombrowsky et al. emphasize that "coordination at process level does not necessarily translate into coordinated outcomes" (Dombrowsky et al., 2022) for it is also critical to redefine objectives, and actions in a dynamic way. It is difficult to determine causality of many problems while providing efficient, cost-effective solutions for surveillance, simulations (Krapez et al., 2020). Collaborative governance, quantitative approach, optimization techniques for supporting management are yet in an initial stage and completely scarce in developing countries. Multifunctionality is rarely implemented in nature-based solutions.

Furthermore, data collection, adaptable design approaches, inter-basin data transfer and analysis are major challenges to be tackled. Although Industry 4.0 tools have enabled more efficient information acquiring, techniques and methodologies for identifying relationships across sectors supported by accurate data are needed alongside with optimization techniques for supporting cross-sectoral approaches.

Simulation and optimization models are valuable tools for obtaining more significant results and can provide instruments for fulfilling the mentioned gaps while offering a wider vision of the systems. Cross-sectoral perspectives can deliver important information that traditional approaches can miss as it can be seen in Fig. 1, where the cross-sectoral analysis not only brings data about each sector but the one pertaining the relationships among sectors.

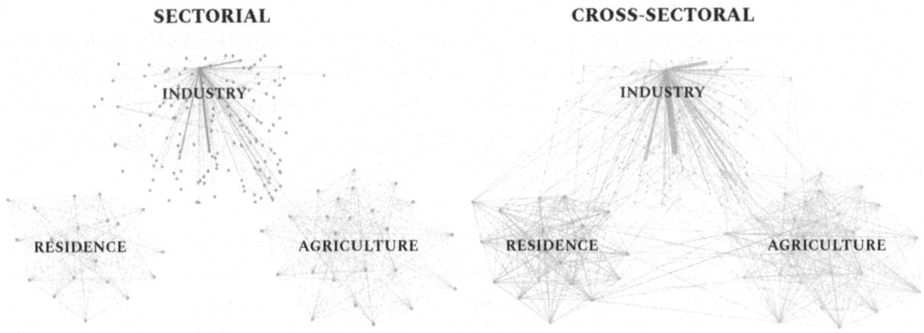

Fig. 1. Sectors analyzed separately vs cross-sectoral analysis.

The sectorial perspective focuses on the collaboration and coordination among the activities or procedures of its own sector, as it can be seen in the left network representation shown in Fig. 1, where the nodes denote the physical spaces, activities or operations and the arcs represent the water inputs and outputs from one node to another. Within this kind of perspective, it is considered that the situation in other sectors can be tackled as a separate problem and that solving those problems individually then bringing those solutions together can constitute a difference on the whole system. While this is in part true, complex problems such as the one addressed here require a special treatment for understanding the aspects that will provide the necessary information to ideate strategies and solutions that bring significant results. On the other hand, the cross-sectoral approach seen in the right network representation takes on account what sectorial perspective is considering plus adding the connections between the nodes pertaining to Residential, Industrial and Agricultural sectors, as it is shown in Fig. 1. Nonetheless, considering these relationships increases the challenges for not only constructing the models but for implementing the solutions given that the computational, mathematical, and operational complexity is increased. Even more robust and flexible tools are required to be created in order to face the challenges associated to a cross-sectoral simulation and optimization.

4 Conclusions

Simulation and optimization models are vital for solving complex problems such as the ones related to water. Those models require a balance between complexity and simplicity. While simpler models contribute to the generation of faster solutions, they can often misrepresent the system leaving vital aspects outside of the analysis. On the other hand, models that take on account more elements tend to increase the level of complexity leading to higher costs of implementation, more computing time, and additional problems associated with data collection and analysis. Nonetheless, due to the nature of the global hydric problem it is required to have more impactful solutions. Solutions that can only be achieved through a better understanding of the actual system that tools like simulation can offer. Cross-sectoral approaches provide information that is commonly set aside because of the additional complexity and work needed. This information bridges considerable gaps that otherwise would remain unattended. For it is important to continue simulating and optimizing from a cross-sectoral systemic perspective.

The present review condenses some of the various cross-sectoral water investigations that have been developed in recent years and highlights the gaps that are still pending to bridge. Future research will aim to empower the use of optimization and simulation models in cross-sectoral water for fulfilling hydric needs and contributing to understand the system dynamics more fully.

This article raises to highlight that cross-sectoral perspectives can change the paradigms of model construction and problem-solving in more than one problem as the one stated in this work, and that this approach can lead to other ways of handling and managing information in problems and in developing more sustainable systems.

References

Gryzwna, A., Nieścioruk, K.: Changes of hydrographic network of uściwierskie lowering according to cartografic materials. J. Ecol. Eng. **17**(4), 148–153 (2016). https://doi-org.pbidi.unam.mx:2443/10.12911/22998993/64705

Baron, S., Schmitt, T.G., Kaufmann Alves, I., Schöffel, S., Schwank, J.: Cross-sectoral optimization and visualization of transformation processes in urban water infrastructures in rural areas. Water Sci. Technol. **72**(10), 1730–1738 (2015). https://doi-org.pbidi.unam.mx:2443/10.2166/wst.2015.378

Birta, L.G., Arbez, G.: Modelling and Simulation: Exploring Dynamic System Behaviour, 3rd edn. Springer International Publishing (2019)

Cook, M., Webber, M.: Food, fracking, and freshwater: the potential for markets and cross-sectoral investments to enable water conservation. Water (Switzerland), **8**(2), 45 (2016). https://doi-org.pbidi.unam.mx:2443/10.3390/w8020045

Chuenchum, P., Suttinon, P., Ruangrassamee, P., Dunn, C.N., Van Weele, B.: Cross-Sectoral Impacts of Water Deficits in Nan River Basin, Thailand (2017)

Deng, B., et al.: Hybrid metaheuristic machine learning approach for water level prediction: a case study in Dongting Lake. Front. Earth Sci. **10**, 928052 (2022). https://doi-org.pbidi.unam.mx:2443/10.3389/feart.2022.928052

Desakota Study Team: Re-imagining the Rural-Urban Continuum. Institute for Social and Environmental Transition. Nepal. ISBN: 978-9937-8019-1-1 (2008)

Dombrowsky, I., Lenschow, A., Meergans, F., Schütze, N., Lukat, E., Stein, U., Yousefi, A.: Effects of policy and functional (in)coherence on coordination – a comparative analysis of cross-sectoral water management problems. Environ. Sci. Policy, **131**, 118–127 (2022). https://doi-org.pbidi.unam.mx:2443/10.1016/j.envsci.2022.01.019

Eberhard, R., Johnston, N., Everingham, J.-A.: A collaborative approach to address the cumulative impacts of mine-water discharge: negotiating a cross-sectoral waterway partnership in the Bowen Basin, Australia. Resour. Policy, **38**(4), 678–687 (2013). https://doi-org.pbidi.unam.mx:2443/10.1016/j.resourpol.2013.02.002

Eriksson, M., Gordon, L., Kuylenstierna, J.: Cross-sectoral approaches help build water resilience – reflections. Aquat. Procedia **2**, 42–47 (2014). https://doi.org/10.1016/j.aqpro.2014.07.007

Harrison, P.A., Dunford, R.W., Holman, I.P., Rounsevell, M.D.A.: Climate change impact modelling needs to include cross-sectoral interactions. Nat. Clim. Chang. **6**(9), 885–890 (2016). https://doi.org/10.1038/nclimate3039

Koutroulis, A.G., Grillakis, M.G., Daliakopoulos, I.N., Tsanis, I.K., Jacob, D.: Cross sectoral impacts on water availability at +2 °C and +3 °C for east Mediterranean island states: the case of Crete. J. Hydrol. **532**, 16–28 (2016). https://doi-org.pbidi.unam.mx:2443/10.1016/j.jhydrol.2015.11.015

Krapez, J.-C., et al.: Assessment of the Triangle Method (T-VI) for Detection of Water Leaks from Airplane and UAV. IGARSS 2020 - 2020 IEEE International Geoscience and Remote Sensing Symposium, Geoscience and Remote Sensing Symposium, IGARSS 2020 - 2020 IEEE International, 4715–4718 (2020). https://doi-org.pbidi.unam.mx:2443/10.1109/IGARSS39084.2020.9323349

Langergraber, G., et al.: Towards a cross-sectoral view of nature-based solutions for enabling circular cities. Water **13**(17), 2352 (2021). MDPI AG. Retrieved from https://doi.org/10.3390/w13172352

Lara-Rosano, F.D.J.: Aplicaciones de las Ciencias de la Complejidad al Diagnóstico e Intervención en Problemas Sociales. Centro de Ciencias de la Complejidad. UNAM, México (2017)

Weber, L., Marchezan, E., Carlesso, R., Marzari, V.: Cultivares de arroz irrigado e nutrientes na água de drenagem em diferentes sistemas de cultivos Irrigated rice cultivars and drainnage water nutrient under differnt managements systems. Ciência Rural **33**(1), 27–33 (2003)

Pronk, G.J., et al.: Increasing water system robustness in the Netherlands: potential of cross-sectoral water reuse. Water Resour. Manag. **35**(11), 3721–3735 (2021). https://doi-org.pbidi.unam.mx:2443/10.1007/s11269-021-02912-5

Quesada, H.B., Baptista, A.T.A., Cusioli, L.F., Seibert, D., de Oliveira Bezerra, C., Bergamasco, R.: Surface water pollution by phar-maceuticals and an alternative of removal by low-cost adsorbents: a review. Chemosphere. **222**, 766–780 (2019). https://doi.org/10.1016/j.chemosphere.2019.02.009

Ran, Y., Deutsch, L., Lannerstad, M., Heinke, J.: Rapidly intensified beef production in Uruguay: impacts on water-related ecosystem services. Aquatic Procedia **1**, 77–87 (2013). https://doi.org/10.1016/j.aqpro.2013.07.007

Schmid, W., Hanson, R.T.: Simulation of intra- or transboundary surface-water-rights hierarchies using the farm process for MODFLOW-2000. J. Water Resour. Plann. Manage. **133**(2), 166–178 (2007). https://doi-org.pbidi.unam.mx:2443/10.1061/(ASCE)0733-9496(2007)133:2(166)

Schöffel, S.: Visualizing Multi-Variate Process Data Concepts, Applications, and Evaluation PhD. Dissertation thesis. Vom Fachbereich Informatik der Technischen Universität Kaiserslautern (2018)

Schoffel, S., Schwank, J., Ebert, A.: A User Study on Multivariate Edge Visualizations for Graph-Based Visual Analysis Tasks. 2016 20th International Conference Information Visualisation (IV), Information Visualisation (IV), 2016 20th International Conference, Information Visualisation (IV), 2014 18th International Conference On, 165–170 (2016). https://doi-org.pbidi.unam.mx:2443/10.1109/IV.2016.41

Seibel, C., Kuhlmann, J.-W.: Dynamic water electrolysis in cross-sectoral processes. Chemie-Ingenieur-Technik, **90**(10), 1430–1436 (2018). https://doi-org.pbidi.unam.mx:2443/10.1002/cite.201800114

The United Nations World Water Development Report. Wastewater: The Untapped Resource. Paris, UNESCO. (2017). www.unesco.org/new/en/natural-sciences/environment/water/wwap/wwdr/2017-wastewater-the-untapped-resource/

Tian, J., Cheng, W., Long, Z., Zhu, Z., Xu, G.: Multilevel partitioning with multiple strategies for complex water distribution network. J. Water Resour. Plann. Manage. **148**(12), 04022064 (2022). https://doi-org.pbidi.unam.mx:2443/10.1061/(ASCE)WR.1943-5452.0001622

UN Water (2017). UN World Water Development Report 2017: WASTEWATER the untapped resource https://www.unwater.org/publications/un-world-water-development-report-2017

United Nations World Water Development Report (2021) https://www.unwater.org/publications/un-world-water-development-report-2021/

University of Oxford (2015). Water Use and Stress. Our World in Data. https://ourworldindata.org/water-use-stress#:~:text=Globally%2C%20approximately%2019%20percent%20of,%2C%20industrial%20and%20domestic%20uses)

Wang, D., et al.: Cross-sectoral urban energy–water–land nexus framework within a multiscale economy: the case of Chinese megacities. J. Clean. Prod. **376**, 134199 (2022). https://doi-org.pbidi.unam.mx:2443/10.1016/j.jclepro.2022.134199

Wimmer, F., et al.: Modelling the effects of cross-sectoral water allocation schemes in Europe. Clim. Change **128**(3), 229–244 (2015). https://doi-org.pbidi.unam.mx:2443/10.1007/s10584-014-1161-9

World Water Assessment Programme/ UN Water (2017). UN-Water. 2018. The United Nations World Water Development Report 2018: Nature-Based Solutions for Water. Paris, UNESCO. www.unesco.org/new/en/natural-sciences/environment/water/wwap/wwdr/2018-nature-based-solutions/. (2017)

Xiao, L., Liu, J., Ge, J.: Dynamic game in agriculture and industry cross-sectoral water pollution governance in developing countries. Agric. Water Manag. **243**, 106417 (2021). https://doi-org.pbidi.unam.mx:2443/10.1016/j.agwat.2020.106417

Yue, Q., Guo, P., Wu, H., Wang, Y., Zhang, C.: Towards sustainable circular agriculture: an integrated optimization framework for crop-livestock-biogas-crop recycling system management under uncertainty. Agric. Syst. **196**, 103347 (2022). https://doi-org.pbidi.unam.mx:2443/10.1016/j.agsy.2021.103347

Estimating Wind and Emission Parameters in an Atmospheric Transport Model

Andres Yarce Botero[1,2(✉)], Santiago Lopez Restrepo[2], Olga Lucia Quintero[2], and Arnold Heemink[1]

[1] Department of Applied Mathematics, TU Delft, 2628CC Delft, The Netherlands
{a.yarcebotero,A.W.Heemink}@tudelft.nl

[2] Mathematical Modelling Research Group, EAFIT University, 050022 Medellín, Colombia
{ayarceb,slopezr2,oquinte}@eafit.edu.co

Abstract. The present study proposes a novel data assimilation (DA) approach for estimating emission and wind direction parameters in an advection-diffusion model. This implementation aims to improve the prediction of a chemical transport model over long distances by updating the emission operator in the model using DA techniques. As a first step, we want to test the method in a small-scale scenario. A low-dimensional advection-diffusion model was utilized to evaluate the effectiveness of the proposed approach under various sampling observation numbers. The model's emission and wind parameters are perturbed as a source of uncertainty. The parameters are sequentially estimated with the adjoint-free Ensemble Kalman filter with an augmented state vector. These sequential DA techniques exploit the ensemble of multiple model realizations to reduce uncertainty in the state and parameter representation. An associated stream function with a divergence-free condition controls the wind fields, and the estimation of this stream function through the assimilation process allows corrections of the wind fields without violating physical laws. The technique's performance was compared against validation observations such as the Root-Mean Square (RMS), and it was found that the number of assimilated observations had a significant impact on the parameter estimations results. This study demonstrates the potential of the proposed DA approach for improving the prediction of transport in the advection-diffusion model through parameter estimation.

Keywords: Air Quality modeling · Parameter estimation problem · Chemical Transport Model · 2D advection-diffusion model · Data Assimilation · divergence-free · LOTOS-EUROS · Ensemble Kalman filter information · atmospheric transport model

1 Introduction

Chemical Transport Models (CTMs) predict the concentration, transport, and deposition of several components emitted from natural and anthropogenic sources. Although

Supported by TuDelft and EAFIT.

M. Mujica Mota and P. Scala (Eds.): EUROSIM 2023, CCIS 2032, pp. 31–43, 2024.
https://doi.org/10.1007/978-3-031-68435-7_3

these atmospheric transport models have been evolving for years, refining the physical, chemical, and mathematical structure representing reality, uncertainty is always there due to the highly nonlinear phenomena these models try to describe. One of the primary sources of uncertainty in CTMs comes from the emission inventories, which could differ significantly from the technique's location and time against reality. We inspired or contributed to a real-life scenario where we have a CTM model in a region with coarse meteorological input information, so the winds are the secondary source of uncertainty between the model and real directions. Uncertainty in the emission inventories or parameters involved in the emission processes may be reduced but never eliminated [1]. With the help of data assimilation (DA), incorporating information in these models alleviates some difficulties.

Emission parameter estimation has been the center of attention for DA around the globe for CTM implementations, mainly because the emission inventories tend to have high uncertainty values, with some different degrees of uncertainty between regions and source sectors [3,4]. In South America, updating emission inventories can be challenging due to historical institutional limitations, making it difficult to have up-to-date information available. This significant variability in CTMs from the emission parameters [5] has been given the principal attention in DA techniques to incorporate observations into numerical models for determining the most appropriate state and parameter sets that produce a new state that follows this observation [6]. DA method is commonly used to incorporate measurements into the system, estimate its states and parameters, and enhance subsequent simulation forecasts, making it a valuable tool for improving model performance.

In this work, the state and model parameters, emissions and winds, are estimated simultaneously using the Ensemble Kalman Filter (EnKF) to estimate the error covariances among several variables with a small scale model of a CTM, an emission advection-diffusion model. EnKF is the DA technique that generates an update at the first moment of the probability density function found in the analysis once new observations are available [8,9]. An ensemble of model propagation is created by perturbing emission factors and wind direction parameters. The method for estimating the parameters using the data provided by the observations uses covariances that include the relationship between states and initial conditions and compliance with the observed stated. The "augmented state space" vector technique, proposed by Jazwinsky in 1970 [10], is used to estimate parameters. Here both the state variables and the model parameters are part of the state vector and can be estimated through the sequential steps of the filter. This approach has been used in different parameter estimation studies, specifically with the LOTOS-EUROS mode" as CTM [11,12].

This chapter is organized as follows: In Sect. 2, we introduce the specific methodology for the EnKF, including the stochastic model for parameter perturbation and propagation, as well as the technical details of the code and stream function formulation. In Sect. 3, we focus on the results of this technique with the advection-diffusion model. In Sect. 4, we present our discussion and concluding remarks. This study offers valuable information on the effectiveness of stream function DA in improving CTM and can contribute to developing more accurate air quality monitoring systems.

2 Methodology

2.1 The Ensemble Kalman Filter for Wind and Emission Parameter Estimation

Ensemble Kalman Filter (EnKF) is a sequential DA mathematical procedure used to incorporate data into a dynamic model to improve its representation of reality [19,20]. One benefit of the such sequential technique is the estimation of the current state $\mathbf{x} \in \Re^{n\times 1}$, with n as the state dimension of a dynamical system that evolves according to some numerical model operator $\mathcal{M}$ from the past to the current time.

$$\mathbf{x}_k = \mathcal{M}_{(k-1)\rightarrow k}(\mathbf{x}_{(k-1)}(\lambda_{k-1}, \gamma_{k-1}))\ , \tag{1}$$

where k denotes the time index. It follows here that a discretization in time takes place within the EnKF DA technique. This model operator transforms the model states from time $k-1$ to time k. λ_k represents the emission parameter and γ_k represents the wind direction parameter for the current time k, with the following distribution:

$$\gamma \sim \mathcal{N}\left(\gamma_n, Q_1\right)\ , \tag{2}$$

$$\lambda \sim \mathcal{N}\left(\lambda_n, Q_2\right)\ , \tag{3}$$

where the mean corresponds to the nominal values γ_n and λ_n of the parameters, and the covariances Q_1 and Q_2 describe the degree of freedom of the magnitude that those to have in the system.

2.2 Stochastic Uncertainty Representation for Wind and Emission Parameters

A stochastic representation of the parameter uncertainty is required to implement the DA method. The emissions employed by the model operator are consequently modeled as a stochastic process with a factor of random variation.

$$\hat{\boldsymbol{e}}_k \ = \ \boldsymbol{e}_k\,(\,1\,+\,\boldsymbol{\gamma}_k\,)\,. \tag{4}$$

Here, $\boldsymbol{e}_k$ is the nominal emission from the emission inventory. The emission deviation is modeled as a colored noise process [10] as follows:

$$\boldsymbol{\gamma}_k \ = \ \alpha_k \boldsymbol{\gamma}_{k-1} + \sigma\sqrt{1-\alpha^2}\ \boldsymbol{w}_\gamma\ , \tag{5}$$

where $\boldsymbol{w}_k$ is a white noise process with zero mean and unity standard deviation:

$$\boldsymbol{w}_\gamma \ \sim \ \mathcal{N}(0,1)\,. \tag{6}$$

The stochastic factors are taken from a normal distribution with zero mean and standard deviation σ. The temporal correlation coefficient $\alpha_k \in [0,1]$ is used to describe the temporal variation, where the value should be set between two extremes: for $\alpha = 0$, the deviation is pure white noise with completely different values for every sample; for $\alpha = 1$ there is no temporal variation at all, and the deviation factor is a single sample out of the normal distribution. In this study, the correlation parameter is described using a temporal length scale τ following [21]:

$$\alpha_k = \exp\left(-|t_k - t_{k-1}|/\tau\right) , \tag{7}$$

where the stochastic model state is formed by augmenting the state vector with a similar model for the wind parameter perturbation. The DA estimation is performed based on a first guess or prior estimate $\mathbf{x}^b \in \Re^{n\times 1}$. In ensemble-based methods, an ensemble of N model realizations and n states is generated:

$$\mathbf{X}_k^b = [\mathbf{x}_k^{b[1]}, \mathbf{x}_k^{b[2]}, .., \mathbf{x}_k^{b[N]}] \in \Re^{n\times N} , \tag{8}$$

which correspond to the different realization of the model generated by perturbing initial conditions assumed to be normally distributed:

$$\mathbf{x} \sim \mathcal{N}(\mathbf{x}^b, \mathbf{B}) . \tag{9}$$

The mean $\overline{\mathbf{x}} = \frac{1}{n}\sum_{i=1}^{n} \mathbf{x}^i$ and the covariance background error matrix are:

$$\mathbf{B} = \frac{1}{n-1}\sum_{i=1}^{n}(\mathbf{x}^i - \overline{\mathbf{x}})(\mathbf{x}^i - \overline{\mathbf{x}})^T \in \Re^{n\times n} . \tag{10}$$

The parameters involved are the emission factor γ and the wind direction λ factor, which has the following distribution for the ensemble method of N model realizations:

$$\Gamma = [\gamma_{[1,1]}, \gamma_{[1,2]}, ..., \gamma_{[1,N]}] , \tag{11}$$

$$\Lambda = [\lambda_{[1,1]}, \lambda_{[1,2]}, ..., \lambda_{[1,N]}] . \tag{12}$$

Each one of the model state, emission factor, and wind deviation matrices ensembles are rest by the mean value of the ensembles:

$$\mathbf{X}_k^b = [\mathbf{x}_k^{b[1]} - \overline{x}, \mathbf{x}_k^{b[2]} - \overline{x}, .., \mathbf{x}_k^{b[N]} - \overline{x}] \in \Re^{n\times N} , \tag{13}$$

$$\Gamma = [\gamma_{[1,1]} - \overline{\gamma}, \gamma_{[1,2]} - \overline{\gamma}, ..., \gamma_{[1,N]} - \overline{\gamma}] , \tag{14}$$

$$\Lambda = [\lambda_{[1,1]} - \overline{\lambda}, \lambda_{[1,2]} - \overline{\lambda}, ..., \lambda_{[1,N]} - \overline{\lambda}] . \tag{15}$$

From these deviation arrays, we create the following augmented vector and calculate the approximate covariance matrix of this augmented system:

$$P_a = \begin{bmatrix} \mathbf{X} \\ \cdots \\ \Gamma \\ \cdots \\ \Lambda \end{bmatrix} . \tag{16}$$

This matrix can be expanded for each one of the components:

$$\begin{bmatrix} \mathbf{x}_k \\ \gamma_k \\ \boldsymbol{\lambda}_k \end{bmatrix} = \begin{bmatrix} \mathcal{M}_{(k-1)\to k}\left(\mathbf{x}_{(k-1)}(\lambda_{k-1}, \gamma_{k-1})\right) \\ \alpha_k\, \gamma_{k-1} \\ \beta_k\, \boldsymbol{\lambda}_{k-1} \end{bmatrix} + \begin{bmatrix} 0 \\ \sigma\sqrt{1-\alpha^2}\boldsymbol{w}_\gamma \\ \eta\sqrt{1-\beta^2}\boldsymbol{w}_\lambda \end{bmatrix} . \tag{17}$$

Using the augmented vector (17), it is possible to estimate both the state and the emission correction factor using a sequential DA approach. The non-linear operator $\mathcal{M}$ propagates the augmented state vector $\boldsymbol{x}$ in time, while the right part of the expression corresponds to the stochastic forcing $\boldsymbol{w}_k$ over the elements of the state. The covariance matrix is calculated as

$$\mathbf{P} = \frac{P_a P_a^T}{N-1} . \tag{18}$$

The observations are also assumed to be normally distributed

$$\mathbf{y} \sim \mathcal{N}\left(\mathbf{H} \cdot \mathbf{x}, \mathbf{R}\right) , \tag{19}$$

where $\mathbf{H} \in \Re^{m \times n}$ is a linear operator that propagates the state space into the observation space, and $R \in \Re^{m \times m}$ is the observation error covariance matrix. In this way, the deviation observation matrix is:

$$\mathbf{Y} = \left[\mathbf{y}_{[1,1]} - \overline{\mathbf{y}}, \mathbf{y}_{[1,2]} - \overline{\mathbf{y}}, ..., \mathbf{y}_{[1,N]} - \overline{\mathbf{y}}\right] . \tag{20}$$

The EnKF computes an optimal variance gain with the expression:

$$\mathbf{K} = \mathbf{P}\mathbf{H}^T [\mathbf{H}\mathbf{P}\mathbf{H}^T + \mathbf{R}]^{-1} , \tag{21}$$

which is applied to each member of the ensemble:

$$\mathbf{x}_i^a = \mathbf{x}^b + [\mathbf{P}^{-1} + \mathbf{H}^T\mathbf{R}^{-1}\mathbf{H}]^{-1}\mathbf{H}^T\mathbf{R}\mathbf{d}, \tag{22}$$

where the innovation $\mathbf{d}$ deviation space is created with the inclusion of noise $v_i \sim N(0, \mathbf{R})$ is added to the observation $y_i = y + v_i$ for each ensemble that is going to be used to perturb the observations before the difference with the correspondent $\mathbf{H}_i\mathbf{x}$:

$$\mathbf{d}_k = [\mathbf{y}_1 - \mathbf{H}_1\mathbf{x}; \mathbf{y}_2 - \mathbf{H}_2\mathbf{x}, ..., \mathbf{y}_n - \mathbf{H}_n\mathbf{x}] . \tag{23}$$

2.3 Stream Function Formulation

Consider the scalar stream function

$$\Psi(x, y) , \tag{24}$$

where

$$u = \frac{\partial \Psi}{\partial y} , \tag{25}$$

$$v = -\frac{\partial \Psi}{\partial x} . \tag{26}$$

This scalar stream function constructs the vector field, satisfying the divergence-free property $\nabla \cdot f = 0$.

$$f(x, y) = \begin{pmatrix} u \\ v \end{pmatrix} \tag{27}$$

The stream function $\Psi(x, y)$ is a scalar function describing fluid flow in a two-dimensional space. The stream function divergence-free property $\nabla \cdot f = 0$ means that the total flow across any closed curve in the fluid domain is zero. This feature is beneficial because it allows physical constraints, such as continuity or conservation laws. The proposed approach involves perturbing the stream function instead of the two wind velocities. The stream function's divergence-free property is characteristic of an auxiliary space in DA used for parameter estimation, which can have some properties that help describe the magnitudes related to the model space. Section 3 provides an illustrative example of this method, where we select a suitable stream function to demonstrate its effectiveness. Moreover, the stream function divergence-free property reduces the dimensionality of the DA problem. The stream function can transform the state vector into a new space with a smaller dimension, reducing computational costs. The stream function can be used as an auxiliary space for DA by transforming the state vector.

3 Results

3.1 Advection-Diffusion Model

Testing an implementation with a simplified model is widely used in the DA field to check the behavior of new techniques designed. This section considers a 2D advection-diffusion model to which synthetic data is fed through data assimilation (DA). The advection-diffusion model for a gas concentration C is described by the following equation in the directions x and y with a constant diffusion coefficients D_x and D_y, and velocities in the two directions U_x and U_y:

$$\frac{\partial C}{\partial t} = -U_x \frac{\partial C}{\partial x} - U_y \frac{\partial C}{\partial y} + D_x \frac{\partial^2 C}{\partial x^2} + D_y \frac{\partial^2 C}{\partial y^2} + \hat{e}_k . \tag{28}$$

Synthetic data in this scenario correspond to the data sampled from the true model states of the advection-diffusion model propagation with different initial conditions. The true model propagation is one propagation of the model that is taken as the reality. This model propagation is perturbed, and then this synthetic data sampled from here is the observation dataset to assimilate. The change in parameters presents a challenge for DA and predictability. The size of this simple test model is 3600 states (60×60 grid) for the concentrations. The emission parameter depends on the number of sources to be updated, and the wind direction parameter the number is two times the number of states because the horizontal wind directions are u and v. In this case, one advantage of the stream function is reducing the number of wind direction parameters in half because once the stream function parameters are estimated, the wind direction is calculated in the model space through the gradient.

For the discretization form, the central step second order finite difference [2] was used to generate the propagating following expression from an initial state. The boundary condition used for solving the experiment was the Dirichlet homogeneous zero or null value fixed in the contour [27].

3.2 Results Advection-Diffusion Model

Figure 1 provides a schematic normally-distributed perturbation for the wind direction along with a comparison of various standard deviation values of the wind direction perturbation for the advection-diffusion model as a grey shade region that describes a set of variations in the direction of the wind from the ideal or true value. In comparison, the different standard deviation values of the wind direction perturbation in the advection-diffusion model Eq. 28 determine the region of uncertainty expanded where the variation in the modeled wind direction occurs. The Figure provides insights into how perturbations in the wind direction can impact the accuracy of the advection-diffusion model. It highlights the importance of accurately estimating wind direction in improving model accuracy. Furthermore, comparing the different standard deviation values of the wind direction perturbation is a useful guide for determining the maximum level of uncertainty or variability that should be considered when modeling wind direction for various applications. The wind fields V_w were homogeneously perturbated in direction assuming that the magnitudes remain constant and equal to $|V_w| = 1$. The noise distribution for generating the distribution is $\mathcal{N}_y \sim \mathcal{N}(\frac{1}{\sqrt{2}}, R)$

$$V\hat{y}_k = V\hat{y}_{k-1} + \boldsymbol{\lambda}_{k-1}\,, \tag{29}$$

$$V\hat{x}_k = \sqrt{1 - V^2\hat{y}_k}\,, \tag{30}$$

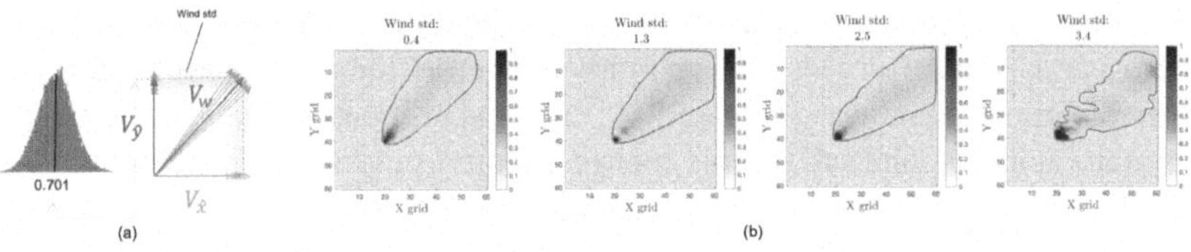

Fig. 1. (a) Schematic perturbation for the wind direction according to the model in Eq. (30) both in one and in two dimensions (b) Comparison for different standard deviation values of the wind direction perturbation for the advection-diffusion model. Colorbar indicates concentration. Different panels correspond to different time steps as indicated.

$$\boldsymbol{\lambda}_k \;=\; \beta_k\, \boldsymbol{\lambda}_{k-1} + \eta\sqrt{1-\beta^2}\, \boldsymbol{w}_{\lambda}. \tag{31}$$

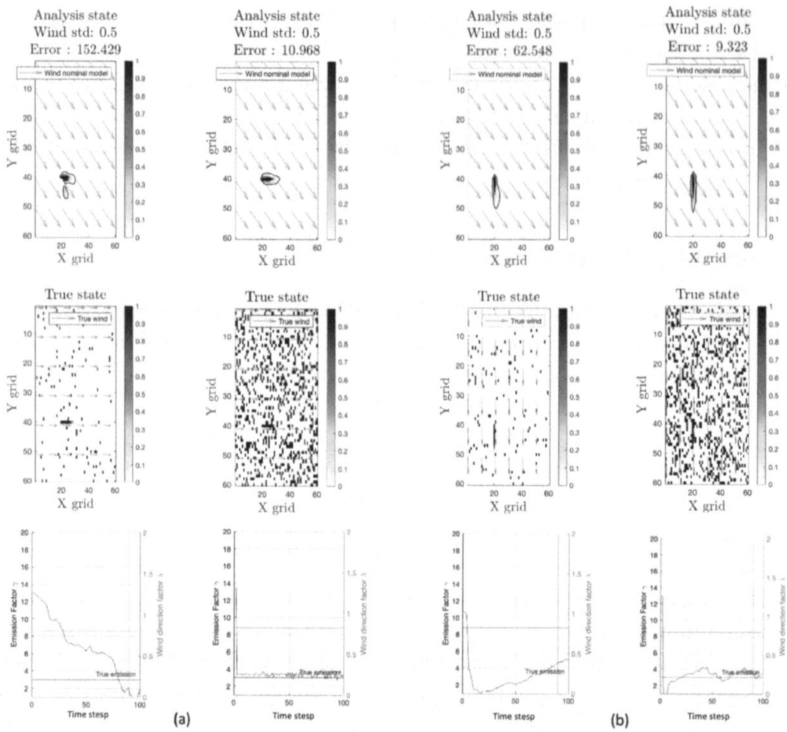

Fig. 2. Panels (a) and (b) show the result of the advection-diffusion emission parameter perturbation and estimation with wind direction not being estimated for the true wind direction pointing to the east (a) and the true wind pointing to the south (b). For each one of these cases, two situations for the dimension of the observation operator $\mathcal{H}$ are shown, one where the number of observations is low and the other where they are high. Colorbars represent the concentration levels. In the last row, the emission and wind direction parameters in time are presented.

The following Figures compare the analysis output between the truth state and three scenarios: emission perturbed, winds perturbed, or emissions and winds perturbed. The number of observations was increased for each experiment to study the discrepancy between the analysis state and the true change. Figure 2 presents the results of an Advection Diffusion emission parameter estimation DA experiment.

The simulation was conducted under two wind directions, one pointing towards the east, and the other towards the south. The top image shows the analysis output, indicating that the wind directions were not updated. Two situations for the dimension of the observation operator H are presented for each case, which has a varying impact on the error reduction. This Figure highlights how fundamental it is to accurately estimate wind directions and how large is the impact of the number of observations on the analysis output.

Figure 3 describes an Advection Diffusion model DA for estimating emission parameters and wind direction. The top image shows the analysis wind field, which results from the DA process. The middle image shows the true wind direction, pointing east in the first case and south in the second case. For each case, two situations with different observation dimensions are shown, where the graphic below shows the estimated parameters. This Figure helps to demonstrate the impact of DA on estimating parameters and wind direction in the Advection Diffusion model. The results suggest that the DA process can improve the accuracy of the estimated parameters, even with fewer observations, and that the true wind direction can be better estimated with more observations.

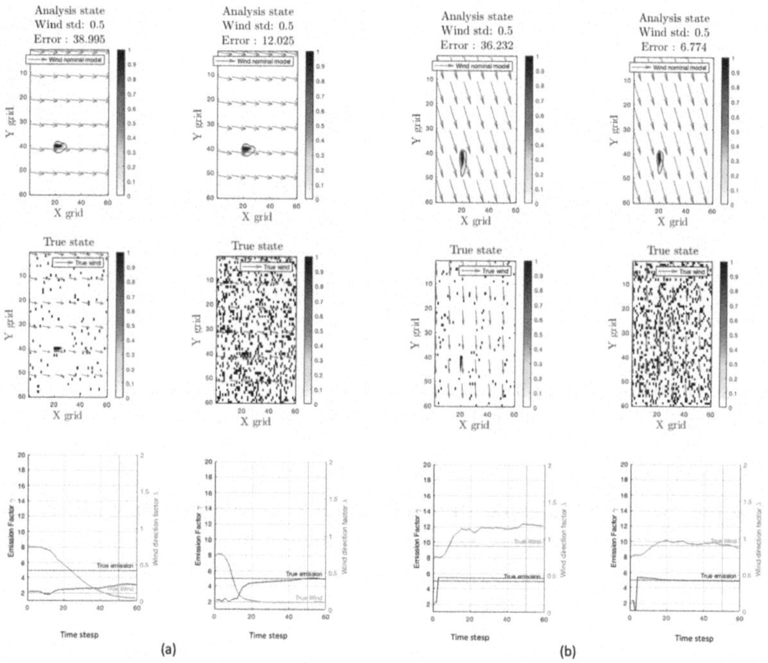

Fig. 3. Panels (a) and (b) show the result of the Advection-diffusion emission parameter perturbation and estimation with wind direction now being estimated for the true wind direction pointing to the east (a) and the true wind pointing to the south. For each one of these cases, two situations for the dimension of the observation operator $\mathcal{H}$ are shown, one where the number of observations is low and the other where they are high. Colorbars represent the concentration levels. In the last row, the emission and wind direction parameters in time are presented.

Figure 4 compares the error between the true state and the analysis state after assimilation and how it relates to the sampled space size (the number of observations as the black pixels in the true state of the images above). The experiment was conducted 50 times under two scenarios: one with a small (2.7%) and another with a large (55%) percent of sampled space size. The Figure shows the density of the group of dots, which confirms the experiment's success in reducing error consistently. The analysis suggests that a larger sample size leads to a more accurate analysis state and lower error. This finding emphasizes the importance of adequately sampling the state space to improve the accuracy of the assimilation process.

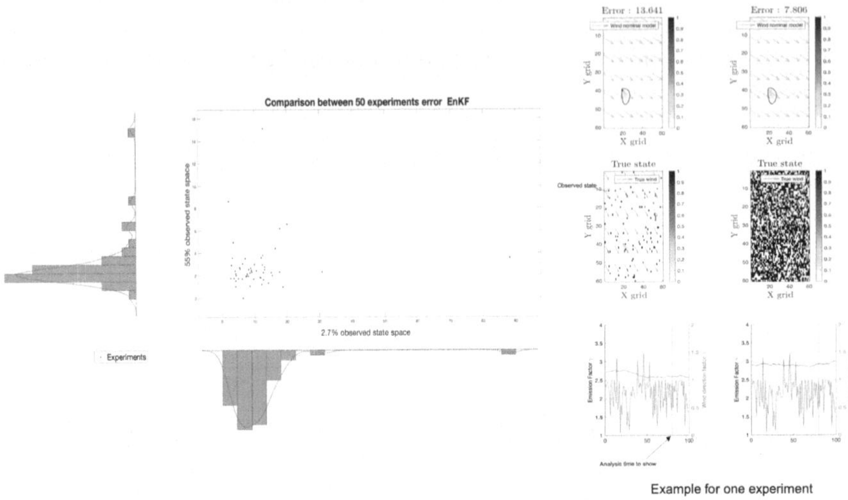

Fig. 4. Statistical comparison for the error for 50 experiments between one scenario with a small (2.7%) and large (55%) size, compared to the state space size, for the observation matrix $\mathcal{H}$.

To test the case with the scalar stream function as an auxiliary space to perform the DA for the parameter estimations, we use the following linear function for a test case.

$$\Psi_b(x, y) = u_b y - v_b x \, , \tag{32}$$

where

$$u_b = \frac{\partial \Psi_b}{\partial y} \, , \tag{33}$$

$$v_b = -\frac{\partial \Psi_b}{\partial x} \, . \tag{34}$$

Figure 5 compares the stream function correction after sequential DA, which involves estimating emission and wind direction parameters. This Figure illustrates the true and analysis wind direction corresponding to the gradient direction, depicted with the red and blue arrows, respectively. Over time, the ends of the arrows align, indicating that the DA process successfully corrected the stream function. This result highlights the potential of DA techniques to improve model accuracy and the importance of accurately estimating emission and wind direction parameters. Overall, the insights from this Figure provide a valuable contribution to the literature on DA and offer guidance to researchers in optimizing their modeling approaches.

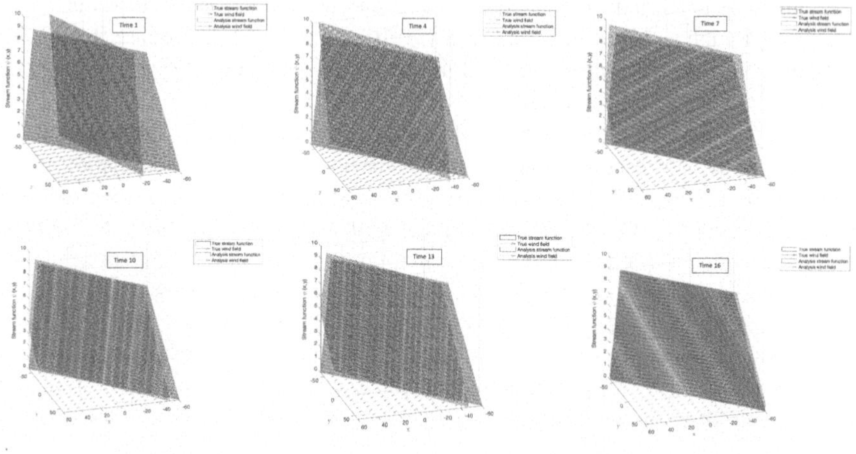

Fig. 5. Comparison of the stream function correction after time through the sequential DA, which estimates emission and wind direction parameters, the true and analysis wind direction correspond to the gradient direction of the stream function, which ends align through time.

4 Discussion

DA techniques have been increasingly incorporated into CTM to improve their accuracy. Most of these models use DA to improve emission estimates by ingesting new data. The next step in our research is to test the stream function DA technique with advection diffusion in a larger model, such as the LOTOS-EUROS, to evaluate its effectiveness. The results of this work can provide valuable insights into the application of DA techniques in different domains and contribute to the ongoing efforts to improve the accuracy of CTMs. Atmospheric trace gasses measurements retrieved from Satellite instruments can perceive different plumes of pollutants emitted from sources and recently different works propose techniques to detect them [22–24], thus estimating the wind direction through the incorporation to the model of concentration plumes can act as a virtual sensor of wind direction.

5 Conclusions

In conclusion, this study presents a novel DA approach that addresses the challenge of estimating emission and wind direction parameters in an advection-diffusion model for improving the accuracy of CTM over long distances. The proposed approach utilizes the ensemble Kalman filter with an augmented state vector to sequentially estimate both parameters while considering additional uncertainties. By perturbing both emission and wind parameters, the proposed approach provides a comprehensive understanding of both parameters' impact on the chemical transport prediction.

The effectiveness of the proposed approach was evaluated using a 2D low-dimensional advection-diffusion model under various sampling observation numbers.

The technique's performance was compared against validation observations using different metrics such as Mean Factorial Bias (MFB) and Root-Mean-Square (RMS). The results indicate that the number of assimilated observations significantly impacts the accuracy of the predicted chemical transport.

One of the significant advantages of the proposed approach is the estimation of the associated stream function that controls the wind fields. The assimilation of this stream function allows for the correction of the wind fields while preserving the physical laws. Considering the uncertainties associated with wind direction, the proposed approach provides a more accurate representation of chemical transport in advection-diffusion models. This study demonstrates the potential of the proposed DA approach for improving the accuracy of CTM in advection-diffusion models. The next crucial step towards advancing our research is introducing spatially variable perturbations to the stream function, thereby generating a diverse range of spatially varying velocity patterns to introduce divergence-free structures on the wind fields to be estimated.

Acknowledgements. Universidad EAFIT and TuDelft University.

References

1. Montoya, O.L.Q., Niño-Ruiz, E.D., Pinel, N.: On the mathematical modeling and data assimilation for air pollution assessment in the Tropical Andes. J. Environ. Sci. Pollut. Res. **27**, 1–20 (2020)
2. Laplace, P.-S., Courant, R.: The finite difference method
3. Park, S., Dash, U., Yu, J., Yumimoto, K., Uno, I., Song, C.: Implementation of an ensemble kalman filter in the community multiscale air quality model (CMAQ model v5.1) for data assimilation of ground-level $PM_{2.5}$. Geoscientific Model Dev. **15**, 2773–2790 (2022). https://gmd.copernicus.org/articles/15/2773/2022/
4. Solazzo, E., Crippa, M., Guizzardi, D., Muntean, M., Choulga, M., Janssens-Maenhout, G.: Uncertainties in the emissions database for global atmospheric research (EDGAR) emission inventory of greenhouse gases. Atmos. Chem. Phys. **21**, 5655–5683 (2021)
5. Bocquet, M., et al.: Data assimilation in atmospheric chemistry models: current status and future prospects for coupled chemistry meteorology models. Atmos. Chem. Phys. **15**, 5325–5358 (2015)
6. Mo, J., Gong, S., He, J., Zhang, L., Ke, H., An, X.: Quantification of SO2 emission variations and the corresponding prediction improvements made by assimilating ground-based observations. Atmosphere **13**, 470 (2022)
7. Sekiyama, T., Kajino, M., Kunii, M.: Ensemble dispersion simulation of a point-source radioactive aerosol using perturbed meteorological fields over eastern Japan. Atmosphere **12**, 662 (2021)
8. Evensen, G., Van Leeuwen, P.: An ensemble Kalman smoother for nonlinear dynamics. Mon. Weather Rev. **128**, 1852–1867 (2000)
9. Evensen, G.: The ensemble Kalman filter: theoretical formulation and practical implementation. Ocean Dyn. **53**, 343–367 (2003)
10. Jazwinski, A.: Stochastic Processes and Filtering Theory. Academic Press, Cambridge (1970)
11. Skoulidou, I., et al.: Changes in power plant NOx emissions over northwest Greece using a data assimilation technique. Atmosphere **12**, 900 (2021)

12. Lopez-Restrepo, S., et al.: An efficient ensemble Kalman Filter implementation via shrinkage covariance matrix estimation: exploiting prior knowledge. Comput. Geosci. **25**, 985–1003 (2021)
13. Menut, L., Bessagnet, B.: What can we expect from data assimilation for air quality forecast? Part I: quantification with academic test cases. J. Atmos. Oceanic Technol. **36**, 269–279 (2019)
14. Hanea, R., Velders, G., Heemink, A.: Data assimilation of ground-level ozone in Europe with a Kalman filter and chemistry transport model. J. Geophys. Res. Atmos. **109** (2004)
15. Lu, S., Lin, H., Heemink, A., Fu, G., Segers, A.: Estimation of volcanic ash emissions using trajectory-based 4D-Var data assimilation. Mon. Weather Rev. **144**, 575–589 (2016)
16. Jin, J., Lin, H., Heemink, A., Segers, A.: Spatially varying parameter estimation for dust emissions using reduced-tangent-linearization 4DVar. Atmos. Environ. **187**, 358–373 (2018)
17. Lopez-Restrepo, S., Yarce, A., Pinel, N., Quintero, O., Segers, A., Heemink, A.: Forecasting PM10 and PM2.5 in the Aburrá Valley (Medellin, Colombia) via EnKF based data assimilation. Atmos. Environ. **232**, 117507 (2020)
18. Milewski, T., Bourqui, M.: Potential of an ensemble Kalman smoother for stratospheric chemical-dynamical data assimilation. Tellus A Dyn. Meteorol. Oceanogr. **65**, 18541 (2013)
19. Evensen, G.: Sequential data assimilation with a nonlinear quasi-geostrophic model using Monte Carlo methods to forecast error statistics. J. Geophys. Res. Oceans **99**, 10143–10162 (1994)
20. Anderson, J., Anderson, S.: A Monte Carlo implementation of the nonlinear filtering problem to produce ensemble assimilations and forecasts. Mon. Weather Rev. **127**(12), 2741–2758 (1999)
21. Barbu, A., Segers, A., Schaap, M., Heemink, A., Builtjes, P.: A multi-component data assimilation experiment directed to sulphur dioxide and sulphate over Europe. Atmos. Environ. **43**, 1622–1631 (2009). https://doi.org/10.1016/j.atmosenv.2008.12.005
22. Finch, D., Palmer, P., Zhang, T.: Automated detection of atmospheric NO 2 plumes from satellite data: a tool to help infer anthropogenic combustion emissions. Atmos. Meas. Tech. **15**, 721–733 (2022)
23. Georgoulias, A., Boersma, K., Van Vliet, J., Zhang, X., Zanis, P., Laat, J.: Detection of NO2 pollution plumes from individual ships with the TROPOMI/S5P satellite sensor. Environ. Res. Lett. **15**, 124037 (2020)
24. Kuhlmann, G., et al.: Detectability of CO 2 emission plumes of cities and power plants with the Copernicus Anthropogenic CO 2 Monitoring (CO2M) mission. Atmos. Measur. Tech. **12**, 6695–6719 (2019)
25. Brandis, A., Centurelli, G., Schmidt, J., Vollmer, L., Djath, B., Dörenkämper, M.: An investigation of spatial wind direction variability and its consideration in engineering models. Wind Energy Sci. **8**, 589–606 (2023)
26. Potts, D., Timmis, R., Ferranti, E., Vande Hey, J.: Identifying and accounting for the Coriolis effect in satellite NO 2 observations and emission estimates. Atmos. Chem. Phys. **23**, 4577–4593 (2023)
27. Cheng, A., Cheng, D.: Heritage and early history of the boundary element method. Eng. Anal. Boundary Elem. **29**, 268–302 (2005)

Impact of Power and CO_2 Emissions on Parking Sustainability: Comparative Analysis Between Automated and Traditional Systems

Felipe Haro(✉), Soheila Antar, and Joaquin Guzman

Noorjax Consulting, Tallinn, Estonia
felipe@noorjax.com

Abstract. Advances in technology and automation are bringing forward vast opportunities for reduction in cost and increase in efficiency in many areas. For example, using autonomous parking systems that rely less on manpower are becoming more popular. However, where some problems are solved, others unfold. This study employs simulations to compare the power consumption and CO_2 emissions between an automated parking system and a traditional one, where both gasoline cars and electrical cars are considered. Results showed that with both types of cars the traditional model uses less energy than the automated model. A performed sensitivity analysis showed that for the automated parking to be more energy efficient, 26% less power should be used by the cars' elevators and 32% by the cars' platforms for the case of gasoline cars. These percentages increase to 43.7% and 55.9% respectively for the case of electrical cars. Different factors combinations for the reduction of CO_2 emission in the automated parking are also presented at the end of the study. The main contribution of the study is a better understanding of the needed efforts to reach an acceptable sustainability level for new automated parking systems.

Keywords: power consumption · C02 emission automated · parking system

1 Introduction

Lack of available parking spaces is a problem that many countries are facing, especially in congested cities in developing countries [1]. In India and Egypt, for instance, 40% of road space is used for parking instead of traffic movement on any average working day [1, 2]. People waste a lot of time searching for available parking spots and this causes traffic congestion [3]. This is why there has been a shift towards utilizing vertical space such as developing multi-level parking systems [1]. Although these systems provide additional parking spots, they still require big land areas due to the need for ramps that consume a lot of space for long slopes. The availability of large sites has become an area of concern, even in developed dense areas. Automatic vertical parking systems, using lifts for transporting cars for example, can aid in solving this problem, since they reduce the need for space to at least 50% when compared to conventional ramp parking [4]. In combination with smart technologies for rapidly allocating available spots [5], these

M. Mujica Mota and P. Scala (Eds.): EUROSIM 2023, CCIS 2032, pp. 44–57, 2024.
https://doi.org/10.1007/978-3-031-68435-7_4

systems, in addition to saving up space can: reduce land and construction cost, reduce waiting times, and require less human intervention, such as patrons [4].

Increased CO2 emissions is considered one of the main factors contributing to global warming [6]. Moreover, the transportation sector is a high contributor to the rise in greenhouse gas emissions. In 1991, 21% of worldwide energy use was attributed to road transport [7], with the major source of energy being fossil fuel [2]. [8] Indicated that 30% of air pollution in road transport is caused by cars searching for spaces to park. Automated parking lot can have a positive impact on the environment by reducing CO2 emissions through reducing mobility from searching for available parking spaces as well as reducing mobility inside parking areas due to the shift to automation. This is countered by the need for a high amount of energy for operating the lift system and any other mechanical components [2]. On the other hand, the use of electrical vehicles is gaining attention in the transportation sector due to their limited CO2 emission [9]. In a future where most of the vehicles on the road become electrical, the above-mentioned tradeoff especially becomes worth investigating.

Previous studies concentrated on finding the amounts of energy consumption and gas emissions related to road transport and on the methods that can be applied to mitigate them. A study by [10] showed that one of the factors that may affect the shift to sustainable transport in Nigeria is the existence of many old cars on the road. [11] Indicated that some of the efforts established in Beijing transportation sector to reduce pollution include driving restriction policies, using electric buses, vehicle emission standards, and eco-driving, among others.

Very few studies have tackled the issue of energy expenditure in parking systems exclusively. [2] Proposed using solar energy to operate the lift in an automatic parking system to reduce CO2 emission. This study provides a sustainability analysis related to the comparison between a traditional car parking used for short stays and a new design of a fully automated car parking that requires no human interaction once the car arrives to the parking.

A business case is used from a private company from Israel, that focuses on the development of automated car parking systems using modern technologies. Included in the analysis we attempt to understand the technological requirements needed in a future that has only electric cars for both automated and traditional underground parking lot. Additionally, comparing the CO2 emissions associated to energy generation at a regional or country level using both types of parking to understand what level of electric generation to CO2 transformation is required as a goal policy in order for automated parking lots to be sustainable in comparison to traditional ones, knowing that currently electric cars are much more efficient to move themselves than automated systems are to move them.

Available energy efficient techniques include using solar and wind energy as a substitute for fossil fuel as proposed by [2] and the employment of novel technologies in traditional power generation systems [6]. [6] Indicated that using integrated gasification combine cycle in coalfired power plants can lead to 45% intensification in thermal efficiency and using combined cycle gas turbine with natural gas can lead to a 60% increase in this efficiency. Utilizing such technologies can help in the reduction of CO2 emissions by up to 50% for each power generation unit.

On this work, we develop simulation models for both car parking structures and we analyze the sustainability impact for each of them.

2 Proposed Simulations

2.1 Simulation of Traditional Car Parking

The base simulation model consists of a subterranean parking building, which has 7 floors, and each floor contains between 15 to 22 parking spaces (depending on the floor) (see Fig. 1). The building has a depth of 35.85 m and has a ramp with an uphill and a downhill track that begins at the entrance of the building and reaches the deepest floor. Cars arrive at the system paying a parking fee, drive to the entrance waiting area where they request entry and are designated a parking space based on the floor with the most unoccupied parking spaces. The model is created in the AnyLogic Software, using the Process Modeling and Road libraries. A hybrid DES (discrete-events simulations) & AB (agent-based) simulation paradigm is used.

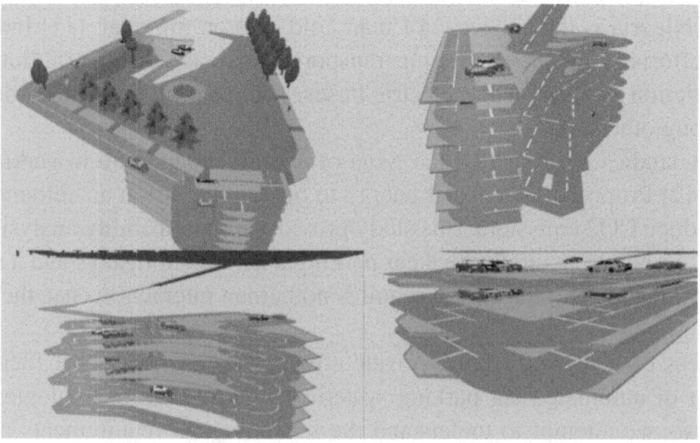

Fig. 1. Baseline car parking system.

There is an elevator that takes car passengers from the parking lot to the exit of the parking lot and vice-versa. It is estimated that one round trip is conducted for each car arriving at the parking lot. A power of 10kW is used for the elevator up and 0kW for the elevator down, since an energy regeneration technology using gravity is employed, which can reduce energy consumption between 20% to 40% [15].

The DES approach is used to simulate the car arrivals to the system, entry to the waiting area, waiting in the queue before entering (See Fig. 2), movement to the parking lot, stopping time in the parking lot (See Fig. 3), and exit from the parking lot (See Fig. 4).

On the other hand, the AB approach is used for the interaction between agents, such as between cars and floors. It is also used to represent the state in which the vehicle is (See Fig. 5), either off or on. In the case of being on, it can be moving or stationary,

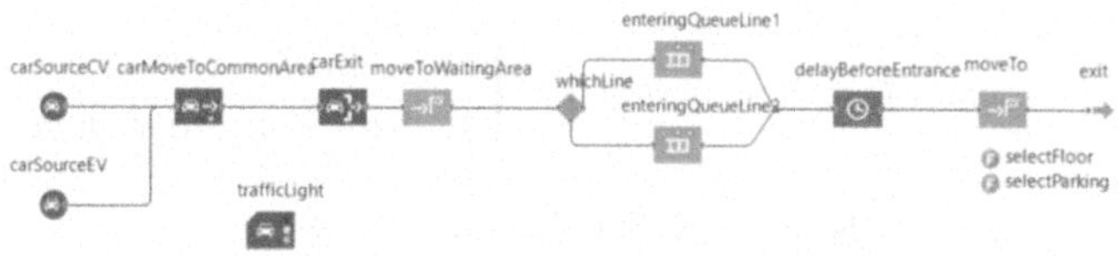

Fig. 2. Arrival at the building and movement to the floor.

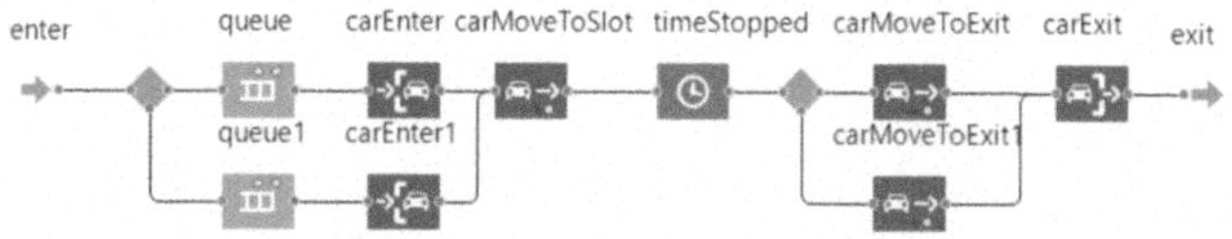

Fig. 3. Entrance to the floor, parked and exit.

Fig. 4. Parking building exit.

and if it is moving it can be on a path with no slope or on a positive or negative slope. The average slope considered in the model is 5° slope. The aforementioned states are used to constantly calculate the energy expenditure of the vehicles, based on the formula presented in the conference paper [12] where the following formula is presented:

$$y = 0.034x_1 + 0.008x_2 \tag{1}$$

where:

$$y : grams\ of\ CO2\ per\ second\,(\frac{g}{s})$$

$$x_1 : speed\ \left(\frac{m}{s}\right)$$

$$x_2 : acceleration\ (\frac{m}{s^2})$$

This is multiplied by another factor depending on whether the vehicle is on a slope or not [13]:

$$y^* = \left(d_1 * f_{5\_Slope}\right) * \left(d_{-1} * f_{-5\_Slope}\right) * y \tag{2}$$

where:

$$y^* : grams\ of\ CO2\ per\ second\ \left(\frac{g}{s}\right) consideringslope$$

$$y : grams\ of\ CO2\ per\ second\,(\frac{g}{s})$$

$$d_1 : dummy\,for\ positive\ slope = \begin{cases} 0, & if\ no\ positive\ slope \\ 1, & if\ it\ has\ positive\ slope \end{cases}$$

$$d_{-1} : dummy\, for\, negative\, slope = \begin{cases} 0, & if\ no\ negative\ slope \\ 1, & if\ it\ has\ negative\ slope \end{cases}$$

$$f_{5\_Slope} : factor\, 5^{\circ}\, positive\, slope$$

$$f_{-5\_Slope} : factor\, 5^{\circ}\, negative\, slope$$

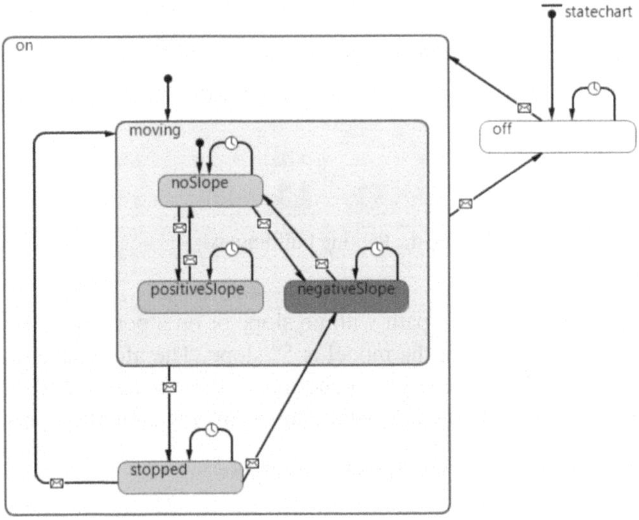

Fig. 5. Cars statechart.

There are two types of cars, which are described below:

Conventional Cars. These vehicles use gasoline for their operation, so their gasoline consumption is calculated every second and, consequently, the emitted CO_2 in grams. The calculation is done using the formula mentioned in [12]. In case where the car is stationary, but with the engine running, the car also consumes gasoline [14], which is 1.8 L/h.

Electric Cars. For electric cars, their energy expenditure is calculated based on the autonomy of a medium-sized electric vehicle which consumes 0.25 kWh for each kilometer traveled [16]. Road gradient factors are also added to this calculation [13]. For these vehicles, it is not considered that they use energy while stopped even if the engine is running.

2.2 Simulation of Automated Car Parking

The proposed model is an automated parking system, which uses an elevator to move the vehicles from the entrance of the parking to any of the underground floors of the parking and vice versa. This works in conjunction with a platform on each floor that moves horizontally. This platform is responsible for moving the car from the elevator to the parking spot designated for the vehicle. All this happens without the need for a driver. Therefore, at the moment of arrival to the entrance of the building, the driver must get out of his car and then the elevator begins the process of parking the vehicle.

The model has a building with similar characteristics to the traditional model, same number of floors, height, number of parking spaces and dimensions. It also has 3 elevators and a platform for each floor, so it does not require a ramp that connects the floors as in the traditional parking model (see Fig. 6).

There is also a common characteristic between this model and the traditional one, which is the selection of the floor and the parking lot. For the floor, the selected one will be the one with the largest current available capacity and then, the car elevator energy will be calculated considering that it only uses energy when going up (people elevator in traditional case). For the selection of the parking lot (See Fig. 7), the car will be in the section of parking with higher capacity in the selected floor (Fig. 8).

```
//select floor based on the floor with max capacity available
Floor selectedFloor=top(floors, floor->floor.currentCapacity);

if (selectedFloor.currentCapacity==0){
    error("No building capacity");
}
INode selectedNode=floorNode.get(selectedFloor);
car.floor=selectedFloor;
selectedFloor.currentCapacity--;

//elevatorEnergy
elevatorEnergy+=general.getElevatorPower()*abs(selectedFloor.levelHeight)
                    /(general.getElevatorSpeed()*3600);
elevatorTrips++;

return selectedNode;
```

Fig. 6. Floor selection algorithm.

```
ParkingLot freeParking=null;
int nFree=0;
SpecialParking specialParking=null;
for (SpecialParking sParking: floor.parkingColl.values()){
    if (nFree<sParking.getNumberOfFreeLots()){
        nFree=sParking.getNumberOfFreeLots();
        specialParking=sParking;
    }
}
car.parking=specialParking.getParking();
specialParking.setNumberOfFreeLots(specialParking.getNumberOfFreeLots()-1);

if (car.parking==null){
    error("no parking available");
}
```

Fig. 7. Parking lot selection algorithm.

The same calculation of energy expenditure and CO_2 for the vehicles is used as in the traditional model, however, for the movement of the vehicle elevator, a power of 15kW is considered for the ascent and 0kW for the descent (given by gravity). The platform moving inside each floor is considered to use 5kW of power in all its paths. For the energy calculation of these components, the following formula was used:

$$E = P * \frac{d}{v} \tag{3}$$

where:

$$E = energy\ (kWh)$$

$$P = component\ power\ (kW)$$

$$d = distance\ traveled\ (km)$$

$$v = component\ speed\ \left(\frac{km}{h}\right)$$

Fig. 8. Automated car parking system

3 Results

Twenty replications were performed for each scenario, with a total of 200 vehicles arriving at the system. The simulation run ends when the last of the 200 vehicles leaves the system. An alpha of 5% is considered as the level of significance.

3.1 Gasoline Cars Scenario

A t-test was performed to compare means with different variance. To compare both models, it is necessary to transform the grams of CO_2 of the vehicles into kWh to have the same unit, considering that all other components are measured in kWh. The results show that the mean of both groups differed by 10.28 kWh. This difference is significant at any significance level, due to the p-value being 5,67E-23, so the traditional model uses less energy than the automated model (see Table 1).

Table 1. t-test System Energy Output Gasoline Cars Scenario.

	Traditional	*Automated*
Mean (kWh)	66,56	76,84
Variance	0,09	1,00
Observations	20	20
Hypothesized Mean Difference (kWh)	0	
p-value two-tail	5,67E-23	

3.2 Electric Cars Scenario

A t-test was performed to compare means with different variance. The results show that the mean of both groups differed by 17.63 kWh. This difference is significant at any significance level, since p-value is 7,01E-24, so the traditional model uses less energy than the automated model (see Table 2).

Table 2. t-test System Energy Output Electric Cars Scenario.

	Traditional	*Automated*
Mean (kWh)	59,61	77,24
Variance	0,14	2,06
Observations	20	20
Hypothesized Mean Difference (kWh)	0	
p-value two-tail	7,01E-24	

4 Sensitivity Analysis

Each country that produces electricity, needs to produce CO_2 to make that electricity available to the general population. This means that to produce x amount of energy (kWh), y amount of CO_2 needs to be used. Each country has an average depending on the power plants they use and how much renewable energy they use. In order to understand the implications of this transformation we need to do a sensitivity analysis to see what level of efficiency in the automated systems is needed in order to be sustainable under the current energy to CO_2 transformation, or what level of improvement in the power plants is needed to generate electricity in a more sustainable way in order to make automated systems more sustainable.

4.1 Gasoline Cars

Car Elevator Power. For the automatic parking to be more energy efficient, the car lift technology needs to use a maximum of 11.16 kW of power. This is considering a scenario where there are only gasoline cars, where there is a conversion of 387.8 g/kWh [17].

This corresponds to 26% less power, compared to the base case (15kW) (see Fig. 9).

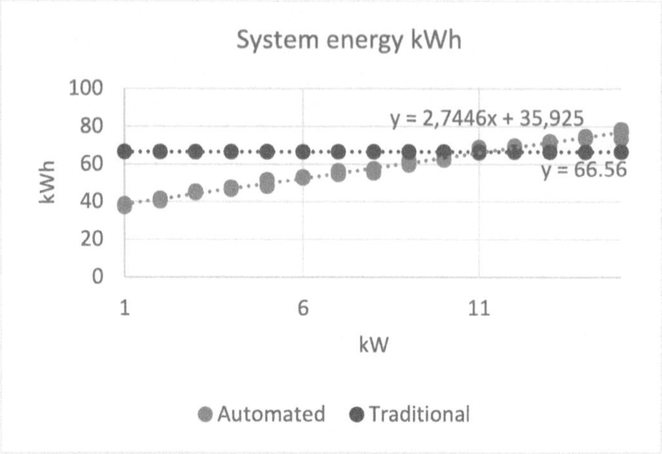

Fig. 9. Car elevator power for gasoline cars comparison.

Car Platform Power. In another scenario, for automatic parking to be more energy-efficient, the horizontally moving platform needs to use a maximum of 3.40 kW of power. This is considering a scenario where there are only gasoline cars, where there is a conversion of 387.8 [g/kWh].

This would be a 32% decrease compared to the value used in the base case (5 kW) (See Fig. 10).

Energy Efficiency Production. For the comparison of the models with 100% gasoline-powered cars, for the automated parking model to be more sustainable, the energy source would need to produce energy more efficiently. Less than 317.05 g of CO_2 would need to be produced for every kWh of energy supplied to the system. This would require a commitment from the government or individuals to create a system that supplies a higher percentage of energy from renewable energy sources (See Fig. 11).

4.2 Electric Cars

Car Elevator Power. When it comes to a system consisting only of electric cars, the decrease in power use for the automated model to consume less energy is larger. The power of the car elevator is required to decrease to 8.43 kW on average for the automated model to consume less energy.

This corresponds to a 43.7% decrease compared to the power used in the base case (15kW) (See Fig. 12).

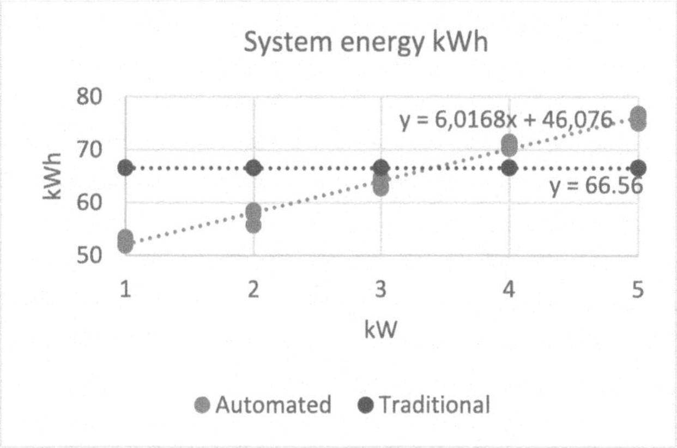

Fig. 10. Car platform power for gasoline cars comparison.

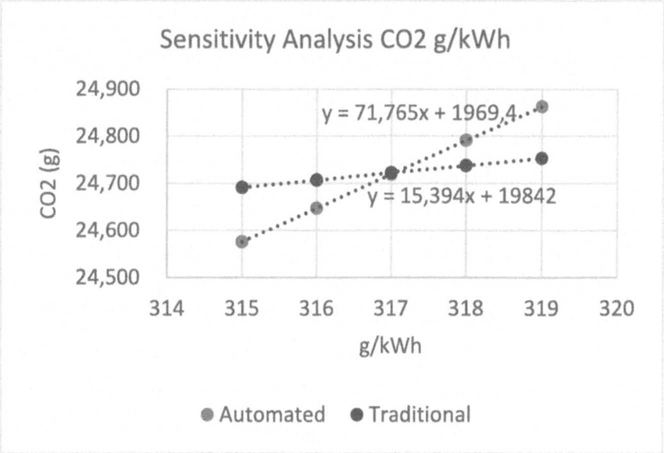

Fig. 11. Energy efficiency production comparison.

Car Platform Power. For the automated model to use less energy than the traditional model, it is necessary to decrease the power of the motor that moves the horizontal platform to 2.20 kW. This represents a 55.9% decrease in power compared to the base case (5kW) (See Fig. 13).

4.3 Carbon Dioxide, Energy and Energy Efficiency

In this experiment we seek to investigate what would be the energy efficiency necessary for the automated parking model to produce a lower amount of CO_2. Therefore, the automated car model was used, with gasoline-powered vehicles.

The response variable is the grams of CO_2 emitted by the system, while the independent variables are energy efficiency, which is the fraction of power efficiency that the

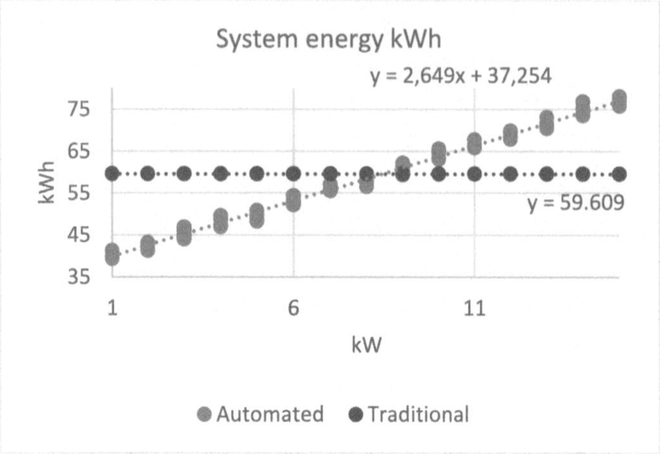

Fig. 12. Car elevator power for electric cars comparison.

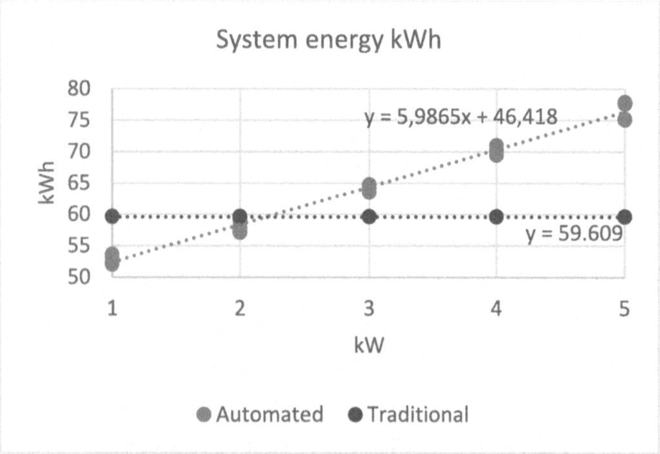

Fig. 13. Car platform power for electric cars comparison.

platform and elevator need to improve overall with respect to the baseline model, and grams of CO_2/kWh.

Considering that the traditional parking model keeps its variables constant (CO_2/kWh and energy efficiency) and uses 25,811.93g of CO_2 in average, there are different combinations that would make the automated system produce fewer grams of CO_2. (See Fig. 14).

As can be seen (See Fig. 15), the feasibility region is given by the combinations of energy efficiency and CO_2/kWh plotted. For example, if CO_2/kWh were 400 g/kWh, an energy efficiency of at least 25% would be required. On the other hand, if CO_2/kWh were 310 g/kWh, there would be no need to improve the energy efficiency of the automated model components. These combinations would result in lower CO_2 emissions compared to the traditional parking model.

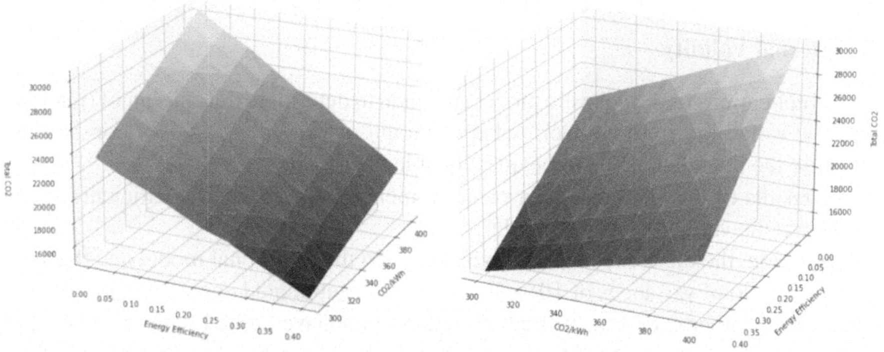

Fig. 14. CO_2 production based on Energy efficiency and CO_2/kWh (Energy Efficiency & CO_2 Perspective).

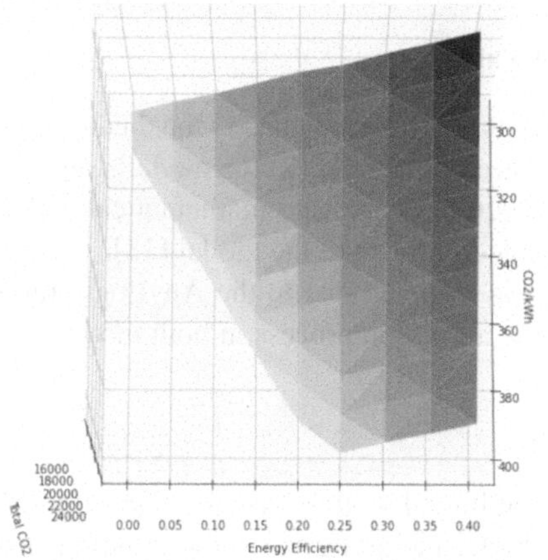

Fig. 15. CO_2 production based on Energy efficiency and CO_2/kWh (Top perspective).

5 Validation

To validate this study, its necessary to mention that all authors declare that they have no conflicts of interest.

The following validation techniques were used to validate the simulation models.

5.1 Face Validity

This is a subjective test of validation, where individuals knowledgeable about the system are asked whether the model and/or its behavior are reasonable [18].

The three contributors to the article cautiously checked that the animations in both 2D and 3D were the same as expected according to the pre-defined algorithms and requirements of the system.

5.2 Internal Validity

Several replications of a stochastic model are made to determine the amount of (internal) stochastic variability in the model. A large amount of variability (lack of consistency) may cause the model's results to be questionable [18].

The results of the models have a variance-mean ratio (variation coefficient) of less than 3%, so the variability is negligible.

In addition, the difference in variability between the traditional model and the automated one (ratio of variances of 10) is because in the automated one, parking spots may be blocked by other cars. For example, a vehicle may have parked with another car in front of it, so if the car wants to leave, the platform will have to relocate the one in front and once that is done, it will finally be able to remove the vehicle that wanted to leave. This situation generates an increase in the energy used by the platform, which is why the variance of energy used is sensitive to the number of relocations required.

5.3 Conceptual Model Validation

Conceptual model validity is determining that (1) the theories and assumptions underlying the conceptual model are correct and (2) the model's representation of the problem entity and the model's structure, logic, and mathematical and causal relationships are "reasonable" for the intended purpose of the model [18].

Independence of data is assumed due to the AnyLogic random number generator. And vehicle arrivals follow a Poisson process in both models.

5.4 Structural Validity

Both the automated and traditional models have the same algorithms for selecting floors and parking for each vehicle, including the same building dimensions, and building entrance. These algorithms were defined by the authors arbitrarily and are the same in both models to have a sense of fairness in the comparisons.

6 Conclusion

The aim of this study was to understand the impact of shifting to the use of automated parking systems on the environment through studying the power consumption and C02 emission in these parkings and comparing the results to those of traditional parkings. The results show, indeed, an increase in power consumption and CO_2 emission in the automated parking system, but this can be improved with better power generation policies at a governmental level.

The results of the study open the door for future sustainable considerations for the design and implementation of such systems. Future studies can expand on this one by investigating methods to decrease energy consumption by taking advantage of the potential benefits of automated parking by limiting the energy use of the technologies. Some of the possible future studies could focus on how to add energy sources to an automated system to make it 100% self-sufficient, or also on a cost-benefit analysis of each model. Another study could be focused on the creation of a semi-automated car parking system in contrast to the alternatives already studied.

References

1. Jog, Y., Sajeev, A., Vidwans, S., Mallick, C.: Understanding smart and automated parking technology. Int. J. u-and e-Serv. Sci. Technol. **8**(2), 251–262 (2015)
2. Elhenawy, Y., Hafez, G., Abdel-Hamid, S., Elbany, M.: Prediction and assessment of automated lifting system performance for multi-storey parking lots powered by solar energy. J. Clean. Prod. **266**, 121859 (2020)
3. Shingare, J., Sarnaik, S., Sangle, Y., Mahindrakar, T., Totla, N.: Design and Development of 4 Level Car Parking Systems (2018)
4. Beebe, R.S., Planning, T.: Automated parking status in the United States–Advantages and Criteria. Parking and transportation planning consulting engineers Group Inc. USA (2001)
5. Kumar, R.S., Surya, B.: Smart parking application using deep learning framework. Int. J. Sci. Res. Netw. Secur. Commun. **8**(4), 1–5 (2020)
6. Al-Ghussain, L.: Global warming: review on driving forces and mitigation. Environ. Prog. Sustainable Energy **38**(1), 13–21 (2019)
7. Paravantis, J.A., Georgakellos, D.A.: Trends in energy consumption and carbon dioxide emissions of passenger cars and buses. Technol. Forecast. Soc. Chang. **74**(5), 682–707 (2007)
8. Wang, F., Liu, Y.: Sweden: Mechanical Parking System. Blekinge institute of Technology Karlskrona, Department of Mechanical Engineering. J. **12**, 14–18 (2017)
9. Weiss, M., Cloos, K.C., Helmers, E.: Energy efficiency trade-offs in small to large electric vehicles. Environ. Sci. Eur. **32**(1), 1–17 (2020)
10. Maduekwe, M., Akpan, U., Isihak, S.: Road transport energy consumption and vehicular emissions in Lagos, Nigeria: an application of the LEAP model. Transp. Res. Interdisc. Perspect. **6**, 100172 (2020)
11. Sun, L., et al.: Reducing energy consumption and pollution in the urban transportation sector: a review of policies and regulations in Beijing. J. Clean. Prod. **285**, 125339 (2021)
12. Oduro, S.D., Metia, S., Duc, H., Ha, Q.P.: Vehicular emission statistical analysis with instantaneous speed and acceleration as predictor variables. In: 2013 International Conference on Control, Automation and Information Sciences (ICCAIS) (2013)
13. National Renewable Energy Laboratory: Contribution of Road Grade to the Energy Use of Modern Automobiles Across Large Datasets of Real-World Drive Cycles (2014)
14. Canada, N.R.: Government of Canada. Natural Resources Canada. https://natural-resources.canada.ca/energy-efficiency/transportation-alternative-fuels/personal-vehicles/fuel-efficient-driving-techniques/21038, last Accessed 14 Feb 2023
15. Al-Kodmany, K.: Tall Buildings and Elevators: A Review of Recent Technological Advances. Buildings (2015)
16. Electric Vehicle Database. https://ev-database.org/car/1471/Kia-EV6-GT, last Accessed 16 Feb 2023
17. U.S. Energy Information Administration: How much carbon dioxide is produced per kilowatthour of U.S. electricity generation? https://www.eia.gov/tools/faqs/faq.php?id=74&t=11#:~:text=This%20equaled%20about%200.855%20pounds,efficiency%20of%20electric%20power%20plants., last Accessed 17 Feb 2023
18. Sargent R.G.: Verification and validation of simulation models. In: Proceedings of the 2010 Winter Simulation Conference, Baltimore, MD, USA (2010)

Sustainable Simulation-Based Digital Twin: An Application Portfolio Management Tool to Minimize Digital Waste and Maximize Useful Life

Antonio Padovano[1] (✉), Francesco Longo[1], Antonio Cimino[2], Martina Cardamone[1], Chiara Sammarco[1], and Pierpaolo Veltri[1]

[1] Modeling & Simulation Center – Laboratory of Enterprise Solutions (MSC-LES), Department of Mechanical, Energy and Management Engineering (DIMEG), University of Calabria, 87036 Arcavacata Di Rende, CS, Italy
antonio.padovano@unical.it
[2] University of Salento, 73100 Lecce, Italy

Abstract. The use of software in industry has become increasingly relevant, although its impact from an environmental perspective is often underestimated because of its intangible nature. Therefore, the need to make technology more sustainable is becoming urgent. This paper investigates this question from a sustainable digital twin technology perspective and points out the importance of analyzing every source of digital waste of a simulation-based digital twin (SBDT). To this end, we adopt Application Portfolio Management (APM) principles, including control charts and TIME analysis, to ensure that the digital twin is always aligned with the real system. We discuss the importance of tracking over time the performance of each Digital Twin component that estimates the Remaining Useful Life in terms of value through an APM dashboard. A proof of concept of a Digital Twin Portfolio Management tool dashboard is developed for monitoring the value as well as the cost-benefit of every component of a SBDT of a smart brewing system.

Keywords: Simulation-based Digital twin · Application Portfolio Management · Sustainable Technology · Lean Digital Waste · Accreditation

1 Introduction

Nowadays, we must address an urgent twofold imperative: we must use technology to make our society and economy more sustainable (*sustainability by technology*) but we should also make technology itself more sustainable (*sustainability in technology*) [1]. This must be done in a responsible way, developing waste-free projects, using resources efficiently, building in proper governance, and building trust [1]. On the surface, Information and Communication Technology (ICT) has a positive impact on the environment as it enables remote working and increases process efficiency. On the other side, the

M. Mujica Mota and P. Scala (Eds.): EUROSIM 2023, CCIS 2032, pp. 58–69, 2024.
https://doi.org/10.1007/978-3-031-68435-7_5

ICT industry contributes to a large percentage of overall energy consumption, CO2 emissions and e-waste. It is well-known that the global IT industry generates as much CO2 as the aviation industry (roughly 2% of the world's global emissions [2]). Emissions from within the IT sector come 50% from the manufacturing of IT equipment and the rest from energy expelled from equipment and data centers [3]. With exponential technologies on the rise and increased relevance of software in industry, the need for more computing power, storage, and energy will continue to surge. Emerging exponential technologies such as simulation, internet of things (IoT), artificial intelligence (AI), augmented reality (AR) and virtual reality (VR), drones, 3D printing, digital manufacturing, and robotic process automation (RPA) increasingly require more computing and storage power through cloud computing.

Sustainability in IT pertains to anything you use internally to run your enterprise, such as owning data centers, managing cloud providers, developing software, managing the life cycle of your hardware, optimizing your energy consumption, or managing your e-waste. This paper investigates this question from a sustainable digital twin technology perspective and points out the importance of analyzing every sources of digital waste of a simulation-based digital twin (SBDT). To this end, we adopt Application Portfolio Management (APM) principles, including control charts and TIME analysis, to ensure that the digital twin is always aligned with the real system. We discuss the importance of tracking over time the performance of each DT component that estimates the Remaining Useful Life in terms of value through an APM dashboard. A proof of concept of a Digital Twin Portfolio Management tool dashboard is developed for monitoring the value as well as the cost-benefit of every component of a SBDT of a smart brewing system.

2 Theoretical Background

2.1 Simulation-Based Digital Twins

Simulation models have been largely used for decision support in production in recent decades [4]. With the growing need for faster decisions and considering the evolution of production processes towards Industry 4.0, simulation has evolved over the last few years towards the DT era [5], thus becoming simulation-based digital twins (SBDTs) [6]. A DT is not a static model, but rather a dynamic, self-evolving entity that reflects the behavior and performance of the physical system it represents [7]. In other words, it is updated with real-time data and new information about the physical system it represents. This data allows the DT to simulate and predict the behavior of the physical system in real-time [8], much like a living organism responds to its environment. Moreover, the term 'living' can also refer to the fact that a DT has the potential to adapt, learn and evolve over time.

In this context, Tao and Zhang [9] report that a DT has four main elements: (i) physical system, (ii) virtual model, (iii) data about both physical and virtual environments, and (iv) integration between physical and virtual environments through their data, allowing synchronism between them. Using simulation as a DT implies several characteristics that differentiate it from traditional approaches. The main difference between a traditional simulation model and a SBDT is its ability to update over time according to changes in physical environments, acting as a virtual mirror of the processes [10]. The connection and synchronization between the digital and physical system are allowed by collecting and sharing data through sensors, smart devices, management systems, databases, among others [11]. This way, a SBDT becomes a daily and efficient tool for decision-making always aligned with the behavior of physical systems. SBDT-enabled decision support tools range from completely autonomous models that operate in real-time to models that only suggest decisions, that is a non-autonomous approach [12]. With the use of advanced data analytics and physical/data-driven simulation methods, a DT allows to support decision-making and to optimize the performance of the physical system it represents [13].

2.2 Sustainable IT and Application Portfolio Management

Although the Fourth Industrial Revolution (4IR) brings positive outcomes for the environment, the increased need for digital services translates into new data centers, computing power, storage, network equipment, and end user IT equipment that are significant cause of greenhouse gases. Today, technology is far from sustainable. Sustainable IT is the production, management, use and disposal of IT to meet our present needs without compromising the ability of future generations to meet their own needs [2, 14]. Therefore, it pertains to anything an organization uses internally to run the processes, such as owning data centers, managing cloud providers, developing software, managing the life cycle of hardware, optimizing energy consumption, or managing e-waste. If a company A intends to implement a SBDT, 3 categories of emissions should be considered, according to the Sustainable IT Greenhouse Gas Emission Taxonomy© [2]:

- direct emissions from owned and controlled sources such as data centers (local and global), enterprise networks, and operations activities;
- indirect emissions that other companies have generated to supply company A with energy, such as electricity, heating, and cooling for their data centers, networks, servers, and end user IT equipment;
- indirect emissions along the entire value chain of company A, which occur outside the organization, such as cloud consumption with Infrastructure as a Service (IaaS), Platform as a Service (PaaS), and Software as a Service (SaaS), ICT equipment such as user devices (computers, smartphones, and tablets), printers, access networks, data centers, enterprise networks, and operator activities.

In this framework, company A will be responsible for sustainability *in* the enterprise, according to the sustainable IT reference model, illustrated in Fig. 1. This means that for a SBDT ecosystem, it would be responsible for IT/OT (Information Technology - Operational Technology) hardware, what they place in the data centers and the cloud, such as hundreds of applications, custom-developed software, data, and information. However, managing the IT/OT ecosystem of a SBDT might be a complex endeavor, from prioritizing hundreds or thousands of applications to the solution architecture and managing the efficiency and effectiveness of a single application. One of the solutions that a Sustainable IT Enterprise must implement is Application Portfolio Management (APM). APM is the process of managing an organization's collection of software applications to ensure they deliver maximum value to the organization throughout their lifecycle.

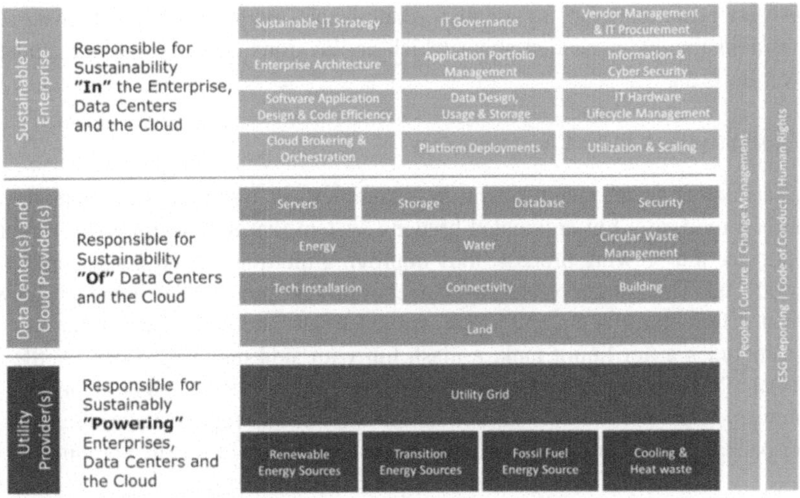

Fig. 1. Sustainable IT reference model [2]. Application Portfolio Management is one of the solutions to be implemented by a Sustainable IT Enterprise

APM includes a range of activities to assess the current technology capabilities, identify potential technology investments to maintain the SBDT, and develop a strategy to allocate resources in the most effective way to achieve the company's goals:

1. *Portfolio assessment and application performance monitoring*, taking stock of software applications, tracking and analyzing their performance and value;
2. *Application rationalization*: identifying redundancies and inefficiencies, and making decisions/adjustments as needed about which applications to retire, replace, update or invest in further;

There are already APM solutions available that provide real-time insights on the performance of apps, infrastructure, and user experience. However, they do not yet include specific functionalities for SBDTs. This study aims at filling this gap by providing a methodology to implement APM for SBDT.

3 Methodology

This study adopts an APM approach to develop a strategy to assess the current technology capabilities of SBDTs, identify potential technology investments to maintain/upgrade the SBDT, and develop a strategy to allocate resources in the most effective way to achieve the company's goals. The present work is based on the following steps:

1. Definition of the case study and scope;
2. Conceptualization and design of an APM tool for SBDTs;
3. Portfolio inventory, to audit all the technology assets used in a SBDT;
4. Portfolio assessment, to assess the performance/value of the technology assets and applications;

5. Portfolio rationalization, to identify redundancies/inefficiencies, and make decisions/adjustments as needed about which applications to retire, replace, update or invest in further.

3.1 Case Study: A Smart Nanobrewery

The case study considered in this work is a nanobrewery system. A smart nanobrewery system is a modern, semi-automated brewing system that uses technology to monitor and optimize the brewing process, thus improving the quality and consistency of the beer produced. The system produces 1 barrel of blonde beer per batch and sits between a dedicated homebrew setup and a professional small-batch commercial brewery. The system includes a hot liquor tank, a mash tun with side manway, a brew kettle with tangential inlet and conical bottom, a temperature-control system and pumping systems. The brewery setup also includes an area for the entry and storage of materials and goods, one for packaging and labeling and another one for storage of finished products.

3.2 Scope Definition: A Simulation-Based Digital Twin of the Nanobrewery

A SBDT of a nanobrewery would typically consists of four components:

- equipment, machinery, sensors and actuators. This would include: smart sensors for measuring various physical parameters such as temperature, pressure, flow rate, pH level, dissolved oxygen level; actuators, such as valves, pumps and motors; controllers to automate various processes such as mashing, boiling and fermentation; communication devices, such as Wi-Fi modules to transmit sensor data or Bluetooth modules for remote monitoring and control.
- the virtual models that include computer-generated 3D models of physical components in the system (e.g. tanks, piping, and other components), as well as models of the brewing process itself. These models can be used to design and optimize the layout of the brewery equipment, as well as to simulate the behavior of the system under different conditions. For example, a CAD model of the mash tun could be used to determine the optimal temperature and mixing conditions for the mash, as well as to identify potential areas of heat loss or inefficiency. Simulation models could be used to predict the behavior of the system under different operating conditions, such as changes in temperature or flow rate. For example, a simulation model of the brewing process could be used to predict the final alcohol content and flavor profile of the beer based on the ingredients and process parameters used.
- real-time data from the physical system collected through sensors, smart devices, management systems, and databases and integrated into the virtual models, allowing for synchronization between the digital and physical systems. Examples are: temperature data, flow rate, pressure data, product data through barcode or RFID, production data, inventory data, etc.
- integration and decision support tools for process optimization, predictive maintenance and risk management. The SBDT can simulate various scenarios and provide insights and recommendations to improve the performance and efficiency of the physical system.

3.3 Portfolio Inventory: Define and Audit the SBDT Technology Components

Based on the scope and case study, the first step of the study is to define the components of the SBDT technology portfolio. These components must be carefully monitored to track their performance as any deviation from optimal functioning might reduce consistently the value (and therefore the lifetime) of the SBDT itself. For the purposes of this work, we consider the following components:

- Data Acquisition: This component is responsible for acquiring data from the physical brewing system. This could include data from sensors, such as temperature and pressure sensors, or data from manual input, such as ingredient quantities.
- Analytics: This component is responsible for analyzing the data that has been acquired and stored. This could include statistical analysis, machine learning, or other analytical techniques.
- Control: This component is responsible for controlling the virtual brewing system based on the data that has been acquired and analyzed. This could include adjusting temperature or pressure settings, or adding ingredients to the virtual brewing system.
- Simulation: This component is responsible for simulating the behavior of the virtual brewing system based on the data that has been acquired and analyzed. This allows the user to see how changes to the physical brewing system might affect the final product.
- Optimization: This component is responsible for optimizing the virtual brewing system based on the data that has been acquired and analyzed. This could include adjusting ingredient quantities or brewing times to achieve a desired flavor profile or quality.

3.4 Portfolio Assessment: Track the Application Performance and Value

In order to keep track of the sustainable performance of the SBDT components over time, it must be clarified the type of waste that can be generated.

E-waste and Digital Waste: The Case of Simulation-Based Digital Twins
SBDT fully embodies the IT-OT convergence by bridging the gap between the physical and digital worlds. For this reason, the SBDT trend undermines the movement to zero-waste economy due to two types of waste: e-waste, digital waste. In the context of this study, waste refers to all those assets (hardware and software) whose value decreases due to a failure to exercise proper care throughout the lifecycle or sound judgment in managing them. Therefore, the first category of waste is the new generation of energy-intensive computing resources and hardware components, such as servers or high-performance computing clusters, IoT devices, sensors, etc., that are needed to run a SBDT and whose future disposal will have detrimental impacts on the environment (thus representing electronic waste, or *e-waste*). The second category of waste is related to the IT nature of SBDTs, such as redundant and/or unnecessary data that is collected, managed, transmitted or stored, or the unnecessary delivery of low-value applications or services. Following the work of Bell [15], who adapted the traditional seven wastes from lean management to IT (coining the *Lean IT* concept), Romero et al. [16] advise on the emergence of a new frontier in lean practices applied to digital manufacturing. In particular, they warn about the risk of losing digital opportunities or abusing digital capabilities, thus defining the

concept of *digital waste* as any non-value adding digital activity. Digital waste produced during SBDT's functioning and maintenance has an indirect environmental impact. For example, running SBDTs implies continuous calculations, data processing and generation, which can lead to increased energy consumption and greenhouse gas emissions from the operation and maintenance of data centers and other IT infrastructure, which are often powered by non-renewable energy sources. Furthermore, SBDT maintenance is a critical point and has also an impact. "Live fast, die young" is the common destiny of any simulation model, which usually serves a specific purpose (e.g. decision problem) after a one-time validation during its building phase [17] and is usually dismissed once the decision has been made, thus representing digital waste. As the SBDT needs to be continuously monitored and updated to ensure it accurately represents the physical system it is modeling, dynamic and online accreditation models are required for SBDTs to extend the SBDT's lifetime [18]. What is new in SBDT is that accreditation methods must be continuously run to ensure the correctness of the model and decisions. However, the majority of accreditation methods is not applicable to simulation models during their operation as DTs [11]. To ensure accreditation of DT models, it is necessary to compare its results with the physical environment periodically [8, 19]. To this end, control charts are one of the main statistical process control techniques, helpful in monitoring output variables in systems under undesirable sources of variability and, consequently, the need for corrective actions [20, 21] if the process is not operating as expected [22]. An inefficient functioning of the DT might lead to negative environmental impacts such as increased energy consumption, e-waste from hardware disposal, and carbon emissions from data center operations. Eventually, if the system is no longer able to accurately represent the physical system it is designed to replicate, it might become digital waste (or 'die') in the sense that it can become obsolete or irrelevant. This could happen, e.g. if the physical system itself undergoes significant changes that are not reflected in the SBDT, if the data sources that the SBDT relies on become outdated or inaccurate, or if the SBDT components fail to mirror the behavior of the physical system.

3.5 Portfolio Rationalization: Maintaining/Upgrading the SBDT

In order to rationalize the SBDT portfolio, an APM dashboard will be conceptualized, designed and developed. The APM tool will serve as a secondary digital twin, also known as a 'meta-model' that monitors and analyzes the performance of the primary digital twin (see Fig. 2).

The meta-system (see the green circle in the Fig. 2) would use the same inputs (from the physical entity) and outputs (produced by the services) as the primary digital twin, and constantly compare the results of the primary digital twin (see the blue circle) with the actual performance of the physical system, to identify any discrepancies or deviations. By analyzing the performance of the primary digital twin, the meta-model could identify waste, i.e. areas where the digital twin is not accurate, where data quality is poor, or where the assumptions used in the model may be incorrect. This information could then be used to improve the accuracy and efficiency of the primary digital twin.

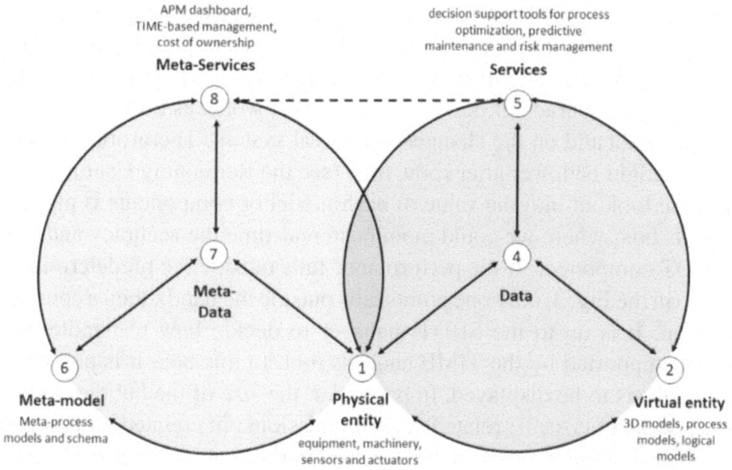

Fig. 2. Conceptual architecture of a Digital Twin meta-model and APM tool

The APM tool includes a TIME analysis, a methodology that helps SBDT managers to identify transformation alternatives and make data-driven decisions [2]. In the TIME analysis quadrant, we set the X-axis with business fit and technical fit on the Y-axis. Each quadrant represents a decision criterion to tolerate, invest in, migrate or eliminate the component. The TIME analysis recommends the following:

- **Tolerate** systems that deliver business value with acceptable technical risk and cost efficiency and meet the sustainability criteria and quality.
- **Invest** in or innovate applications that increase business processes' effectiveness and value, deliver sustainability outcomes, or leverage cross-application stovepipes, service levels, or data volumes that preclude conversion.
- **Migrate** system applications that are critical to providing value to the business but at high risk and high cost, including high sustainability-related costs to maintain them.
- **Eliminate** systems that no longer deliver business value with acceptable technical risk, cost efficiency, or risk and don't meet the sustainability criteria and quality.

Each component will be represented in the quadrant with a bubble whose size can depend on the component/application cost, carbon emission or carbon emission cost. This method will allow enterprises to monitor the SBDT portfolio to identify different transformative alternatives, both from a strategic value and from a sustainable IT perspective.

4 Results

4.1 APM Tool for SBDT: A Dashboard Mock-Up

A mock-up of the web app representing the APM tool for the SBDT of the nanobrewery is presented. The app allows to monitor all the components of the digital twin and give alerts (see the Overall Status Summary box) to the technical team whenever a discrepancy between the real system and the digital counterpart is found. The app keeps track of the response time of the SBDT and provides an estimate of the remaining useful life of every DT component (not only the physical assets, like sensors or hardware) but also virtual models. In fact, the accuracy of the models might get worse as time goes by depending on the received data and on the changes in the real system. Therefore, the actual value of each model might be lower after some time (see the Remaining Useful Life Estimate box). A precise lookout into the value of each model or components is provided in the Control Charts box, where we could monitor in real-time the accuracy and functioning of every SBDT component. If the performance falls outside the predetermined band of acceptability (in the Fig. 3, only one points falls outside the band), then a countermeasure could be taken. It is up to the SBDT manager to decide how to handle the specific situation, also supported by the TIME analysis tool. In this box, it is possible to select the type of analysis to be displayed. In particular, the size of the bubbles might depend on the: i) component cost; ii) related carbon emissions; iii) related carbon costs. Here we can see that five components have a critical carbon emission cost per component. Ten applications in this example may not look like as much cost. Still, if you are a large organization with hundreds or even thousands of components in a complex SBDT, this will add to accumulated costs due to your carbon dioxide emissions.

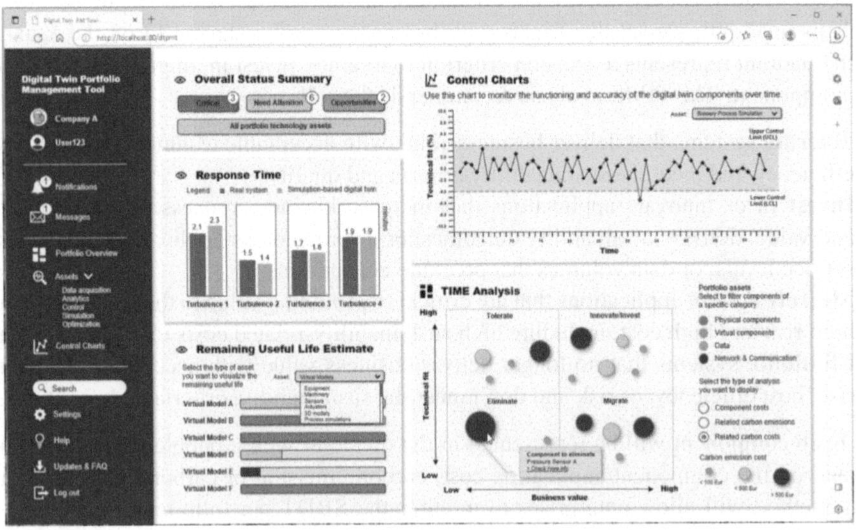

Fig. 3. Mock-up of a Digital Twin Portfolio Management Tool

We have replaced the application costs with carbon emission costs. The business value and the technical fit are the same. For simplicity's sake, we have set the carbon emission cost to €100 per metric ton of carbon dioxide emission in this example. Hence, carbon emissions per application are multiplied by the carbon emission tax.

4.2 Sustainable SBDT: A Preliminary Analysis

Due to their complexity and need for high performance computing requirements, SBDTs generally run on computing resources available on the cloud. In the context of this study, we assume to use an Amazon EC2 C5 instance (model: c5d.12xlarge) to run the simulation model of the nanobrewery. This instance model is well suited to SBDTs as well as high performance web servers, high performance computing (HPC), scientific modeling, dedicated gaming servers, machine learning inference and other compute intensive applications (for full data, see [23]). The model is expected to run continuously as it has to simulate the brewing process and analyze/process data through its machine learning engine. We assume for the purposes of this study that the instance load of the instance (i.e. the utilization/stress rate of the instance considering the average computing needs) is 50%. We estimate then the power consumption according to its load level. To do so, we assume that the digital twin should be up and running 24/7 in order to keep monitoring the brewing process. After estimating the kilowatt hours for compute, storage and networking, we need to multiply this by the cloud provider Power Usage Effectiveness (PUE). PUE is a score of how energy efficient a data center is, with the lowest possible score of 1 meaning all energy consumed goes directly to powering the servers and none is being wasted on cooling. This is based on publicly available information provided by the cloud providers and estimated as PUE: 1.2. Once we have the estimated kilowatt hours for usage of a given cloud provider, we then convert that into estimated CO2e using publicly available data on emission factors for a given electricity grid based on the mix of local energy sources. We provide here a location-based cloud emissions estimate of energy (Watt-hours) and greenhouse gas emissions expressed as carbon dioxide equivalents (metric tons CO2e) caused by the application (computing, storage, networking, etc.). In particular, we assume that the AWS region is eu-south-1 Europe (Milan), which is characterized by an electricity carbon intensity of 233 gCO_2eq/kWh [24]. Based on this data, the carbon footprint of the EC2 workload for an entire year is $1'056'696.7$ gCO_2Eq. ($835'944.7$ gCO_2eq for running the instance and $220'752$ gCO_2eq for its manufacturing). This carbon footprint is a preliminary and rough estimate based on simplified assumptions: the actual energy consumption and greenhouse gas emissions will depend on a variety of factors such as the specific simulation software used, the configuration of the data center, the type of machinery and sensors.

Considering that the price of emissions allowances (EUA) traded on the European Union's Emissions Trading System (ETS) was 85.22 euros per metric ton of carbon dioxide on December 1, 2022 [25], the current carbon emission cost for running the SBDT is 90.05 Eur.

5 Discussion and Conclusions

As a digital 'living' representation of a physical system, the question that stimulates this study is: "can a DT die?". "All that lives must die", Gertrude tells her son in Hamlet, thus inspiring an interesting perspective on DT lifecycle. In this study, we differentiate between the life cycle of the physical entity and the lifecycle of the DT. In a literal sense, a DT cannot 'die' because it does not have a physical body or existence. However, a DT can die in the sense that it can become obsolete or irrelevant if it is no longer able to accurately represent the physical system it is designed to replicate. This could happen, e.g. if the physical system itself undergoes significant changes that are not reflected in the DT, or if the data sources that the DT relies on become outdated or inaccurate. In this sense, the 'death' of a DT is rather a failure to evolve and adapt to changes in the physical system. Digital waste is generated when the DT is not capable to represent accurately the real system. In a sustainable technology perspective, an APM tool for a SBDT portfolio would be very helpful to monitor the Total Cost of Ownership of a Digital Twin. While the design and development stage of a SBDT is an incredible effort, companies shall start taking care on the waste, operational costs and efforts in keeping a hyper-converged infrastructure like a SBDT fully functioning, accurate and always aligned. *Twinning the digital twin* seems to be an interesting solution for managing the technology portfolio of components and assets the DT is made of. We not only need to monitor and optimize the performance of the real system, but we must also keep track of the performance of the SBDT in a novel way and provide recommendations about what to tolerate, what to invest in, what to eliminate and what to migrate (TIME Analysis). This raises another question related to the data sampling frequency. This is an essential aspect for the meta-model to monitor the performance, accuracy and sustainability of a DT. The sampling frequency should be carefully chosen to strike a balance between the information considered for the purpose, the computational resources and the dynamics of the real system.

Finally, this approach will protect the enterprise IT from obsolescence, but it will also keep track of the environmental footprint of such a complex digital solution. Researchers and the scientific community should start focusing on the complexity of running an enterprise-wide digital twin that is also efficient in terms of *technical fit*, *business value* and *environmental performance*. Future work will be carried out to design and develop this approach in a specific industrial case study.

References

1. Accenture – Sustainable technology, https://www.accenture.com/us-en/services/sustainability/sustainable-it-technology, last accessed 2023/03/24
2. Sundberg, N., Barnekow, H.: Sustainable IT Playbook for Technology Leaders: Design and implement sustainable IT practices and unlock sustainable business opportunities. Packt Publishing Ltd, Birmingham, UK (2022)
3. The Economic Times, Reuse and recycle: Google, Microsoft & Dell join forces to tackle e-waste crisis by 2030. Accessed on April 4th, 2023. Available at: https://economictimes.indiatimes.com/magazines/panache/reuse-and-recycle-google-microsoft-dell-join-forces-to-tackle-e-waste-crisis-by-2030/articleshow/81585724.cms?utm_source=contentofinterest&utm_medium=text&utm_campaign=cppst
4. Mourtzis, D.: Simulation in the design and operation of manufacturing sys-tems: state of the art and new trends. Int. J. Prod. Res. **58**(7), 1927–1949 (2020)
5. Rodič, B.: Industry 4.0 and the new simulation modelling paradigm. Organi-zacija **50**(3), 193–207 (2017)
6. Reed, S., Löfstrand, M., Andrews, J.: Modelling cycle for simulation digital twins. Manuf. Lett. **28**, 54–58 (2021)
7. Tao, F., Zhang, H., Liu, A., Nee, A.Y.: Digital twin in industry: state-of-the-art. IEEE Trans. Industr. Inf. **15**(4), 2405–2415 (2018)

8. Rojek, I., Mikołajewski, D., Dostatni, E.: Digital twins in product lifecycle for sus-tainability in manufacturing and maintenance. Appl. Sci. **11**(1), 31 (2020)
9. Tao, F., Zhang, M.: Digital twin shop-floor: a new shop-floor paradigm to-wards smart manufacturing. IEEE Access **5**, 20418–20427 (2017)
10. Wright, L., Davidson, S.: How to tell the difference between a model and a digital twin. Adv. Model. Simul. Eng. Sci. **7**(1), 1–13 (2020)
11. dos Santos, C.H., Montevechi, J.A.B., de Queiroz, J.A., de Carvalho Miranda, R., Leal, F.: Decision support in productive processes through DES and ABS in the digital twin era: a systematic literature review. Int. J. Prod. Res. **60**(8), 2662–2681 (2022)
12. Santos, C.H.D., de Queiroz, J.A., Leal, F., Montevechi, J.A.B.: Use of simulation in the industry 4.0 context: creation of a digital twin to optimise decision making on non-automated process. J. Simul. **16**(3), 284–297 (2022)
13. Qiu, H., Chen, Y., Zhang, H., Yi, W., Li, Y.: Evolutionary digital twin model with an agent-based discrete-event simulation method. Appl. Intell. **53**(6), 6178–6194 (2023)
14. Harmon, R.R., Auseklis, N.: Sustainable IT services: assessing the impact of green computing practices. In: PICMET'09–2009 Portland International Conference on Management of Engineering & Technology (pp. 1707–1717). IEEE (2009)
15. Bell, S.: Lean enterprise systems: using IT for continuous improvement. John Wiley & Sons (2005)
16. Romero, D., Gaiardelli, P., Powell, D., Wuest, T., Thürer, M.: Digital lean cyber-physical production systems: the emergence of digital lean manufacturing and the significance of digital waste. In: Advances in Production Management Systems. Pro-duction Management for Data-Driven, Intelligent, Collaborative, and Sustainable Manufacturing: IFIP WG 5.7 International Conference, APMS 2018, Seoul, Korea, August 26–30, 2018, Proceedings, Part I (pp. 11–20). Springer International Publishing (2018)
17. Sargent, R.G.: Verification and validation of simulation models. J. Simul. **7**(1), 12–24 (2013). https://doi.org/10.1057/jos.2012.20
18. Meng, C., Nageshwaraniyer, S.S., Maghsoudi, A., Son, Y.J., Dessureault, S.: Data-driven modeling and simulation framework for material handling systems in coal mines. Comput. Ind. Eng. **64**(3), 766–779 (2013)
19. Cho, M., Song, M., Yoo, S., Reijers, H.A.: An evidence-based decision sup-port framework for clinician medical scheduling. IEEE Access **7**, 15239–15249 (2019)
20. Apsemidis, A., Psarakis, S., Moguerza, J.M.: A review of machine learning kernel methods in statistical process monitoring. Comput. Ind. Eng. **142**, 106376 (2020)
21. Zwetsloot, I.M., Woodall, W.H.: A review of some sampling and aggrega-tion strategies for basic statistical process monitoring. J. Qual. Technol. **53**(1), 1–16 (2021)
22. dos Santos, C.H., Campos, A.T., Montevechi, J.A.B., de Carvalho, M.R., do Amaral, J.V.S., de Queiroz, J.A.: Simulation-based digital twins monitor-ing: an approach focused on models' accreditation. Int. J. Adv. Manuf. Technol. **124**(7–8), 2423–2435 (2023)
23. AWS EC2 Carbon Footprint Dataset. Available at: https://docs.google.com/spreadsheets/d/1DqYgQnEDLQVQm5acMAhLgHLD8xXCG9BIrk-_Nv6jF3k/edit#gid=504755275
24. European Environment Agency. Greenhouse gas emission intensity of electricity generation. Available on: https://www.eea.europa.eu/data-and-maps/daviz/co2-emission-intensity-12#tab-googlechartid_chart_11. Accessed on: April 4th, 2023
25. Statista. European Union Emission Trading System (EU-ETS) carbon pricing in 2022. Available at: https://www.statista.com/statistics/1322214/carbon-prices-european-union-emission-trading-scheme/#:~:text=The%20price%20of%20emissions%20allowances,dioxide%20on%20December%201%2C%202022

Evaluation of Government Incentive Policy on Industrial Waste Utilization by Agent-Based Simulation of Industrial Symbiosis Network

Shufang Xie(✉), Tao Zhang, Tobias Uhlig, and Oliver Rose

Universität Der Bundeswehr München, 85577 Neubiberg, Germany
shufang.xie@unibw.de

Abstract. With the increasing concern about the limitation of available resources, industry symbiosis (IS) has emerged as one of the mechanisms to achieve sustainable industrial ecology. However, the economic benefit is always the factory's primary consideration, few factories will consider implementing IS unless they see certain economic benefits. Governmental policies are the main factors to facilitate and stimulate IS development. Unfortunately, how governmental policy actions influence the development of IS is not completely clear. This article investigates the government subsidy policy and proposes an agent-based simulation study to investigate the relationship between this kind of policy and the industry symbiosis practice from the government's perspective. We found the expected results, which is that the subsidy policy can positively influence the emergence of the IS relationships which will eventually reduce the waste disposal to the landfill, beneficial to the environment. Moreover, from the simulation experiments, we also find that when incentives reach a certain level, they no longer work. Even if the value of the subsidy is increased, the amount of waste utilization will not increase anymore. We think that our research can serve as a reference for the policymakers who are interested in creating incentive policies to stimulate the IS practice and achieve more sustainable industrial development.

Keywords: agent-based simulation · industrial symbiosis · government incentive policy

1 Introduction

Industry symbiosis (IS) is one of the mechanisms to achieve sustainable industrial ecology which studies the physical exchange of material, energy, and services to engage separate enterprises in a collective approach[1]. It brings together companies from all business sectors with the goal to add value, reduce costs and benefit the environment. The kernel of IS is the association between companies in which the waste or by-products of one become raw materials for another. Because of the increasing concern about the limitations of the available resources and the need for more sustainable industrial development, in the past few years, the topic of IS has gained significant attention in practice, as well as in academic communities. A recent review by [2] highlights that cases of

M. Mujica Mota and P. Scala (Eds.): EUROSIM 2023, CCIS 2032, pp. 70–85, 2024.
https://doi.org/10.1007/978-3-031-68435-7_6

IS have been growing over the past years and are disseminated at the global level. In academic communities, researchers focus on basically one of three stages: 1) identifying IS opportunity and finding potential business partners [3–5]; 2) assessing the underlying economic, environmental, and social benefits brought by the identified IS [6]; 3) implementing IS by paying particular attention on the emerging process of IS and finding out factors that promote, hinder, or ignite the emerging process [7]. Stage 3) is the most important step in IS practice but is also often omitted due to the lack of research tools.

In most cases for waste utilization, the emergence of IS is triggered only after government incentive policies are implemented. The policies can promote IS through regulations, laws, and economic instruments. Regulations and legal frameworks forbidding the disposal of some specific wastes to landfills, thereby compel factories to implement IS practices for their disposals. Besides regulations and laws, the government can also stimulate the IS practice through economic instruments because economic profits are always the company's primary concerns. Some examples of economic instruments are building facilities, fiscal incentives for firms adopting IS, and rising landfill taxes. Several studies have mentioned the importance of government incentive policy for the emergence of IS [8–10]by analyzing a certain amount of real IS cases. However, a deep and detailed analysis of a specific IS under incentive policies is still a research gap.

In this study, we will conduct our research from the policymakers' perspective and analyze the policies' effect on the emergence of the IS which will eventually reduce the emission of waste. As policymakers, we will observe the overall situation, and improvements not only in the economic but also in environmental and social dimensions. Specifically, we try to find how the economic subsidies from the government policy influence the economic activities within the IS network, what is the best scenario for the policy maker and what are the limits of the government policies. The rest of the paper is structured as follows: Sect. 2 gives more details about IS networks and policies. Section 3 introduces agent-based simulation to model IS networks. Simulation results and discussion are given in Sect. 4. Conclusions and future work are provided in the last section.

2 IS Network and Incentive Policy

An industrial symbiosis network is a network of industries among which IS relationships exist. Boons, Frank, et al. [11] define IS network as a process that revolves around the emergence, development, and decline of networks of by-product synergies and utility synergies. IS network can be spontaneously self-organized or planned. The planned model includes a conscious effort to identify companies from different industries and locate them together so that they can share resources across and among them [12]. In this model, the policymakers play a critical role. Most of the Chinese eco-industrial parks are examples of the planned model. The Chinese policy on circular economy is one of the central factors of influence on Chinese industrial symbiosis development [13, 14]. In the self-organizing symbiosis model, an industrial ecosystem emerges from decisions by private actors motivated to exchange resources to meet goals such as cost reduction, revenue enhancement, or business expansion [12]. The early exchanges in the IS network at Kalundborg are the demonstrations of self-organization, those exchanges were business opportunities driven by social networking, and the later expansion will also require support from the local policymakers. Moreover, the policymakers can stimulate the creation of an environment favorable to the spontaneous development of IS relationships among firms, by means of policy measures [15]. In conclusion, in either a self-organized or planned IS network, the government policy relevance is remarkably high.

Compulsory legal requirements to recycle specific materials, higher taxes for waste disposal, and support from government policy for waste reuse projects are the inevitable influencing factors that exist during the development of IS network. In addition, companies are all profit-driven and focus mainly on the economic benefits instead of the environmental commitment which is usually not considered to be beneficial in increasing their profits and competitiveness. It is even argued that IS principles can be implemented on a large scale, but only if there exists the political will to support them, if the institutional structure allows them to be implemented, and if the technological principles are in place [16]. Therefore, it is quite necessary to explore the influence of government policy on the development of IS networks. Some works of literature such as [13, 14] have already clearly demonstrated the positive effect of government policy on the planned IS network. Thus, in this study, we only focus on the self-organized IS network, because it is considered as the most promising one, thanks to its ability to be more resilient to perturbations such as changes in production levels, in symbiotic flows, in the dimension and the number of the actors involved [1].

As mentioned before, we consider one government policy: economic subsidies. There are different types of subsidies, for example, a loan with interest below the market rate, the free use of equipment or office space, a tax break, cash subsidies, and so on. A tax break reduces the amount of taxes that a business or individual must pay, and a cash subsidy is a payment from the government made to a business or individual to encourage companies to invest in or expand their operations. In this study, we only focus on cash subsidies. The cash subsidy prescribes that, for each unit of waste reuse, the government pays the subsidy to the factories which are willing to reuse waste to replace the raw material. Such a measure can improve the economic benefits for the factories implementing the IS practice, so motivating them to establish the IS relationships, thereby reducing waste emissions.

The government usually decides to reach some climate targets within certain years and a certain amount of cash is planned for those years in advance. To reach the final climate target, the government needs to figure out what is the best suitable cash grant plan. Continuous subsidies without proper controls can cause economic harm or distortive effects. Subsidies can give recipients an unfair advantage over their competitors or be an inefficient use of public money if they do not bring about net positive change. Therefore, usually, the government will not subsidize enterprises indefinitely, they will make appropriate adjustments according to fiscal revenue, thus, most financial assistance measures are time-limited and degressive by design.

In addition, with the government's incentive policy to motivate the IS practice, some factories may choose to use waste to replace the original raw materials as input materials for the sake of interest. In this case, the factory's original supplier which usually provides the raw material will be affected and lose many customers, at the end they may not make a profit anymore and even go bankrupt finally. Moreover, the improvement of social benefits is another topic for the development of IS practice. On the social side, more employment opportunities can be created because of waste preprocessing or other reasons, at the same time some workers in the factories which produce the original raw materials will lose their jobs due to demand decline. Because the study is just at the beginning, the social performance of IS will not be involved for now.

3 Agent-Based Simulation of is Network

In this section, we will introduce the agent-based simulation to model the generic IS network. We will give details on the design of the agent itself, the behaviors of agents, the interaction among agents, and so on.

3.1 IS Network and Agent-Based Simulation

Self-organized IS networks are recognized as complex adaptive systems (CAS) [17], where companies are agents who interact with each other in exchanging wastes, sharing resources, and so on, without a central agent deliberately managing the system. Agent-based modeling (ABM) is one of the most suited techniques to study CASs [18]. The basic unit in the agent-based simulation is the agent. Wooldridge and Jennings [19] define an agent as a computational system that interacts with its surroundings. The agent can have autonomy, social ability, reactivity, and proactiveness. These characteristics are represented by the properties, objectives, and behavioral rules of the agent. The main advantages of adopting ABM in an IS network lie at the core of understanding the learning processes and complex macrostructures by studying the interactions among individual agents [20]. Agent-based simulation has already been used to study the complexity and dynamics of IS practice. Albino, Fraccascia, et al. [21]use an agent-based model to investigate the role of contracts to support the emergence of self-organized industrial symbiosis networks. Gabriel and Daniel [22] present a comparative evaluation of the IS indicators available in the literature. The indicators were simulated through an agent-based model. Luca, Devrim, et al. [23] designed an agent-based model to explore the redundancy strategy on the performance of companies involved in IS. Because failures of IS networks may be caused by actor behavior that leads to unanticipated economic losses, Kasper, Gijsbert, et al. [24] explore the effect of these behaviors on IS network robustness by using an agent-based model. We are also going to use agent-based simulation in this study.

3.2 Agents and Populations

We consider the IS network to be made up of many factories and customers. Two types of agents are considered in our agent-based model: one agent is one factory that needs some raw materials, purchasing from its suppliers to produce a single main product, along with some wastes which are destined to be disposed of in landfill sites. Instead, through creating IS relationships wastes produced by one factory can be used as input materials for another factory. The other agent type is the final customer who buys the main product produced by one factory. The demand of the customer is stochastic over time. Factory agents in the same industry make up one industry population in which all factory agents have the same types of input and output but different capacities and production plans. Likewise, customer agents which require the same product but with different demands make up one customer population. The IS is built between the populations.

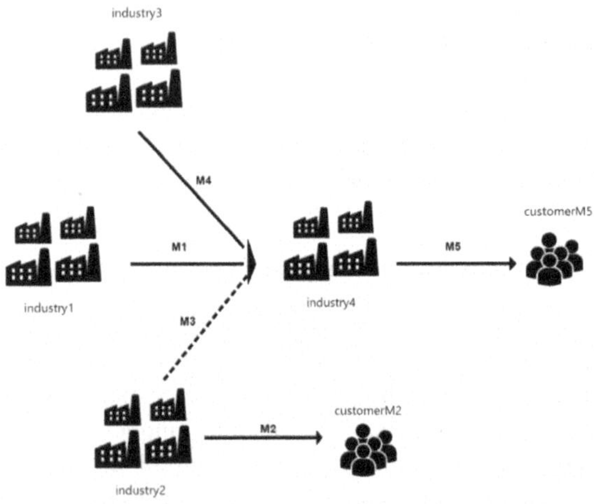

Fig. 1. An example of IS network

An example of IS network is shown in Fig. 1. Totally we considered four arbitrary industries in the example model, we refer to them simply as Industry1, Industry2, Industry3, and Industry4. Each industry is made up of a certain number of factories. Industry 3 and 1 are the original suppliers for Industry 4. Factories in Industry 3 and Industry 1 provide two different types of raw materials M4 and M1 respectively for the factories in Industry 4 as the inputs. Through IS practice, the wastes M3 produced by the factories in Industry 2 can be used as input materials for the factories in Industry 4 to substitute the raw materials M1 which are produced by the factories in Industry 1. Two final customer populations consume the product M2 and M5 from Industry 2 and 4 respectively.

3.3 Behaviors of Agents

Some important behaviors of the factory agents are shown in Fig. 2. Each factory produces products according to a production plan, subject to varying capacities. For many different kinds of reasons like equipment outages, staffing and materials shortages, and so on, the production capacity in each factory changes uncontrollably. Each factory determines the price of the main product accordingly based on the full cost of running the business, customer demand, competition, and some other factors that may affect the price. The full cost of running the business could be the labor cost, marketing cost, manufacturing costs, and so forth. Each individual factory has a natural tendency to grow and go through similar life cycles: establishing, expanding/shrinking, and closing. If the demand far exceeds the capacity of all factories, new factories are built. If the factories cannot make any profit, they will be closed.

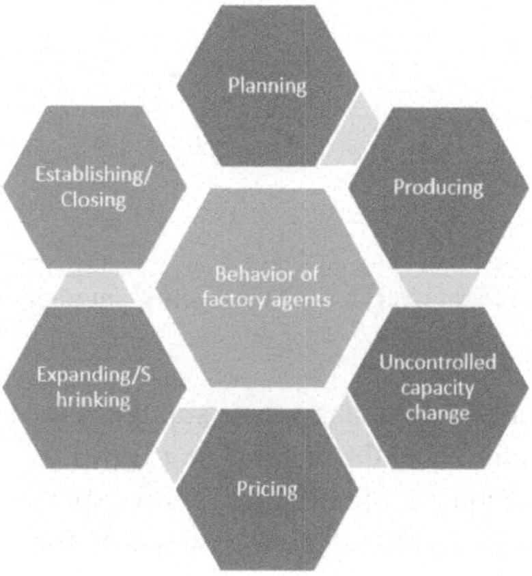

Fig. 2. Behaviors of the factory agent

3.4 Interactions of Agents

Figure 3 shows a brief overview of the interaction pattern among agents. Each factory agent can be a supplier or customer for other factories, customer can be a factory agent or the final customer agent. Different from some research reports in which the researchers, for the sake of simplicity, assume that each factory can send its output materials only to one factory and can receive materials only from one factory [15], in our agent-based simulation model, the agents have many to many relationships which can describe the interaction among factories in the real world more appropriately. One factory can have more than one customer and multiple suppliers as well.

After a purchase order is created on the customer side, the customers will send out the information including the total amount of demand and the latest response time to the suppliers to request the quotation. The suppliers will handle all the requests at regular intervals, within each interval, the suppliers will wait and collect all the demands from the customers. And then based on the customer trust level, the suppliers give the required quotation and available quantity only to the customers whose latest reply deadlines have not been exceeded. Then the customers will analyze and select the suppliers based on the price offers from all their suppliers and send the purchase confirmation to the selected suppliers. The purchase order ends when the selected suppliers deliver the products to the customers.

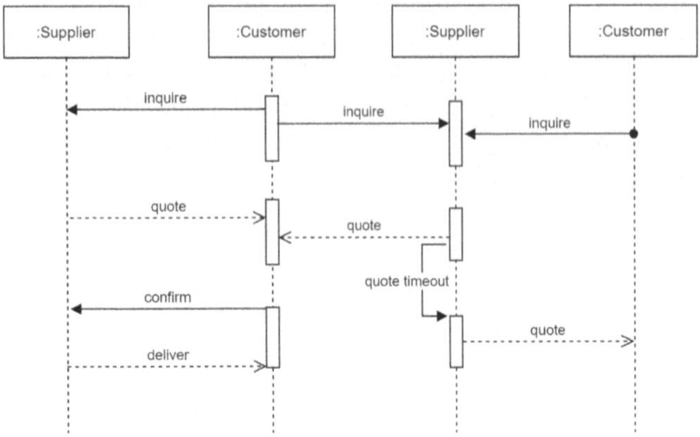

Fig. 3. An overview of the agents' interaction

In the real life, customers may change their suppliers for various reasons. It could be because the supplier cannot meet the deadlines or fulfill the orders, the quality of the product or service is not adequate, the product is too expensive, and so on. Similarly, the customers of the suppliers are also changing over time for certain reasons. Some old customers will be lost, while some new customers will be attracted. Each supplier or customer in the IS network is modeled as an agent, who decides its supplier or customer. Two types of trust level functions are defined, supplier trust level and customer trust level, which measures the willingness of two agents to build the supplier and customer relationship respectively.

The supplier trust level measure depends on three factors: 1) the price of the product.2) delivery satisfaction. 3) the order accuracy. In this model, Customers are more inclined to choose suppliers with cheaper quotations, the trust level decreases with higher quotes. Moreover, getting orders into customers' hands just in time is a big part of customer satisfaction, both too late and too early can lose the trust of the customers. Besides, order accuracy is also considered as one influencing factor in our model. Higher order accuracy will boost customer loyalty, if the customers almost always receive the exact amount of items they ordered, customers are less likely to switch their suppliers. Therefore, the update function of the supplier trust level is defined as follows, the offered amount of the supplier is always less than or equal to the required amount of the customer:

$$\begin{aligned} \textit{trust Level} =& \textit{trust Level} \\ &+ a * (\textit{last Price} - \textit{current Price})/(\textit{max Price} - \textit{min Price}) \\ &+ b * (1 - 2 * \mathrm{ABS}(\textit{due Date} - \textit{arrival Date})/\mathrm{ABS}(\textit{max Delay})) \\ &+ c * (2 * \textit{offered Amount}/\textit{required Amount} - 1) \end{aligned} \tag{1}$$

The customer trust level measure depends on two factors:1) customer demand and 2) changes in customer demand. Usually, the suppliers tend to choose customers with high and increasing demand. Thus, in this model, the update function of the customer trust level is defined as follows, the ordered amount of the customer is always less than or equal to the offered amount of the supplier:

$$\begin{aligned} \mathit{trust\ Level} &= \mathit{trust\ Level} \\ &\quad + d * (2 * \mathit{ordered\ Amount}/\mathit{offered\ Amount} - 1) \\ &\quad + e * (\mathit{ordered\ Amount\ CurP} - \mathit{ordered\ AmountPreP})/\mathit{max\ Differ} \end{aligned} \tag{2}$$

In trust level functions (1) and (2), a, b, c, d, and e are the weights of factors that influence the trust level. In this study, we consider all the factors to have the same influence. If the value of the trust level is below a certain bound, the agent will lose its existing suppliers or customers. If this is the case, the agent will randomly choose another one to replace the lost supplier or customer.

3.5 Rules of Agent Behavior and Interactions

Agents behave and interact with other agents according to some predefined rules. In this section, we will describe all rules designed for agent behaviors and interactions.

Rules of Agent Behavior

In the last section, we have introduced all behaviors of each agent. Now we are going to give detailed information about the rules. The explicit rules regarding planning, production, capacity change, pricing, and establishing and closing a factory are listed below.

Rule of Planning

- Make-to-stock planning
- 80% capacity loading
- 8 weeks planning horizon

Rule of Producing

- The production tries to stick to the production plan
- Producing starts at the planned time
- Production will stop if one of the output inventories is full
- If raw materials are not enough, the planned amount can be adjusted
- The queue adopts the FIFO rule
- Processing times are amount-dependent

Rule of Uncontrolled Capacity Change

- Capacity loses randomly but follows a distribution
- The lost capacity will be recovered after some time
- The capacity cannot reduce anymore if reaching the minimal capacity

Rule of Establishing a Factory

- If the difference between total demand and capacity of the industry from last year is greater than 5 times the maximum of all factories' capacity, new factories will be established this year

- The establishment of the factory follows the uniform distribution between a minimum which is given in advance and the end of the year.
- The capacity of the new factories is random and between min and max, the sum of all newly built factories fills the capacity gap
- If 20% of factories in an industry are closed last year, no new factories will be established this year
- If the return on investment is below 10% in the last two consecutive years in the industry, no new factories will be established this year

Rule of Closing a Factory

- If a factory doesn't make any profit for two years, the factory will be closed
- If only one factory is left in the industry, the factory will not be closed

Rule of Pricing Each factory can price its product, in our model the price is defined based on the inventory, the demand, and the trust level of the customer.

$$price = f(lowestPrice,\ demand,\ inventory,\ trustLevel)$$

The rules for the pricing behavior are listed below:

- Prices are updated in a fixed period
- If the inventory is too high, the price will decrease
- If the demand is greater than the capacity, the price will increase
- The customer with a higher trust level can get a lower price

Rule of Agent Interactions

In our model, the agents interact with each other by buying input materials from the suppliers and selling the main products to the customers. The detailed rules for buying and selling are listed as follows:

Rule of Buying

- Start to buy materials for a production batch when the batch is within the planning horizon
- Only buy materials if the inventory is not enough
- Send inquiries to all suppliers in its supplier list
- Sort quotes by the prices and trust levels
- Drop expired quotes
- Order materials from the top quote to the bottom, until the required amount is ordered, and reject other quotes

Rule of Selling

- Sort inquiries by the trust level
- Drop expired inquires

- Send quotes from the top to the bottom, until the inventory is empty, and discard the other inquires
- Put back the rejected quotes into the inventory

4 Simulation Experiments and Results

4.1 Experiment I

We define several different subsidy scenarios and use simulation to analyze their efficacy. All designed subsidy scenarios are divided into three categories: no grant, continuous grant, and discrete grant. No grant means that the government doesn't provide any subsidy to push the IS establishment. Another type is a continuous grant, in this type, the government will continuously support and provide grants to the factories which implement the IS practice to reuse the waste. In the discrete grant type, considering that most of the financial supports are time-limited, instead of a steady cash grant, the government gives the grants in a discrete way: the grants are offered for a certain period, then terminate, and after a while, the support may start again. The government makes a subsidy plan every ten years, in every ten years, 8 different scenarios are defined in a string shown in Table 1.

Table 1. Scenarios in Experiment I

Id	Scenario	Years	Id	Scenario	Years
1	All	10	6	1–1-1–1-1–1-1–1-1–1	5
2	4-1-4–1	8	7	1-2-1-2-1-2-1	4
3	2-1-2-1-2-1-1	7	8	1-3-1-3-1-1	3
4	3-3-3-1	6	9	1-4-1-4	2
5	2-2-2-2-2	6	10	None	0

The number in odd positions represents the number of years the government provides a continuous subsidy for the factories which are willing to develop the IS relationship with other factories and reuse their wastes, while the number in even positions represents the years in which government terminates the subsidy policy. For example, in the second scenario, the government provides funding incentives for waste recycling plants first for four consecutive years, then stops for one year, after that, offers incentives again for another four consecutive years, then stops for another year. All together with the no-grant type and the continuous grant type, we designed 10 scenarios for the experiments, from the aspect of the government grant model and sorted them by the total grant years in descending order.

The scenarios are evaluated using the example model mentioned in Sect. 3.2. This model was implemented in AnyLogic. The initial number of factories for industry1, industry2, industry3, and industry4 are 102, 304, 115, and 123 respectively, and the number of customers in two customer populations (customerM5 and customerM2) is

212 and 330 separately. We simulated each scenario for a simulation run of 10 years and perform 10 replications. We investigate the economic and environmental influence to measure the success of the government subsidy policy. We use waste usage to judge the impact of government policies on the environment and the contributions to GDP and profits to reflex the economical influences. The contributions to GDP and profits include all four industries in the model. The contribution to GDP is the total income of all factories, and the profit is usually the difference between total income and total expenses. In our simulation model, each factory obtains revenues from selling its main product. Sales revenues, together with the subsidy which is granted by the government based on the used waste amount are the income of the factory. The total expenses are the production cost which includes material disposal cost, material preprocess cost, direct material cost, direct labor cost, and factory overhead. The material disposal cost is the cost factories pay for the waste disposed of in the landfill. Material Pre-processing costs are also taken into account in the model because usually, the wastes discharged from one factory cannot be directly used, and additional pre-processing steps are always required. Direct material cost is used to buy the input raw material from the supplier, and the direct labor cost is paid in the form of salaries that are paid to the employees. Additionally, the overhead costs are all indirect costs like rent expenses of the factory, management costs, etc.

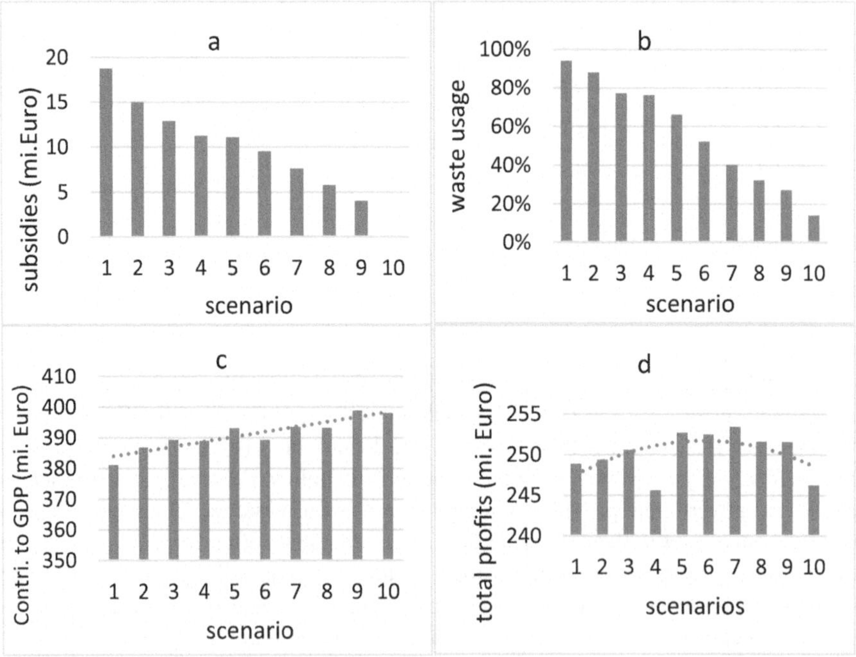

Fig. 4. Simulation results from Experiment 1

The results are shown in Fig. 4. Total cash subsidies, waste usage, contribution to GDP, and profits in ten years are depicted in charts a, b, c, and d respectively for all scenarios. From charts a and b, we can see that the total cash subsidies and waste usage go down as the grant years are reduced. The interesting finding is that the contributions to GDP are slightly increasing while the grant years are declining, and the total profits reach the maximum at around half of the study horizon. This means that the subsidies harm the contributions to GDP. Meanwhile, a good subsidy scenario can help the industries to make more profits, but a bad scenario may prevent the industries from making profits. Another interesting finding comes from Scenarios 4 and 5 both with 6 years of granting. The Total cash subsidies are almost the same, but the waste usages are quite different because of the distribution of 6 granting years in ten years. The following experiments are done to closer examine this observation more.

4.2 Experiment II

In this experiment, we still consider ten years period but only 6 subsidizing years and investigate the influences of 6 distribution scenarios listed in Table 2 on the example model of IS network. All the model and simulation configurations remain the same as in the last section.

Table 2. Scenarios in Experiment II

Id	Scenario	Years	Id	Scenario	Years
11	6-4	6	14/5	2-2-2-2-2	6
12	3-2-3-2	6	15	3-2-2-1-1-1	6
13/4	3-3-3-1	6	16	4-3-2-1	6

The results are shown in Fig. 5. The total cash subsidies are very close in all scenarios from chart a. However, the waste usage in chart b is varying a lot. This further confirms that the distribution scenario of cash subsidies has a very big impact on the performance of IS network. According to chart b, scenario 15 is the best one to utilize the waste. If looking into the details of this scenario, 3-2-2-1-1-1, we can easily figure out the scenario's pattern in which the subsidizing and non-subsidizing periods are slowly decreased in sequences of 3-2-1 and 2-1-1, the non-subsidizing periods are always shorter than the previous subsidizing period. We can consider this to be a promising pattern of subsidy distribution. In addition, from charts b and c, we can also conclude that the subsidies are harmful to GDP.

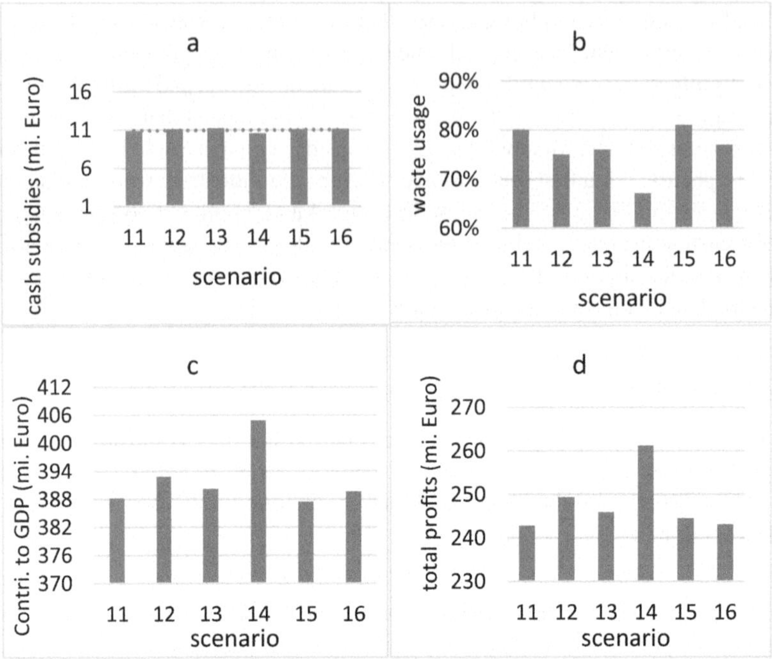

Fig. 5. Simulation results from Experiment II

4.3 Experiment III

In this experiment, we still consider ten years period but with varying subsiding periods from 0 to 10 years. The subsidizing years are distributed according to the promising patterns found in Sect. 4.2. The 11 scenarios are listed in Table 3. We investigate the influences of total cash subsidies on waste usage only. All the model and simulation configurations still keep the same as in the last section.

Table 3. Scenarios in Experiment III

Id	Scenario	Years	Id	Scenario	Years
17	10-0	10	23	4–6	4
18	9-1	9	24	3–7	3
19	8-2	8	25	2–8	2
20	4-2-3-1	7	26	1–9	1
21/15	3-2-2-1-1-1	6	27	0–10	0
22	3-3-2-2	5			

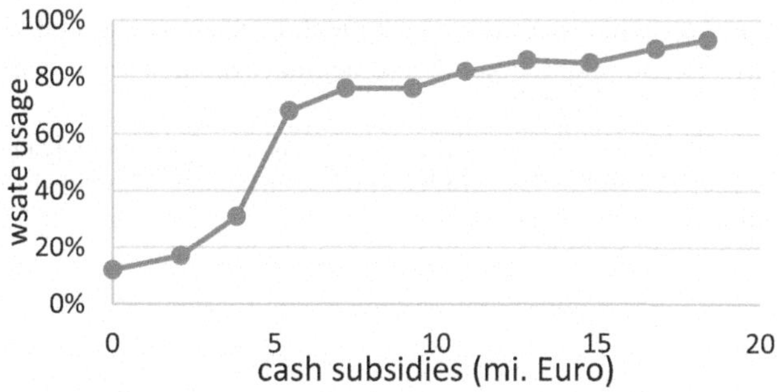

Fig. 6. Simulation results from Experiment III

The results are drawn in Fig. 6. A certain amount of waste is utilized without cash subsidies. This is caused by the shortage of the original raw material. The factories have to use the waste in this situation, otherwise, they will stop production. Also, the waste cannot be 100% utilized even though cash subsidies are always there. The reason is the surplus of the generated waste. If the storage of waste is full, we have to dispose of the waste. From Fig. 6, waste usage goes up dramatically at the beginning of the subsidy increase, and this rapid increase slows down after the subsidy reaches a certain amount. The policymakers can determine an operating area in Fig. 6. For example, the subsidy from 5.5 to 7.7 million and the waste usage from 68%–76%. The policy can be implemented according to this area.

5 Conclusion and Future Work

Industrial symbiosis is an important approach to achieving sustainability and reducing waste significantly. In this paper, we proposed an agent-based simulation model to investigate how government policy can encourage this practice and influence economic activities within the IS network from the government's perspective. Different scenarios are created by designing several various fiscal incentive strategies. Our results show the expected results, which is that the subsidy policy can positively influence the emergence of the IS relationships which will eventually reduce the waste disposal to the landfill. Besides, the experiment results also show that subsidy grant patterns influence the policy's efficacy. Moreover, we find that the total amount of waste reuse will not continue to increase, even if the value of the subsidy is increased, which means the subsidy policy no longer works after the subsidy reaches a certain level. Therefore, we think that our research can serve as a reference for the policymakers who are interested in creating incentive policies to stimulate the IS practice and promote the sustainable development of the industry.

Even though the study demonstrated a general way to evaluate government policies by agent-based simulation, the study is just carried out on some artificial data and has no way to validate the built model. Additional research is still needed to build and validate the model with real data before the application. In the future, we will also research how government policy influences the social side because policymakers usually focus on a broad range, not just the economic and environmental, but also the social side. Moreover, the rules of behaviors and interactions should be improved and more realistic.

Acknowledgment. This work is part of the project "EnICO – Energieeffiziente IndustrieCluster Optimierung", project number 439187891, which is financially supported by the Research Foundation of Germany.

References

1. Chertow, M.R.: Industrial symbiosis: literature and taxonomy. Annu. Rev. Energy Environ. **25**(1), 313–337 (2000)
2. Godina, R., Azevedo, S.G., Matias, J.C.O.: A comprehensive review of industrial symbiosis. J. Clean. Prod. **247**(20), 119113 (2020)
3. Low, J.S.C., et al.: A collaboration platform for enabling industrial symbiosis: application of the database engine for waste-to-resource matching. Proc. CIRP **69**, 849–854 (2018)
4. Cecelja, F., Trokanas, N., Raafat, T., Yu, M.: Semantic algorithm for industrial symbiosis network synthesis. Comput. Chem. Eng. **83**, 248–266 (2015)
5. van Capelleveen, G., Amrit, C., Yazan, D.M.: A literature survey of information systems facilitating the identification of industrial symbiosis. From Science to Society, 155–169 (2018)
6. Yazan, D.M., Fraccascia, L.: Sustainable operations of industrial symbiosis: an enterprise input-output model integrated by agent-based simulation. Int. J. Prod. Res. **58**, 392–414 (2020)
7. Henriques, J.D., et al.: Implementing industrial symbiosis incentives: an applied assessment framework for risk mitigation. Circ. Econ. Sustain. **2**, 669–692 (2022)
8. Azevedo, J., et al.: Guidelines for industrial symbiosis—a systematic approach for content definition and practical recommendations for implementation. Circ. Econ. Sustain. **1**(2), 507–523 (2021)
9. Madsen, J.K., Boisen, N., Nielsen, L.U., Tackmann, L.H.: Industrial symbiosis exchanges: developing a guideline to companies. Waste Biomass Valoriz. **6**, 855–864 (2015)
10. Lybæk, R., Christensen, T.B., Thomsen, T.P.: Enhancing policies for deployment of Industrial symbiosis–What are the obstacles, drivers and future way forward? J. Clean. Prod. **280**, 124351 (2021)
11. Boons, F., Spekkink, W., Jiao, W.: A process perspective on industrial symbiosis: theory, methodology, and application. J. Indust. Ecol. **18**(3), 341–355 (2014)
12. Chertow, M.R.: "Uncovering" industrial symbiosis. J. Indust. Ecol. **11**(1), 11–30 (2007)
13. Su, B., Heshmati, A., Geng, Y., Yu, X.: A review of the circular economy in China: moving from rhetoric to implementation. J. Clean. Prod. **42**, 215–227 (2013)
14. Yong, R.: The circular economy in China. J. Mater. Cyc. Waste Manage. **9**, 121–129 (2007)
15. Fraccascia, L., Giannoccaro, I., Albino, V.: Efficacy of landfill tax and subsidy policies for the emergence of industrial symbiosis networks: an agent-based simulation study. Sustainability **9**(4), 521 (2017)
16. Evans, S., Wen, Z., Ma, M.: The influence of policy on industrial symbiosis from the firm's perspective: a framework. J. Clean. Prod. **213**, 1172–1187 (2019)
17. Chertow, M.R., Ehrenfeld, J.: Organizing self-organizing systems toward a theory of industrial symbiosis. J. Ind. Ecol. **16**(1), 13–27 (2012)
18. Axelrod, R.: The complexity of cooperation. Agent-based models of competition and collaboration Robert Axelrod. Princeton studies in complexity. Princeton University Press, Princeton N.J., Chichester (1997)

19. Wooldridge, M., Jennings, N.R.: Intelligent agents: theory and practice. Knowl. Eng. Rev. **10**, 115–152 (1995)
20. Demartini, M., Tonelli, F., Govindan, K.: An investigation into modelling approaches for industrial symbiosis: a literature review and research agenda. Clean Log. Supply Chain **3**, 100020 (2022)
21. Albino, V., Fraccascia, L., Giannoccaro, I.: Exploring the role of contracts to support the emergence of self-organized industrial symbiosis networks: an agent-based simulation study. J. Clean. Prod. **112**, 4353–4366 (2016)
22. Mantese, G.C., Amaral, D.C.: Agent-based simulation to evaluate and categorize industrial symbiosis indicators. J. Clean. Prod. **186**, 450–464 (2018)
23. Fraccascia, L., Yazan, D.M., Albino, V., Zijm, H.: The role of redundancy in industrial symbiotic business development: a theoretical framework explored by agent-based simulation. Int. J. Prod. Econ. **221**, 107471 (2020)
24. Lange, K., Korevaar, G., Nikolic, I., Herder, P.: Actor behaviour and robustness of industrial symbiosis networks: an agent-based modelling approach. JASSS **24**(3), 8 (2021)

Emission-Based Production and Inventory Control System: A Stability-Based Analysis for System Cost and Service Level

Rishav Deval(✉) and Jayendran Venkateswaran

Industrial Engineering and Operations Research, Indian Institute of Technology Bombay, Mumbai, India
{deval.r,jayendran}@iitb.ac.in

Abstract. A discrete-time stability analysis is performed on an Emission-based Production and Inventory Control System (*EPICS*) to identify stability boundaries for control parameters. An initial simulation-based analysis is performed to determine short-term system performance on sustainability dimensions: service level (social measure) and system cost (economic measure) for stability boundaries of Production Inventory (*PI*) control parameters and Emission-control parameters. Negative Item Fill Rate (*IFR*) as an undesired state from order cancellation/return due to high Emission-control weightage results in high system costs. Further, as variability in the system increases, system performance for both cost and service level under higher Emission weightage is unaffected but for higher *PI* weightage system performance is significantly impacted.

Keywords: EPICS · APIOBPCS · Perceived Emission Rate · Emission Permit · Average System Cost · Carbon Cost

1 Introduction

Sustainability has emerged as a critical consideration in supply chain management, with research efforts increasingly focused on reconciling environmental concerns with economic and social factors. This emphasis on sustainable supply chain management has given rise to Green Supply Chain Management (*GSCM*), which seeks to minimize the environmental impact of supply chain operations while enhancing economic performance and social well-being. However, current research suggests that *GSCM* may not be sufficient to address the complex sustainability challenges facing supply chain management, as it tends to prioritize environmental concerns over other sustainability dimensions, such as economic viability and social equity.

To address this limitation, Emission-based Production and Inventory Control Systems (*EPICS*) have emerged as a promising approach to achieving sustainable supply chain management. *EPICS* is designed to maintain inventory and supply lines while simultaneously regulating system emissions to reduce the carbon footprint of a production system. *EPICS* accomplishes this by tuning parameters to control the system's

M. Mujica Mota and P. Scala (Eds.): EUROSIM 2023, CCIS 2032, pp. 86–100, 2024.
https://doi.org/10.1007/978-3-031-68435-7_7

flow and ensure emissions are maintained at permissible levels in real time. Under this setup, environmental regulations mandated by regulators ensure that production systems comply with commitments towards sustainability at an individual level, which can have significant implications for system performance.

Despite the potential of *EPICS* to contribute to sustainable supply chain management, there still needs to be a significant knowledge gap regarding the short-term impact of sustainability considerations on service levels and systems costs. This knowledge gap is critical since supply chain management is characterized by a complex interplay of multiple players, each with almost identical operations, who must collaborate to ensure sustained growth of the individual subsystem and overall supply chain. An important question arises of how much to order and ensure that the subsystem operates per desired expectations without compromising sustainability objectives. Therefore, this paper aims to conduct a stability-based cost and service level analysis of an EPICS using a stock management structure.

The paper seeks to offer valuable insights into the trade-offs between sustainability considerations, service levels, and system costs, thereby providing decision-makers with a comprehensive framework for balancing competing objectives in supply chain management. Specifically, the paper adopts a two-pronged approach to the analysis. Firstly, a stability analysis is carried out to determine a stable region for control parameters. The stability analysis is critical since it allows decision-makers to understand how system parameters impact the system's dynamics. Secondly, the paper investigates the short-term impact of change in environmental regulations on the system performance concerning system cost and service level, thereby shedding light on the relationship between sustainability considerations and system performance. This analysis is significant since it allows decision-makers to understand how environmental regulations impact system performance and try to identify optimal policies that balance sustainability objectives with economic and social considerations.

The paper's main contribution is that it offers the first step towards a deeper understanding of the role of *EPICS* in achieving sustainable supply chain management. The paper seeks to provide decision-makers with an initial framework for balancing competing objectives in supply chain management by conducting a stability-based cost and service level analysis of *EPICS*. Specifically, the paper seeks to provide insights into the short-term impact of sustainability considerations on service levels and system costs, enabling decision-makers to make informed decisions about balancing environmental, economic, and social objectives. Overall, the paper contributes to the emerging body of literature on sustainable supply chain management and highlights the importance of incorporating a multidimensional approach to sustainability that considers economic, environmental, and social dimensions.

1.1 Literature Review

Recent literature suggests using Economic Order Quantity (*EOQ*) techniques to order optimal quantities with emission as a system constraint. *EOQ* model includes an emission cap mechanism for emission reduction and a total cost trade-off between costs and emissions [1]. An analytical support emission reduction without significant cost increment is observed for these tradeoffs [2]. Heavily relied on an optimization approach with emission caps [3], these static approaches fail to account for the dynamic nature of environmental considerations. A transition from a linear approach to a dynamic approach of systems thinking is required as sustainability requires a process-based approach instead of a goal-oriented view [4].

GSCM focuses on incorporating environmentally friendly practices into operations [5] by adopting green practices from environmental pressures (regulations, consumer preferences, and stakeholder expectations) [6]. Pressure toward green practices broadly includes adopting advanced technologies, sustainable design, and introducing new development of green metrics [7]. A positive relationship can be observed between adopting green practices and system (supply chain) performance for reduced cost, improved quality, and enhanced sustainability. Among multiple stakeholders, two relevant pressure groups: customers (as demand creators) and government (as regulators) [5], significantly assert environmental pressure for green practices. Emission regulations like strict emission caps, taxes on emissions, cap-and-offsets, and cap-and-trade can significantly increase the value of supply chain collaboration [1]. An Emission Permit System (*EPS*) is an overall class of various market-based systems that certify firms for compliance based on mandated government regulations [8]. *EPS* keeps the firm under stakeholders' (governmental and public) scrutiny and requires dealing with institutional pressure from these stakeholders [9].

The stock management structure has been immensely helpful in understanding various supply chain dynamics, such as the bullwhip effect, the impact of delays, ordering stability, etc. [10]. The popular (well-studied) version of the same is the Inventory and Order-based production control systems (*IOBPCS*) [11] family of models. These models include the supply line, the level of available inventory, and the expected incoming demand [12]. Feedback in *IOBPCS* from desired inventory creates an automated pipeline inventory and order-based production control system (*APIOBPCS*) [13] by considering discrepancies in work-in-process or supply line providing stability. *APIOBPCS* is considered for a fixed desired inventory target, and substantial literature is available for system stability and controllability [14–18], the effect of information update [17].

Emission-based Production and Inventory Control System (*EPICS*) integrates an emission module based on Emission Permit Scheme (*EPS*) into the Production and Inventory Control System (*PICS*). The prevailing literature focuses on the dynamics of emission feedback, namely, Perceived Emission Rate (*PER*) and Cumulative Net Emission (*CNE*) [19]. Also, the impact of emission consideration on system stability in *EPICS* for the continuous domain is available in [20]. Herein, the term stability of the system refers to the ability to overcome the effects of disturbances and attain a state of equilibria. Based on the Bounded Input Bounded Output (*BIBO*), a system can respond into stable, unstable, or critically stable for certain parameter setups from the state of equilibria. Since service level and costs are affected by the control parameters and smoothing factor of demand forecasting thus, *APIOBPCS* performs variably within the stable region [10]. Using Laplace transformation, an initial analysis of *EPICS* is already available in literature where a full-weightage (refer to Sect. 3.1 for more details) to emission feedback shifts the stable regions slightly (insignificant compared to *APIOBPCS*) on $(\beta, \alpha)-$plane but significantly reduces system emission [20].

In this article, discrete-time (z-transforms) is performed on *EPICS* to identify simulation-based stability bounds and the system's performance on sustainability dimensions. Motivated by *EPICS* [20], this paper aims to study the short-term system performance of *EPICS* on sustainability dimensions, Average System Cost (*ASC*) as system economic objective, and Item Fill Rate (*IFR*) i.e., service level as system social objective. This paper is organized as follows: Sect. 2 describes the *EPICS* model with model description, equations, and performance measures. Section 3 deals with Stability Analysis for *EPICS* in the discrete-time domain and related concepts, followed by Sect. 4 for Simulation Experimentations and Results on performance measures with some discussions. Finally, conclusions are presented in Sect. 5 related to the simulation study, future work, and some limitations.

2 Emission-Based Production and Inventory Control System

2.1 Model Description

The Emission-based Production and Inventory Control System (*EPICS*) model analyzed in this article is adapted as an extension of *APIOBPCS* with an emission module [20]. Initially introduced in [19] with non-linearity constraints on *SR* (lost sales instead of system backlogs), dynamics are analyzed for Perceived Emission Rate (*PER*) and Cumulative Net Emission (*CNE*) Feedback. It assumes a continuous review policy (immediate information update in the system) and fails to discuss system performance against costs and service level. The classical assumption of an infinite capacity of storage and processing stages is assumed to avoid the non-linearity and saturation effect. A periodic-review approach is applied due to its wide applicability in inventory management problems. For the simulation setup in this analysis, customer demand (*CD*) is an independent and identically distributed (*i.i.d*) as normal demand with mean μ_D and standard deviation σ_D. Note that any unfilled demand for the system (stock-out) is assumed to be a system backlog and satisfied in the future period(s).

From [19], *CNE* Feedback helps to achieve a net-zero target, whereas, *PER* Feedback deals with some operational adjustment based on a change in emission perception. Interestingly, *PER* Feedback is sandwiched between *CNE* feedback and no feedback. A combination of feedback from PER and CNE provides flexibility to reach net zero. This net zero target requires a long-term investment using capacity upgradation [20]. But choice control parameters (adjustment rates) in *EPICS* lead to a better balance system tradeoff and prevent undesired system instability. The choice of adjustment rate reduces the system's environmental footprint but compromises the system service level [20]. Further, incorrect choice of adjustment rate can bring undesired havoc to the system. Identifying stable parameters is necessary to measure the system's performance.

2.2 System Model and Equations

The EPICS structure in Fig. 1 below is a stock-flow diagram representing information and material flow with accumulation as stocks. As a single manufacturing firm with a build-to-order strategy, work-in-process (*WIP*) and inventory (*INV*) are two stocks maintained as per desired level in the production process. Production Release (*PREL*) as an inflow and outflow as Production Completion Rate (*PCR*) drives the production stage referred to as *WIP* in the system. Similarly, *INV* for the system is an accumulation of net flow between the input *PCR* and Sales Rate (*SR*). The flow of material in the system is represented from placing orders to eventual sales, i.e., *PREL* to *SR*.

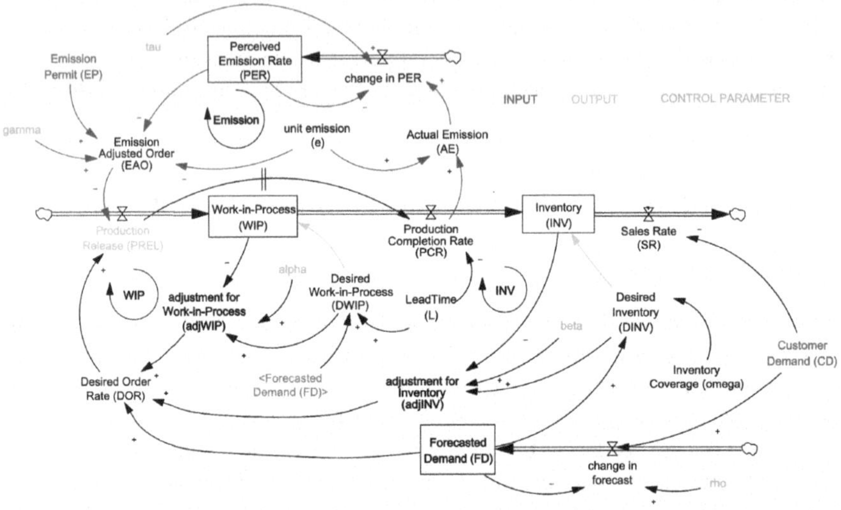

Fig. 1. Emission-based Production and Inventory Control System (*EPICS*) in Stock-Flow.

As an information flow to the system, demand is forecasted as *FD* for the future using first-order exponential smoothing with a smoothing rate 'ρ'. The push approach of customer demand, and *CD* drives the Production-Inventory (*PI*) system. *FD* with adjustment to *WIP*, *adjWIP;* and adjustment to *INV*, *adjINV* contribute as information to compensate by placing an order at *PREL* for the system moving from the desired state due to *SR*. *AdjWIP* and *adjINV* both use an anchor and adjustment heuristic to adjust the discrepancy between desired level and the current level with some adjustment rate [21]. For the desired level of stock, Little's Law provides insight into the level of desired throughput for a given lead time (for *DWIP*) and inventory coverage (for *DINV*) [22]. In addition to the above information feedback from system emission, Emission Permit (*EP*) as a soft constraint provides additional feedback to adjust *PREL*. Based on the assumption [20], *PCR* only contributes to the system's actual emission with an inventory coverage as ω.. First-order exponential smoothing is used to smoothen Actual Emission (*AE*) and update the Perceived Emission Rate (*PER*) with smoothing rate τ. *PER* is the forecasted emission used to adjust for orders and not account for the past emission excess. Finally, the discrepancy between *EP* and *PER* with adjustment rate is adjusted to *PREL* as emission feedback.

The above system has three balancing loops; two as per the classical model: the *WIP*-balancing loop and the *INV*-balancing loop. Furthermore, a delay involved in decision-making (as well as due to lead time) in the system introduces oscillations (or bullwhip to a system with multiple players). Intending to minimize the environmental footprint, the *PER*-balancing loop (as a third balancing loop) tries to balance *PER* level to the current *EP* in the system.

Equation (1) below represents the Production Release (*PREL*) of a period as the sum of forecasted demand (*FD*), adjustment to work-in-process (*adjWIP*), adjustment to inventory (*adjINV*), and emission-adjusted orders (*EAO*) to move the system back to the desired state. Forecasted Demand (*FD*) denoted in Eq. (2) is a first-order exponential smoothing with smoothing factor ρ. . Adjustment to work-in-process (*adjWIP*) and inventory (*adjINV*) signifies the fractional rate of adjustment with adjustment rate, α and β in Eqs. (3) and (4), respectively. Adjustments are required to achieve the desired level from the current level of *WIP* and *INV*. Equation (5) and (6) below gives *WIP* and *INV* accumulation based on net flow for a period. Production Completion Rate (*PCR*) is modeled as a fixed pipeline delay in Eq. (7) with lead time *L*.

Emission module feedback is presented in Eqs. (8–10). An Emission Permit (*EP*) as a constraint to the *PI* system is the desired level for the Perceived Emission Rate (*PER*). Emission Adjusted Orders (*EAO*) in Eq. (8) is a rate of adjustment for orders based on the discrepancy between system *EP* and *PER* level. Equation (9) below uses a first-order exponential smoothing of Actual Emission (*AE*) with smoothing rate τ to update *PER* level. Further, Eq. (10) represents *AE* as a function of *PCR* and is directly proportional to *PCR* with a constant unit emission *e*.

$$PREL_n = FD_n + adjWIP_n + adjINV_n + EAO_n \tag{1}$$

$$FD_n = FD_{n-1} + \rho(FD_{n-1} - CD_{n-1}) \tag{2}$$

$$adiWIP_n = \alpha(DWIP_n - WIP_n) \tag{3}$$

$$adjINV_n = \beta(DINV_n - INV_n) \tag{4}$$

$$WIP_n = WIP_{n-1} + PREL_{n-1} - PCR_{n-1} \tag{5}$$

$$INV_n = INV_{n-1} + PCR_{n-1} - SR_{n-1} \tag{6}$$

$$PCR_n = PREL_{n-L} \tag{7}$$

$$EAO_n = \gamma \frac{EP_n - PER_n}{e} \tag{8}$$

$$PER_n = PER_{n-1} + \tau(PER_{n-1} - AE_{n-1}) \tag{9}$$

$$AE_n = e \cdot PCR_n \tag{10}$$

Figure 1 above provides an overview of the proposed *EPICS* model with black causal links per the classical *APIOBPCS* model and blue causal links per the proposed extension from the emission module. It is to be noted that the system with green variables represents control parameters, whereas variables labeled red are input to the system and considered exogenous variables.

2.3 Performance Measure Modeling: Service Level, System Cost, and Environment Cost

Order Fill Rate (*OFR*); Item Fill Rate (*IFR*); Average Holding Cost (*AHC*), and Average Backlog Cost (*ABC*) [10] and Average Emission Cost (*AEC*) are five performance measures selected to analyze the system behavior under the stable tunning parameter of *EPICS*. Below are set of equations used to model these performance measures. Equation (12) represents a measure for *OFR*; Eq. (13) denotes *IFR* and Eqs. (14) and (15) cost associated with the system (holding and backlog); *AEC* (carbon cost) is represented in Eq. (16) and Eq. (17) denotes Average System Cost (*ASC*).

$$OFR_n = \frac{1}{n}\sum\nolimits_{i=1}^{n} OF_i \tag{11}$$

$$OF_i = \begin{cases} 1, \ INV_i \geq CD_i \\ 0, \ \ otherwise \end{cases} \tag{12}$$

$$IFR_n = 1 - \frac{\sum_{i=1}^{n} INV_i^-}{\sum_{i=1}^{n} CD_i} \tag{13}$$

$$AHC_n = \frac{1}{n}\sum\nolimits_{i=1}^{n} C_h \cdot INV_i^+ \tag{14}$$

$$ABC_n = \frac{1}{n}\sum\nolimits_{i=1}^{n} C_b \cdot INV_i^- \tag{15}$$

$$AEC_n = \frac{1}{n}\sum\nolimits_{i=1}^{n} C_e \cdot (EP_i - PER_i)^- \tag{16}$$

$$ASC_n = AHC_n + ABC_n + AEC_n \tag{17}$$

where,

$$INV_i^- = \begin{cases} -INV_i, INV_i < 0 \\ 0, \qquad\qquad otherwise \end{cases}$$

$$INV_i^+ = \begin{cases} INV_i, INV_i \geq 0 \\ 0, \qquad\qquad otherwise \end{cases}$$

$$(EP_i - PER_i)^- = \begin{cases} PER_i - EP_i, PER_i - EP_i > 0 \\ 0, \qquad\qquad\qquad\qquad otherwise \end{cases}$$

3 Stability Analysis for Emission-Based Production and Inventory Control System (EPICS)

To determine stability boundaries in the discrete-time domain, z-transforms are used on Eqs. (1–10). Since, the performance measures from Eq. (11–17) in this study do not impact system feedback, thus ignored for stability consideration. The difference equations transformed to z-domain and given in Eq. (18–27) below, respectively.

$$PREL(z) = FD(z) + adjWIP(z) + adjINV(z) + EAO(z) \tag{18}$$

$$FD(z) = \frac{\rho}{\rho + z - 1} CD(z) \tag{19}$$

$$adiWIP(z) = \alpha(DWIP(z) - WIP(z)) \tag{20}$$

$$adjINV(z) = \beta(DINV(z) - INV(z)) \tag{21}$$

$$WIP(z) = \frac{(PREL(z) - PCR(z))}{z - 1} \tag{22}$$

$$INV(z) = \frac{PCR(z) - CD(z)}{z - 1} \tag{23}$$

$$PCR(z) = \frac{PREL(z)}{z^L} \tag{24}$$

$$EAO(z) = \gamma \frac{EP(z) - PER(z)}{e} \tag{25}$$

$$PER(z) = \frac{\tau}{\tau + z - 1} AE(z) \tag{26}$$

$$AE(z) = e \cdot PCR(z) \tag{27}$$

EPICS under consideration has two exogenous variables, Customer Demand (*CD*) and Emission Permit (*EP*) (assuming $CD(z) = u(z)$ and $EP(z) = k(z)$). Above, Eqs. (18–27) in 'z' are solved to obtain the transfer function between system output *PREL* and two input *CD* and *EP*. The resulting system is a Multiple-Input and Single-Output (MISO) system and can be solved independently as two Single-Input and Single Output (SISO) systems to gain a transfer function. Equation (28) below represents the transfer function for the MISO system when solved independently.

$$C(z)PREL(z) = \begin{bmatrix} C_u(z) & C_k(z) \end{bmatrix} \begin{bmatrix} u(z) \\ k(z) \end{bmatrix} \tag{28}$$

where, $C(z)$: coefficient of $PREL(z)$; $C_u(z)$: coefficient of input *CD*; $C_k(z)$: coefficient of *EP*. Equation (28) coefficients are represented below in Eq. (29–31) in terms of control parameters on which stability is dependent.

$$C(z) = 1 + \frac{1}{z^L} \left\{ \frac{\gamma\tau}{\tau + z - 1} + \frac{1}{z - 1} \left(\alpha z^L + \beta - \alpha \right) \right\} \tag{29}$$

$$C_u(z) = \frac{\rho(1 + L\alpha + \beta\omega)}{\rho + z - 1} + \frac{\beta}{z - 1} \tag{30}$$

$$C_k(z) = \frac{\gamma}{e} \tag{31}$$

Further, the system is assumed to have a fixed pipeline delay *L*- time units with an inventory coverage ω. Equation (28) above is a general expression for the transfer function with α as the adjustment rate for work-in-process, β as the adjustment rate for inventory, and γ as the adjustment rate for emission feedback. 'τ' as the smoothing rate for Perceived Emission Rate (*PER*) and intuitively represents the rate of *PER* update for emission.

3.1 Characteristic Polynomial

Using the above generic transfer function system is *Bounded-Input Bounded Output* (BIBO) stable *if and only if* the denominator polynomial (also called characteristic polynomial) has poles within the unit circle of the complex plane [23]. Equation (32) below gives the general expression for characteristic polynomials.

$$p(z) = e(\rho + z - 1)(z - 1)\left[z^L(\tau + z - 1)(z - 1) + \tau\gamma(z - 1) + (\tau + z - 1)\left(\alpha z^L + \beta - \alpha\right)\right] \quad (32)$$

In Eq. (32), e represents a constant rate per unit emission. Thus, for any $e > 0$, the polynomial $p(z)$ is stable and insignificant to stability boundaries. Also, the factor $(\rho+z-1)$ in characteristic polynomial will represent a stable point only if $0 \leq \rho < 2$. But the literature suggests $0 \leq \rho \leq 1$ as the smoothing rate of demand and cannot exceed 1. So, factor $(\rho+z-1)$ also represents a stable pole of the system [10]. Furthermore, factor $(z-1)$ in Eq. (32) is also stable and insignificantly contributes to stability conditions as it has a unit root.

$$p'(z) = z^L(\tau + z - 1)(z - 1) + \tau\gamma(z - 1) + (\tau + z - 1)\left(\alpha z^L + \beta - \alpha\right) \quad (33)$$

Note that under a scenario when $\gamma = 0$ (no emission feedback), the above transfer function is reduced to APIOBPCS, available in classical literature [10]. Furthermore, $\tau = 0$ represents no update in the system *PER*-level i.e., a constant adjustment to *PREL* (completely exogenous) to the state of the system as feedback, $EAO_n = \gamma \frac{EP_n - PER_0}{e}$. Therefore, system stability boundaries are intact with classical setup for any value of $\gamma \in R^+$ under $\tau = 0$. Moreover, for $\tau = 1$ system has an immediate change in the perception of emission. This scenario (assuming $\gamma = 1$) under the continuous domain is already discussed in [20] where stability boundaries are slightly shifted from the original setup of APIOBPCS on $(\beta, \alpha)-$ plane.

3.2 Stability Boundaries for EPICS

Polynomial $p\prime(z)$ in Eq. (33) is reduced to study stability boundaries for specific parameter setup. The stability boundaries are independent of the smoothing demand forecast factor ρ and a fixed inventory coverage ω but they significantly contribute dynamics of system as well as service level and cost [10]. So, to identify the stable region in terms of control parameters, $(\alpha, \beta, \gamma, \tau)$ system is assumed to have fixed lead time, $L = 3$.

$$p'(z) = z^5 + (\alpha + \tau - 2)z^4 + (\alpha\tau - \alpha - \tau + 1)z^3 + (\gamma\tau + \beta - \alpha)z + ((\beta - \alpha)(\tau - 1) - \gamma\tau) \quad (34)$$

Section 4 performs a simulation study using Exploratory Modeling Analysis (*EMA*) under stability boundaries. Service level (Order Fill Rate and Item Fill Rate) against System Cost (Average Holding Cost, Average Backlog Cost, and Average Emission Cost) as a system metric used for short-term sustainability. Control parameters from $(\gamma, \theta)-$ plane in Fig. 2 is used to understand the impact of PI-control ($\alpha = \beta = \theta$) and Emission-control (γ) on system performance.

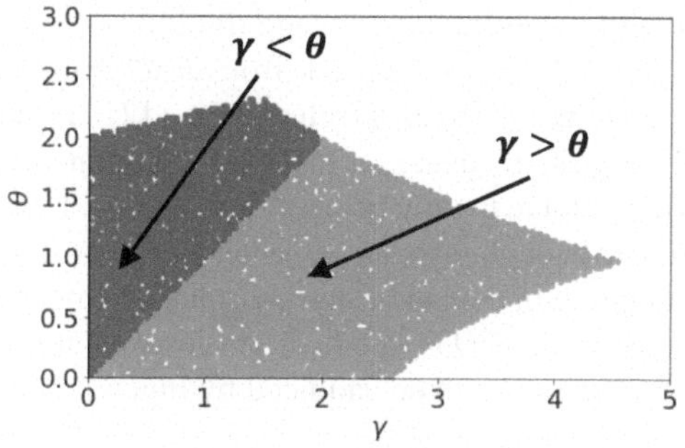

Fig. 2. Sampled points from $(\gamma, \theta)-$ plane to test system performance.

4 Simulation Experiments and Results

4.1 Initial Setup

EPICS is modeled into Venism® using Eq. (1–10) above with an initial state of dynamic equilibrium (inflow and outflow for any stock are equivalent). A randomly normal distribution with mean $\mu_D = 100$ and standard deviation $\sigma_D \in \{1, 5, 10, 20\}$ *i.i.d.* customer demand (*CD*) is considered in setup. Further, a step-decrease in Emission Permit (*EP*) by 20% from the initial *EP* level is assumed. A build-to-order strategy in production strategy to the system ensures no inventory coverage to the system i.e., $\omega = 0$ to reduce holding cost. Cost coefficients, $[\, C_h \; C_b \; C_e \,]$ corresponding to this simulation study are $[\, 1 \; 2 \; 8 \,]$ based on [10] and \$9.29/*mtCO2e* [24].

This simulation study is restricted to understanding the impact of *PI*-control parameters ($\alpha = \beta = \theta$) and Emission-control parameters (γ) on system performance without policy intervention. Both smoothing rates, ρ and τ are set to 0.2 to simplify computational experiments. An open-source toolkit, PySD [25] is used for the simulation at a one-time step under Euler integration by integrating the system model into Python. A total of 24321 sample points were randomly generated in $\gamma \times \theta$ in $[0, 5] \times [0, 3]$ for $(\gamma, \theta)-$ plane.

4.2 Simulation Result

Using Eq. (34), stable region is identified on $(\gamma, \theta)-$ plane under two operational regimes where $\gamma > \theta$ implying higher weightage to emission considerations (refer to the green region in Fig. 2) and for $\gamma < \theta$ where *PI*-control for higher weightage in the decision process (refer to the red region in Fig. 2). This segregation is relative to the position of γ and θ against $\gamma = \theta$ and an analysis is presented in below to study the impact of coefficient of variation, $\nu \in \{0.01, 0.05, 0.1, 0.2\}$ for system performance. Note that in Fig. 2 concentration of stable point is relatively higher for $\gamma > \theta$ implying higher flexibility in exploring system performance under this operational regime.

Figure 3 below represents the concentration of stable points to *OFR* metric under varying ν such that grey bins representing $\nu = 0.01$ (base case) against $\nu \in \{0.05, 0.1, 0.2\}$. *OFR* is defined as a fractional instance at which demand is fully satisfied.

Therefore, the system with no inventory coverage (build-to-order strategy) is expected to have a shortfall in demand due to feedback from *EAO* resulting in reduced *PREL* than desired order rate causing fall in *OFR* (refer to Eq. (12)). From Fig. 3 below in the base case of $\nu = 0.01$ once the feedback of *EAO* is incorporated into *PREL* following subsequent demand will continue to only be partially satisfied. Low variability of 1% in demand cannot accumulate enough inventory to cater *CD* for a 20% decrease in *EP*. On the contrary, higher variability in demand causes surplus inventory to build up and have higher *OFR* as in some instances *CD* is satisfied. This argument for an increase in *OFR* with increased demand variability can be validated by Fig. 3.

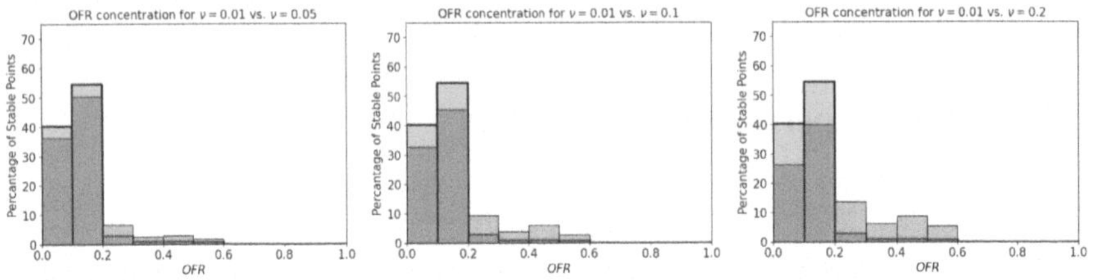

Fig. 3. Histogram representing frequency stable points for Order Fill Rate (*OFR*) under the varying $\nu \in \{0.01, 0.05, 0.1, 0.2\}$

IFR for the system is defined as a fraction of cumulative demand satisfied against cumulative system demand by the customer. For visualization, *IFR* is labeled into $\{(-\infty, 0), [0, 0.2), [0.2, 0.5), [0.5, 0.75), [0.75, 0.9), [0.9, 1)\}$ under stable region as per Fig. 2. Classical PICS refrains the system from building up the system backlog by ordering sufficiently. But, under EPICS setup a significantly higher weightage to emission control against negligible weightage to *PI*-control causes cumulative backlog to be more than cumulative demand. This dominating feedback causes *IFR* to be negative (for more detail refer to Eq. (13)) and can be validated from Fig. 4 in label $(-\infty, 0)$.

As expected, when weightage to *PI*-control is higher than Emission-control, higher *IFR* is observed and can be validated from the below figure. The system assumes a 20% reduction in *EP* i.e., an available 80% as an emission allowance. Hence, ensuring *IFR* more than 75% under $\gamma < \theta$ for the stable region. Furthermore, as variability increases, *IFR* slightly reduces under the same operational regime for increased γ (refer to the box in Fig. 4). Under $\gamma > \theta$, an accumulation of backlog due to higher Emission-control weightage causes *IFR* for the system to significantly reduce.

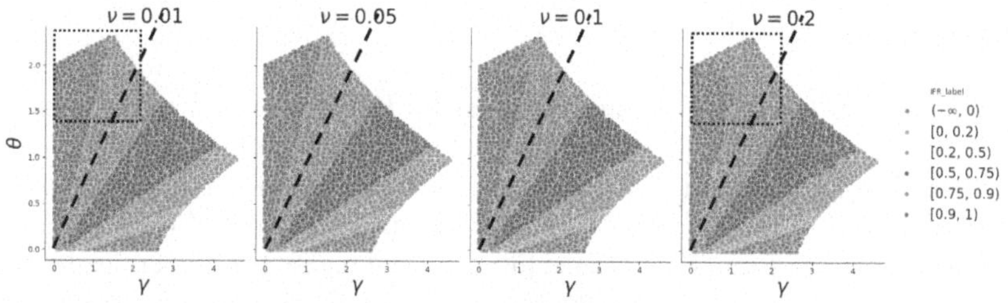

Fig. 4. Item Fill Rate *(IFR)* performance in stable region for varying $\nu \in \{0.01, 0.05, 0.1, 0.2\}$

Figure 5 below is the average system cost, *ASC* considering holding, backlog, and carbon cost labeled into $\{(0, 1750), [1750, 2500), [2500, 2750), [2750, 3000), [3000, \infty)\}$. For $\gamma < \theta$ operational regime, a higher weightage to *PI*-control helps the system to reduce *ASC*, ensuring it is almost below 1750. As variability in demand increases, *ASC* also increases for higher weights of control parameters (refer to boxes in Fig. 5). More importantly, there is no significant change in the cost in $\gamma > \theta$ for varying demand. *IFR* observed in Fig. 4 as negative implies the system has high backlogs due to higher adjustments from *EAO*. Therefore, simulation results are expected to have high system costs for backlog which can be validated with $[3000, \infty)$ label in Fig. 5.

A higher weightage to *PI*-control implies the system incurs higher carbon (emission) costs instead of insignificant holding costs and low backlog costs. Under such a setting, a significant contribution to *ASC* is from emission (carbon) cost for short-term adjustment. Interestingly, for $\gamma > \theta$ regime, higher Emission-control weightage causes the system to incur a higher backlog cost. From Fig. 5, a clear conclusion can be drawn about *ASC* as the system has a higher cost when Emission-control weightage is higher because of higher backlog cost. Thus, a clear trade-off is visible in choosing control parameters as $\gamma > \theta$ can cause higher backorder cost, and $\gamma < \theta$ contribute higher carbon cost to *ASC*.

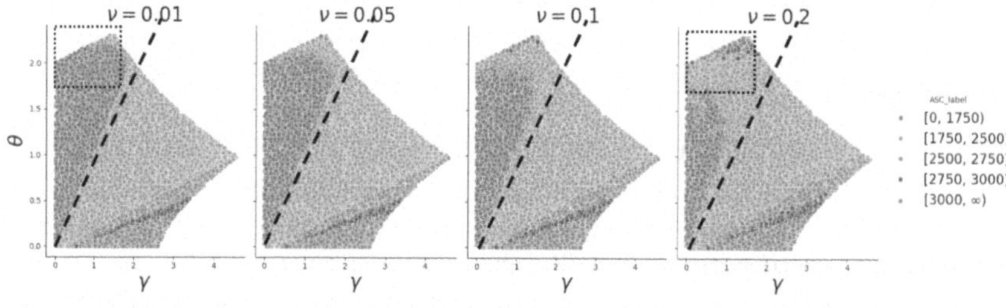

Fig. 5. Average System Cost *(ASC)* in stable region for varying $\nu \in \{0.01, 0.05, 0.1, 0.2\}$

5 Conclusions

Three tunning parameters and the smoothing rate for *PER* define the stability boundaries of *EPICS*. On the contrary, these four parameters along with the smoothing rate for demand and inventory coverage define dynamics in the system. In this study, stability analysis is performed for two control parameters namely, *PI*-control ($\alpha = \beta = \theta$) and Emission-control (γ). From Eq. (34) and Fig. 2, $\gamma < \theta$ region has higher flexibility as a stable region concentration is higher than $\gamma > \theta$ region. Further, as discussed above for system performance on service level, *OFR* is inefficient in providing anything concrete to balance trade-off against system cost. Interestingly, *IFR* as a key performance index to the system provides a better understanding in Fig. 4. Also, a negative *IFR* in the stable region is an undesired state to the system due to higher cumulative backlog than actual cumulative demand. A counterintuitive argument about negative *IFR* can be validated because higher Emission-control results in order cancellation/return (negative *PREL*).

Service level and System cost are two conflicting objectives for a system to be optimized where service level must be maximized with minimal cost. Classical setup involves two complementary costs, holding and backlog cost with a tradeoff between two costs determined depending on the relative positioning of α and β [10]. But emission consideration introduces emission (carbon) cost to system cost which further requires a detailed analysis based on the relative position of α, β, and γ. Negative *IFR* due to high backlog (order cancellation/return resulting in negative inventory from higher Emission-control) and further incurring high backlog cost. Under this scenario, no emission or holding cost is incurred; thus, *ASC* is driven by backlog cost only. An additional constraint on parameter selection to refrain from negative *IFR* is further required. With low holding cost (due to zero inventory target) and backlog cost, a significant contribution from emission (carbon) cost to total cost from violation of emission permit causes varying dynamics in Fig. 5 (refer to boxes). Higher weightage to γ causes system to perform inefficiently on performance measures. More importantly, in short term measures $\alpha = \beta = \theta < \gamma$ performs better and a part of future work a detailed analysis is required from α, $\beta < \theta$. This study tries to understand impact of sustainability considerations on short-term dimensions without any policy interventions.

5.1 Limitations

EPICS relies on proportional to the Production Completion Rate (*PCR*) as actual emission and depends solely on the production process as an emission source. Such a setup introduces carbon leakages at various levels and limits the above setup from proposing more robust and generic policies. Also, the above proposed aggregate box model fails to find support for applicability due lack of realistic data to validate and model calibration. Importantly, the sustainability of a system requires dealing with the long-term consequences of the system. This study is restricted to understanding the short-term impact on sustainability considerations in *EPICS* and fails to provide managerial insights.

5.2 Future Work

A stable system with a negative *IFR* is the undesired state for the system. *PI*-control and Emission-control are insufficient and require a detailed analysis to study system performance for (α, β, γ) parameters and ensure positive *IFR* with a better trade-off between service level and system cost. A long-term impact of policy interventions is required to determine system tradeoffs on proposed sustainability dimensions for sustainability in the system. A possible future direction is integrating capacity upgradation and technology transition modules and applying the *EMA* technique (both involve high uncertainty and delays) to develop robust policies and managerial implications. Furthermore, system boundaries in *EPS* can be stretched to incorporate a generic carbon market.

References

1. Benjaafar, S., Li, Y., Daskin, M.: Carbon footprint and the management of supply chains: insights from simple models. IEEE Trans. Autom. Sci. Eng. **10**(1), 99–116 (2012)
2. Hua, G., Cheng, T.C.E., Wang, S.: Managing carbon footprints in inventory management. Int. J. Prod. Econ. **132**(2), 178–185 (2011)
3. Chen, X., Benjaafar, S., Elomri, A.: The carbon-constrained EOQ. Oper. Res. Lett. **41**(2), 172–179 (2013)
4. Hjorth, P., Bagheri, A.: Navigating towards sustainable development: a system dynamics approach. Futures **38**(1), 74–92 (2006)
5. Seuring, S., Müller, M.: From a literature review to a conceptual framework for sustainable supply chain management. J. Clean. Prod. **16**(15), 1699–1710 (2008)
6. Lee, S.M., Kim, S.T., Choi, D.: Green supply chain management and organizational performance. Indust. Manage. Data Syst. **112**(8), 1148–1180 (2021)
7. Hervani, A.A., Helms, M.M., Sarkis, J.: Performance measurement for green supply chain management. Benchmark. Int. J. **12**(4), 330–353 (2005)
8. Zhou, J., et al.: A review of development and reform of emission permit system in China. J. Environ. Manage. **247**, 561–569 (2019)
9. Sun, Y.H.: How to improve and implement the pollution discharge license system. Environ. Protect. **14**, 16–17 (2014)
10. Bijulal, D., Venkateswaran, J., Hemachandra, N.: Service levels, system cost and stability of production–inventory control systems. Int. J. Prod. Res. **49**(23), 7085–7105 (2011)
11. Towill, D.R.: Dynamic analysis of an inventory and order based production control system. Int. J. Prod. Res. **20**(6), 671–687 (1982)
12. Disney, S.M., Towill, D.R.: Eliminating drift in inventory and order based production control systems. Int. J. Prod. Econ. **93**, 331–344 (2005)
13. Simon, J., Naim, M.M., Towill, D.R.: Dynamic analysis of a WIP compensated decision support system. Int. J. Manuf. Syst. Des. **1**(4), 283–297 (1994)
14. Dejonckheere, J., et al.: Measuring and avoiding the bullwhip effect: a control theoretic approach. European J. Operat. Res. **147**(3), 567–590 (2003)
15. Disney, S.M., Towill, D.R., Van de Velde, W.: Variance amplification and the golden ratio in production and inventory control. Int. J. Prod. Econ. **90**(3), 295–309 (2004)
16. Ortega, M., Lin, L.: Control theory applications to the production–inventory problem: a review. Int. J. of Prod. Res. **42**(11), 2303–2322 (2004)
17. Venkateswaran, J., Son, Y.-J.: Effect of information update frequency on the stability of production–inventory control systems. Int. J. Prod. Econ. **106**(1), 171–190 (2007)

18. Disney, S.M., Towill, D.R.: A discrete transfer function model to determine the dynamic stability of a vendor managed inventory supply chain. Int. J. Prod. Res. **40**(1), 179–204 (2002)
19. Deval, R., Venkateswaran, J.: Production and inventory control system dynamics under emission feedback. In: Proceedings of International System Dynamics Conference (2022), Frankfurt, Germany URL for PaperID_#1298 at System Dynamics Conference (2022)
20. Deval, R., Venkateswaran, J.: Stability analysis of emission-based production and inventory control systems (EPICS). In: 2022 IEEE International Conference on Industrial Engineering and Engineering Management (IEEM). IEEE (2022)
21. John, S.: *Business dynamics.* Irwin/McGraw-Hill c2000.., (2010)
22. Venkateswaran, J., Hasti, C.: Stability of production-inventory control systems considering inventory shortages. *Industrial Engineering & Operations Research Indian Institute of Technology Bombay, Powai, Mumbai–400076, Extended Abstract* (2006)
23. Jury, E.I.: Theory and Application of the z-Transform Method. Wiley, New Jersy, USA (1964)
24. https://www.spglobal.com/commodityinsights/en/market-insights/latest-news/energy-transition/042522-china-data-carbon-emission-allowance-price-at-929mtco2e-april-22#article0
25. Houghton, J., Siegel, M.: Advanced data analytics for system dynamics models using PySD. revolution, 3(4) (2015)

Healthcare

Discrete-Event Modeling of Human Behavior for Spread of Diseases on University Campuses

Hazel Griffith, Cristina Ruiz-Martin(✉), and Gabriel Wainer

Carleton University, Ottawa, ON, Canada
cristinaruizmartin@sce.carleton.ca

Abstract. The COVID-19 pandemic has highlighted the importance of defining sound policies to make attending workplaces safe. Sometimes, deciding on different policies is challenging as this highly depends on the behavior of the individuals. We introduce a Discrete Event based method to study such policies, including human behavior along with information about the workplace layout and building characteristics such as ventilation rate or room capacity. We exemplify how to use this method using the case study of Carleton's University Campus. We introduce a case study focusing on the effect of the ventilation policy on the number of disease cases on campus.

Keywords: DEVS · disease spread · human behavior

1 Introduction

The COVID pandemic has revealed that crowded events or locations are hot spots for the spread of airborne viruses. When functioning at normal capacity, workplaces, including university campuses, can be considered crowded places where people congregate for a long period of time, usually for eight hours or more. For example, Carleton has over 30,000 students attending lectures and traveling the halls. Some lectures have over 200 students sitting in a closed space for at least 90 min.

On top of complying with the regulations, workplaces and university campuses cannot afford to get the disease spread among their employees or students because that will highly impact the daily activities of the organization. Therefore, a safe environment should be provided.

To provide a safer environment, we need to collect data on how different policies affect the spread of the disease. Doing this with real-life experiments is not viable as we may risk the health of the employees/students, seeing the effect of policies takes time, and many variables may affect the result in unpredictable ways. We may collect this data by running experiments using simulation models that realistically model how COVID (or other diseases) spread across the workplace or university campuses. Using these models, we can run experiments to study the effect of different parameters on the spread of the disease. For example, different ventilation rates in rooms or offices, the number of infected people attending work, or different room capacity limits. The data

M. Mujica Mota and P. Scala (Eds.): EUROSIM 2023, CCIS 2032, pp. 103–118, 2024.
https://doi.org/10.1007/978-3-031-68435-7_8

obtained from these experiments can be used to inform new strategies, policies, and identify hotspots in the workplace.

Although there are already many simulation models to study the spread of COVID-19, there is limited work focusing on workplaces, including human behavior, workplace layout, and building characteristics in the study.

In this research, we approach this limitation by building a model that includes the above-mentioned aspects. More specifically, we developed a method to study the spread of diseases in workplaces considering human behavior and workplace characteristics. We base our method on the architecture to study diffusion processes in dynamic multiplex networks (ADPM) presented in [1]. We adapted the architecture to study the spread of diseases in a workplace and exemplify the use with the case study of the Carleton University Campus.

2 Background

Different simulation models have been built and executed to study the spread of disease during the COVID-19 pandemic. Such simulations were useful for determining the impact of COVID-19 depending upon many factors, which include control policies, the area's physical layout, citizens' mobility, etc. Many of these models are based on the well-known Susceptible-Infected-Recovered models [2, 3]. These SIR models have evolved, and now they include other states (e.g., exposed, deceased) as well as geographical level transmission dynamics [4–8].

Other studies include the use of a novel Monte Carlo simulation procedure for modeling the spread of COVID-19 over time [9]. This study focuses on simulating the rate at which cases will appear in Australia and the UK based on knowledge of the virus and the initial number of cases, over a series of arbitrarily created scenarios. They calculated the day when the number of new cases per day would peak for both countries. Their results were found to be accurate, indicating that their model could be applied to other nations and pandemics.

In [10], the authors focused on the impact of urban structure on the spread of COVID-19. They focused on individual cities and how their physical layout affects the capacity of citizens to respond quickly to mobility-related policies. Such policies include stay-at-home orders, social distancing, etc. They found that while densely packed cities experience a greater initial infection rate, they are also easier to enforce and transmit mobility policies to citizens. They use a SIR-type compartmental model where simulated individuals flow between compartments that may be labeled as susceptible, infectious, or recovered. Their results proved that investing resources in early monitoring and prompt ad-hoc interventions in more vulnerable cities could make future pandemics easier to contain.

Many other works, as presented by Loring et al. in their literature review [11], consider human behavior and social relations such as workplace, profession, or household, among others. However, they do not focus on developing specific models to evaluate the risk of getting covid in the workplace. Some other studies have focused on addressing the risk of getting covid at the workplace by developing high-level probabilistic models [12].

In this work, we focus on addressing the above-mentioned limitation by developing a method to study the spread of diseases in workplaces considering human behavior and workplace characteristics.

To do so, we consider the spread of a disease as a diffusion process. Diffusion processes are phenomena that represent how an element may spread within a given medium. For example, you could consider the diffusion of traffic across a network of roads, or how information travels between people in an emergency response system. The connections over a multiplex dynamic network (i.e., a network with different types of connections) determine the medium through which the element travels and how it travels.

In [1], the authors propose ADPM (Fig. 1) to model these types of processes. There are components to model the nodes, the direct and indirect connections between nodes, how the connections between nodes change over time, how the behavior of the nodes changes over time, and how the diffusion process starts. They also proposed a development process to instantiate this architecture, as detailed in [1]. We base our method on this architecture. We adapt the architecture to study the spread of diseases in a workplace and exemplify the use with the case study of the Carleton University Campus. Following the original research in [1], we also define the models in DEVS [13] and implement them using Cadmium Simulator [14].

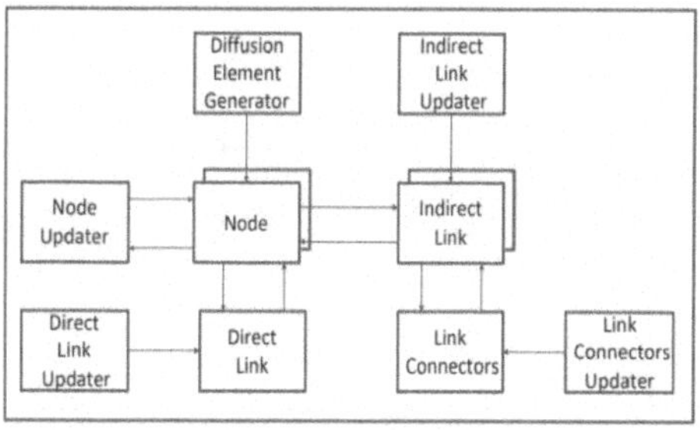

Fig. 1. Complete ADPM architecture.

3 An Architecture to Simulate the Spread of Diseases in Workplaces

In this section, we explain how we adapted ADPM [1] to the case study of the spread of diseases.

Figure 2 shows the elements of the ADPM used to study the spread of an airborne virus. We selected three of the components: Node, Direct Link and Node Updated. The Node model is instantiated with a Person model that will get and transmit the disease. The Node Updater model is instantiated with a Weather model as the changes in the weather will influence the behavior of people. Finally, the Direct Links model is instantiated with several Room models because interactions between people will happen in rooms. In the case of an airborne virus being spread across a population, an interaction in a Room could represent sneezing near another person.

In this case, Indirect links would represent an infected person handling an item and passing it to another person. With airborne viruses, the odds of a person becoming sick through indirect contact with a sick person are very low. Therefore, we have not included the part of the architecture that handles indirect interactions.

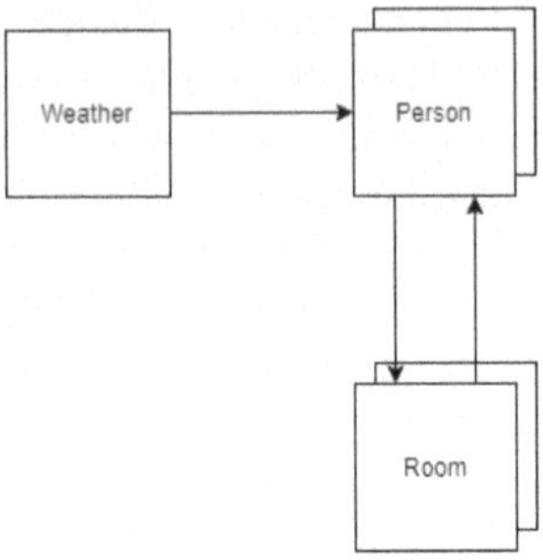

Fig. 2. The adapted ADPM architecture

Each model present in Fig. 2 is further expanded as shown in Fig. 3, and detailed in the rest of the section. Each component is defined as parameterized models using the DEVS formalisms.

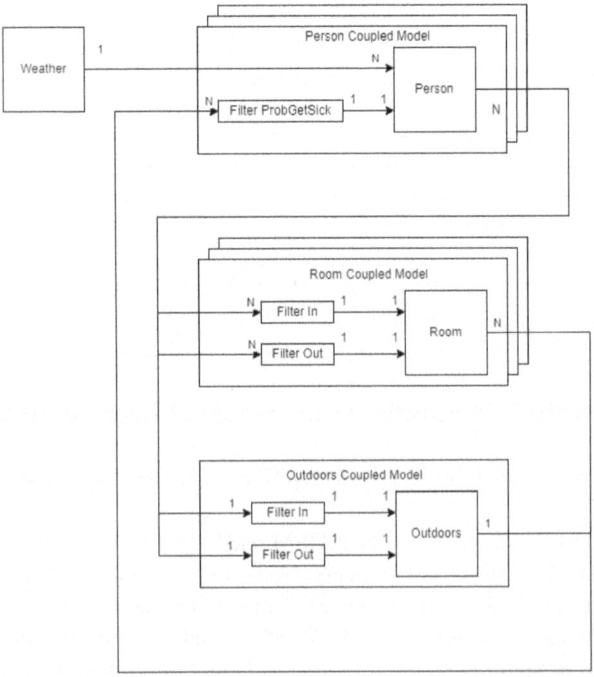

Fig. 3. Decomposition of the components in the adapted ADPM architecture

3.1 Person Model

The Person component is defined as a coupled model with two components: a filter to determine which messages are for that specific person and a person atomic model to represent the person's behavior.

The behavior of a person includes their daily schedules, interpersonal relationships, personal choices in terms of masking, social distance, how to move around the workplace, and their probability of being infected after exposure. The explained behavior is a function of the weather. For example, when the weather is mild, a person may choose a safer path to move around the workplace, but if it rains or snows, they will favor being inside.

Each person model will have a general relationship (e.g., family, acquittance, stranger, etc.) established with each other person to realistically model how they engage in safe behaviors or not.

In this case, the main behavioral factors we consider are if people would wear masks, social distance, and comply with maximum occupancy policies.

Although we have focused on those behavioral factors, the model can be extended to include others. The behavioral components to be included will depend on the disease under study and the experiments to be performed.

The Person model is defined as a parameterized model whose parameters are specified in an XML file, as shown in Fig. 4.

Each person has the following attributes:

- *ID*: a unique ID identifying the person
- *location*: the location of the person at the beginning of the simulation
- *currStartTime*: the time the person becomes active in the simulation
- *timeInFirstLocation*: the time the person will spend in the current location
- *isSick*: identifies if the person has the disease or not
- *exposed*: identifies if the person has been exposed to the disease or not
- *vaccinated*: identifies if the person is vaccinated or not
- *wearingMaskCorrectly*: identifies if the person, when wearing a mask, wears the mask properly fitted or not
- *socialDistance*: identifies if the person is prone to keep social distance or not
- *weatherThreshold*: identifies how prone the person is to engage in unsafe behavior when the weather conditions are not favorable. It is a value between 0 and 10. Zero represents low tolerance to bad weather.
- *relationship*: identifies the type of relationship with other persons in the model.
- *behaviorRulesPerson*: identifies the behavior patterns in terms of maintaining a safe social distance, wearing a mask, and entering a room at a maximum occupancy based on the type of relationship with a given person
- *locationPlan*: determines the person's schedule

```xml
<DecisionMakerBehavior>
<ID>3</ID>
<location>home</location>
<currStartTime>1280</currStartTime>
<timeInFirstLocation>760</timeInFirstLocation>
<isSick>False</isSick>
<exposed>False</exposed>
<vaccinated>False</vaccinated>
<wearingMaskCorrectly>True</wearingMaskCorrectly>
<socialDistance>True</socialDistance>
<weatherThreshold>6</weatherThreshold>
<relationship>
    <relationship ID="1" type="friends" />
    <relationship ID="4" type="acquaintances" />
    <relationship ID="5" type="acquaintances" />
    ...
</relationship>
<behaviorRulesPerson>
    <personRelations status="acquaintance" safeDistanceProb="50"
        maskWearingProb="50" enterMaxOccRoomProb="50" />
    <personRelations status="friends" safeDistanceProb="20"
        maskWearingProb="20" enterMaxOccRoomProb="20" />
  ...
</behaviorRulesPerson>
<locationPlan>
    <locationPlan room="38-VS - 2285" timeinroom="90" startTime="610 />
    <locationPlan room="Outdoors" timeinroom="10" startTime="700" />
    <locationPlan room="home" timeinroom="490" startTime="710" />
  ...
</locationPlan>
</DecisionMakerBehavior>
```

Fig. 4. Person XML file

3.2 Room Model

The Room component is defined as a DEVS coupled model with three atomic models, two filters to determine if the persons are coming or leaving that specific room, and a room atomic model modeling the room itself. Note that we are distinguishing a specific type of room called outdoors to represent outdoor locations where the probability of getting infected with an airborne virus is very low.

Each room atomic model determines the probability of a person being exposed to the virus based on several factors. These factors include whether the person and other people in the room wear a mask and maintain social distance, and how many sick people are in the room. We also use the current concentration of CO2 in the room as CO2 is usually used as a proxy for the number of viral particles in a room when there are sick people.

The Room model is also defined as a parameterized model whose parameters are specified in an XML file, as shown in Fig. 5. The XML specifies the following characteristics: the room's ID, its ventilation rate, social distance threshold, maximum occupancy, several adjustable factors that contribute to the probability of exposure calculations (mask-wearing, social distance, vaccination, etc.), the respiratory increase per minute, the size in square meters and the height.

```xml
<RoomParameters>
<ID>5,RH - 1100</ID>
<ventilationRating>3</ventilationRating>
<socialDistanceThreshold>35</socialDistanceThreshold>
<maxOccupancy>26</maxOccupancy>
<wearsMaskFactor>1</wearsMaskFactor>
<socialDistanceFactor>1</socialDistanceFactor>
<vaccinatedFactor>-2</vaccinatedFactor>
<sickPeopleCO2Factor>2</sickPeopleCO2Factor>
<highCO2FactorThresholds>1001,1840,1857,1904</highCO2FactorThresholds>
<highCO2Factors>1,2,3,4</highCO2Factors>
<respIncreasePerMin>340000</respIncreasePerMin>
<squareMetres>71.019997</squareMetres>
<height>2.438000</height>
</RoomParameters>
```

Fig. 5. Room XML file

Modeling CO2 Concentration

To model the CO2 concentration in the room, we first calculate the increase of CO2 because of breathing. According to Environmental Health Perspectives, "*Although typical outdoor CO2 concentrations are approximately 380 ppm, outdoor levels in urban areas as high as 500 ppm have been reported (Persily 1997).*"[15]. For our model, we define the outdoor concentration as 400 ppm. In [16], the authors state that "*an average adult exhale contains 35,000 to 50,000 parts per million (ppm) of CO 2 on each breath*"; in [17], the authors define the volume of a breath at 500 mL per breath; and in [18], the authors state that a person breaths between 12 and 20 times per minute. Using these values, we calculate the respiratory increase per minute as shown in the following equation. We get a respiratory increase of 340000 mg of CO2 per minute.

$$
respIncPerMin\left(\frac{mg}{min}\right) = \left(500\left(\frac{mL}{Breath}\right) * \left(\frac{35000ppm + 50000ppm}{2}\right)\left(\frac{\frac{1\,mg}{L}}{1ppm}\right) * \left(\frac{12\left(\frac{breaths}{min}\right) + 20\left(\frac{breaths}{min}\right)}{2}\right)\right) = 340000\frac{mg}{min}
$$

We then calculate the decrease in CO2 because of ventilation. The air changes per hour (ACH) are usually anywhere between 1 and 8, but this will be a parameter of our room model since different buildings and workplaces can choose different rates. The ventilation rate per minute is calculated as follows:

$$ventilationRatePerMin\left(\frac{L}{min}\right) = \frac{ACH * volume\ Of\ Room\left(m^3\right)}{60\ minutes} * \frac{1000L}{m^3}$$

The new CO2 concentration is calculated linearly based on the CO2 concentration flow as follows:

$$CO2\ Concentration\left(\frac{mg}{L}\right) = CO2\ Concentration\ Flow\left(\frac{mg}{L}\right) + \Pr evious\ CO2\ Concentration\left(\frac{mg}{L}\right)$$

CO2ConcentrationFlow represents the change in the CO2 concentration since the previous CO2 concentration was calculated. The CO2ConcentrationFlow is calculated by adding the CO2 entering the room from its ventilation per minute, the CO2 being breathed out by everyone in the room per minute, and subtracting the CO2 leaving the room per minute due to ventilation.

$$CO2ConcentrationFlow\left(\frac{mg}{L}\right) = \frac{ElapsedTime(min)}{volumeOfRoom\left(m^3\right) * \frac{1000L}{m^3}} * \left(CventilationRatePerMin\left(\frac{L}{min}\right) * averageCO2\left(\frac{mg}{L}\right) + respIncPerMin\left(\frac{mg}{min}\right) * numberOfPeople - ventilationRatePerMin\left(\frac{L}{min}\right) * PreviousCO2Concentration\left(\frac{mg}{L}\right)\right)$$

The limitation of this equation is that the longer the elapsed time variable is, the less accurate it becomes. We have tested that for intervals of 10 min or less, and the formula is accurate. Therefore, we are sticking to those intervals.

The CO2 concentration equations were validated using historical data from a room at Carleton University Campus. We collected the CO2 concentration values at half-hour intervals over 24 h, the volume of the room, and the number of people in the room. The ACH value was unknown. We calibrated the model to find the ACH value. Then, we validated our formula by conducting a t-test to judge if there were any inconsistencies between the historical data set and the data set the model produced.

3.3 Outdoors Model

The Outdoors is an atomic model that represents outdoors locations where the probability of being exposed to the virus is really low. This model is similar to the room model but with a really small probability of being infected as long as there is no close contact.

3.4 Weather Model

The Weather atomic model simulates general weather patterns such as rain and snow. In this version of the model, it generates a random weather pattern. This model can be adapted to generate weather patterns that matches the obverted weather of given locations. This model is important because the behavior of a person is usually influenced by the weather. For example, in cold climate weather, where the temperatures can reach -30C, people will prefer to move around through indoor corridors that may be crowded and not well ventilated instead of going outdoors.

4 Implementation

The model was implemented in Cadmium [14] and it is available at the following link: https://github.com/SimulationEverywhere-Models/COVID_Campus_Simulation.

The model is parameterized with the number of people and number of rooms. That way, we can run a simulation with any number of people, over any duration, and have any number of rooms.

Each instance of the Person atomic model is created the XML file described above that contains the person's schedule, relationships with each other person, and several traits that determine their behavior. These traits include their ID, whether or not they are vaccinated, are sick, have been exposed, will wear a mask correctly if they are alone in a room, and the severity of the weather necessary for them to choose to travel via the tunnels. Each instance of the room is also created from an XML as described in the previous section.

In order to run different scenarios, we just need to update the XML describing the behavior of the people and the rooms.

4.1 Visualization of the Simulation Results

Once the simulation is complete, some data logs are generated. This data is then parsed by a separate program implemented in Python, and different graphs and metrics are generated based on the data obtained. These graphs will allow showing statistical analysis done on the results of the chosen policies. It will help identify areas where there is an increased risk of infection or safer policies.

Currently, the Python graph generator generates some graphs such as the rooms grouped by average exposure probability vs ventilation rate, or probability of exposure and occupancy vs time for each room, among others. These graphs give insights into which rooms pose the greatest risk for exposure and what virus mitigation policies may be effective.

5 Case Study

We show how to use the model with the case study of Carleton's University Campus.

In this case study, the rooms accurately resemble Carleton University's campus. We got the dimensions for 4352 rooms from a Building Information Model (BIM) of the campus. This model did not contain ventilation data. Therefore, this will be a parameter for our experiments. We did not consider all the rooms. We focused on the rooms where students would congregate, so only rooms with an area greater than 40 square meters were considered. Also, many rooms were missing the height of the room and this value is important to calculate the CO2 concentration in the room. This brought it down to 434 usable rooms with over 40 square meters and complete dimensions. With the information extracted from the BIM model, XML files for every room were generated.

Carleton Campus has a special feature; students can move from one room to the other going out or through a set of tunnels. We have modeled the tunnels as an instance of the Room atomic model where there is a chance of being exposed to the virus.

Our simulation can also consider a varied number of users, and we define the schedules of each person to accurately resemble the schedules of students and faculty on Carleton University's campus. We were not able to gather actual data from scheduling services. Therefore, we had to implement our own scheduling.

We implemented an event-based scheduling system for students by obtaining real data on rooms at Carleton's campus. To implement the event-based scheduling system, we created events (e.g., lectures, tutorials, etc.) for every room. The events were labeled as being a tutorial, lab, or lecture, depending upon the size of the room. Lectures were scheduled in the largest rooms, followed by labs, and tutorials were assigned to the smallest rooms. The events were given durations according to their type. Lectures are 90 min long, labs are 180 min, and tutorials are 60 min long. Events are created to fill a regular school day from 8:30 am to 9:00 pm. These events were then passed to the students being generated, where students are assigned schedules for events on a first-come, first-serve basis. A boolean is used to determine whether an event should be booked over its max occupancy or if it should only book at most a number of students equal to the room's max occupancy.

A possible limitation of this approach could be if there are too many students and not enough rooms with events for them to be given schedules. For example, if we were going to consider the first 30 rooms from our selection, we could only generate a population of 1350 people. This can change depending on what rooms we choose to use in a simulation. Despite this limitation, if we focus on a set of people, we can simulate the interactions within a faculty or the students in a given set of programs.

We use this scenario to study the effect of ventilation on the spread of COVID on campus.

5.1 Ventilation Experiments

Although we already know that ventilation has a strong impact on the transmission of COVID-19, we use the above mention case study to determine, through simulation data, how strong is the effect of ventilation on the probability of exposure to COVID-19. We define the probability of exposure as the chance of a person being exposed to COVID-19 particles. After exposure, different people have different probability of becoming sick, and this depends on many factors such as the person's medical history, but it cannot be influenced by any COVID-19 safety protocols. In this analysis, we only focus on the person's probability of being exposed to the virus.

We will consider two scenarios over a two-week period as we assume that in two weeks a person infected/exposed to COVID will most likely identify that has been exposed and would test for the disease. Both scenarios include 30 rooms, where 10 are lecture halls, 10 are student's labs, and 10 rooms are for tutorials. Both scenarios consider 1200 people, as this is the maximum number of people that can be generated with the chosen rooms.

The parameter that changes between both scenarios is the ACH (Air Changes per Hour). In the first scenario, we set the number of ACH to 1 for every room, while in the second scenario, we set ACH to 8 for every room. The room representing the tunnels will remain the same, and no change in the ventilation rate will be applied.

Table 1 shows the data we have used to initialize the rooms, and Table 2 the rest of the parameters used in our study.

Table 1. Data of Rooms used in Experiments

Lecture Halls		Laboratories		Tutorials	
Area (m3)	Height (m)	Area (m3)	Heigh (m)	Area (m3)	Height (m)
313.46	3048	121.98	3200	54.77	3048
185.27	3200	143.65	2438	45.91	3048
198.31	3200	122.46	2438	53.67	3048
238.53	3200	134.33	2438	72.91	3048
286.38	2438	150.28	2438	52.48	3048
568.02	2438	120.25	2438	65.27	3048
208.14	2438	112.86	2438	52.36	3200
323.87	2438	132.82	2438	54.92	3200
339.39	2438	132.82	2438	55.48	3200
209.48	2438	132.84	2438	53.68	3200

Simulation Results

From each simulation scenario, we ran five replications and obtained the CO2 concentration at each half-hour interval for every room, and the probabilities of exposure at each half-hour for every person. Five replications were enough to achieve a 95% confidence interval.

We calculated the mean, sample variance, and half-width confidence intervals with 95% confidence for the probability of exposure values for all 1200 people. Then an across-replication, we also calculated the sample mean, sample variance, and half-width

Table 2. Parameters of the Simulation

Max occupancy	¾ of the social distance threshold
Social distance threshold	½ of room's area
Wears mask factor	Uniform distribution that is either 1 or 2
Social distance factor	Uniform distribution that it is either 1 or 2
Vaccinated factor	Uniform distribution from −1 to −4
Sick People CO2 Factor	Uniform distribution from 1 to 3
CO2 Factor Thresholds	3 values with uniform distribution between 1000 and 2000 in order from lowest to highest
CO2 Factors	1, 2, and 3 assigned to thresholds from lowest to highest
Respiratory CO2 Increase per Minute	340.000mg of CO2
Room over capacity	30% probability that a room is assigned more students than its max occupancy allows
Relationships	20% Friends; 30% Acquaintances; 50% Strangers
People sick at start	0%
People vaccinated	50%
People wears mask when alone	50%
Weather threshold	Uniform distribution from 0 to 10
Probabilities of wearing a mask, social distancing, and entering a room at max occupancy with Friends	20%, 20%, and 80%
Probabilities of wearing a mask, social distancing, and entering a room at max occupancy with Acquaintance	50%, 50% and 50%
Probabilities of wearing a mask, social distancing, and entering a room at max occupancy with Strangers	80%, 80%, and 20%
Probability of becoming sick after exposure	30%
Time to become sick	Uniform distribution from 0 min to 10080 min (two weeks)
Abandons schedule and stays home when sick	50% chance

confidence interval with 95% confidence. We followed this same approach to calculate the average CO2 concentration among rooms.

Table 3 summarizes this data for the scenario with ACH = 1 and Table 4 the data for the scenario with ACH = 8.

Table 3. Data obtained for Experiment with ACH = 1

ACH = 1	Exposure Probability (%)			CO2 Concentration (mg/L)		
Experiment Replications	Mean	Variance	Conf. Int	Mean	Variance	Conf. Int
E1R1	29.9771	259.0067	± 0.9115	889.2212	88433.4321	± 111.0426
E1R2	29.1751	250.9616	± 0.8972	879.8302	83292.0315	± 107.7664
E1R3	28.868	237.5691	± 0.8730	880.686	88638.6487	± 111.1714
E1R4	30.1844	266.403	± 0.9244	894.845	87977.8172	± 110.7562
E1R5	28.5901	229.0607	± 0.8572	880.77	92257.5773	± 113.4181
Overall	29.3589	0.48236	± 0.8624	885.0705	44.4870	± 8.2817

Table 4. Data obtained for Experiment with ACH = 8

ACH = 8	Exposure Probability (%)			CO2 Concentration (mg/L)		
Experiment Replications	Mean	Variance	Conf. Int	Mean	Variance	Conf. Int
E2R1	23.4795	154.3914	± 0.7037	445.8642	1179.5802	± 12.8246
E2R2	23.7178	161.6669	± 0.7201	444.9765	1070.3994	± 12.2167
E2R3	24.2372	171.1704	± 0.7410	445.4648	1052.7961	± 12.1158
E2R4	23.6137	153.7463	± 0.7023	445.7014	1131.4423	± 12.5602
E2R5	24.561	178.8042	± 0.7573	444.4135	1101.6124	± 12.3936
Overall	23.9218	0.2100562	± 0.5691	445.2841	0.34897	± 0.7335

In Fig. 6, we show the above data graphically. We plot every room's ventilation rate compared to its average probability of exposure values. The graph represents the scenario with an ACH value of 8, and the graph on the right the scenario with an ACH value of 1.

The green dot in the bottom right that is the same for both graphs represent the tunnels on Carleton's campus which are very large and well ventilated so the odds of becoming exposed to COVID-19 remain low. From the graph, we can see that when more points have a lower ventilation rate, they tend to have a higher average probability of exposure. The green, yellow, and red dots represent rooms with an average probability of exposure smaller than 10%, between 10 and 20% and over 20% respectively.

The results of first scenario (ACH = 1) show an overall average probability of exposure equal to 29.3589 ± 0.8624, and an overall average CO2 concentration of 885.0705 ± 8.2817 mg/L. The results of the second scenario (ACH = 8) show an overall average probability of exposure equal to 23.9218 ± 0.5691 and an overall average CO2 concentration of 445.2841 ± 0.7335 mg/L. Comparing these averages, we can conclude

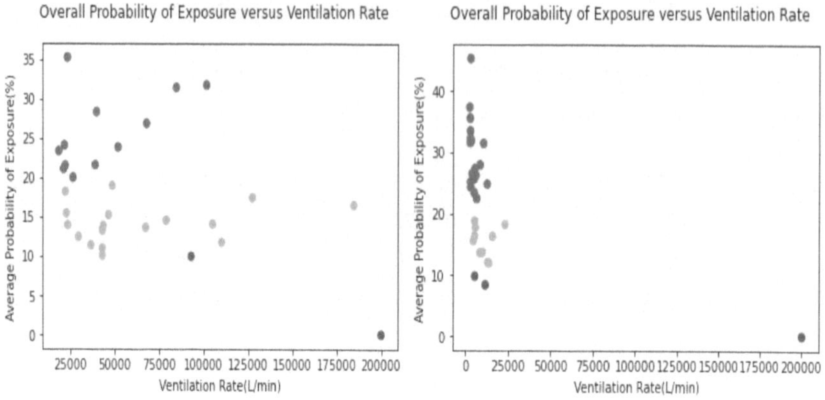

Fig. 6. Average Probability of exposure in each room versus ventilation rate per room. (Left) ACH = 8 (Right) ACH = 1

that when the ventilation of all rooms on campus are improved by 7 ACH, we would see the average probability of becoming exposed to COVID reduce by 5.4371 ± 1.0346.

Although around 5% is not a lot, the results prove that implementing a policy where the ventilation rates of rooms on Carleton's campus are standardized at a high ACH value would improve safety and limit the spread of the virus.

It is important to note that this improvement if for the population defined in table 2, where only 50% of people are vaccinated where the probability of wearing a mask in the presence of friend and acquaintance is relatively low.

With this model, we can also examine the probability of exposure of a specific person (Fig. 7) or the CO2 concentration on specific rooms (Fig. 8). This is relevant because we can analyze the behavior of people with high risk of exposure and the locations they have visited to give further insights on the factors that contribute to the exposure of COVID on campus.

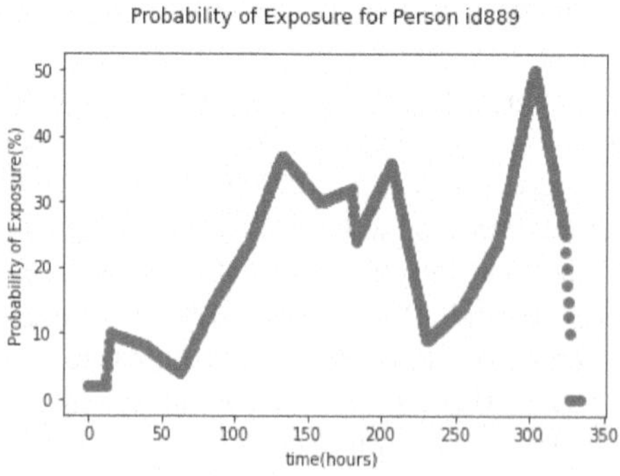

Fig. 7. Probability of Exposure over 2 weeks for Person 889 when ACH = 1

In Fig. 8, we can see large spikes then linear dips. During the day, when people are using the room and breathing CO2 into the air but with poor ventilation, the CO2 levels only keep increasing. Then once the room is empty for the night the ventilation can bring CO2 back down to normal levels.

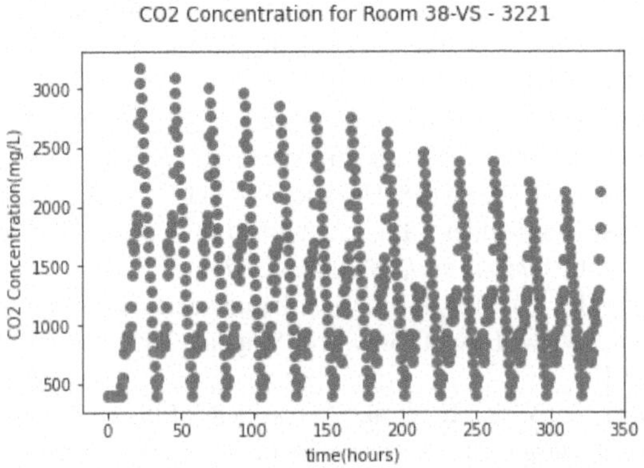

Fig. 8. CO2 Concentration for Room 38-VS - 3221 over two weeks with ACH = 1

6 Conclusions

Being able to define policies to keep workplaces safe is critical as they cannot afford to have many employees sick at the same time. These policies should be supported by experimental data. However, doing real life experiments to collect data is not feasible.

In this paper, we have presented an architecture and a model to study the spread of a disease in a workplace and analyze the effect of different policies. We have shown how to use this model through a case study using real data from Carleton University Campus to simulate the workplace environment.

Some limitations of this research include the validation of the human behavior implemented in the model as well as the factors used to correct the probability of infection based on the different behaviors such as social distance or mask wearing.

References

1. Ruiz-Martin, C., Wainer, G., Lopez-Paredes, A.: Discrete-event modeling and simulation of diffusion processes in multiplex networks. ACM Trans. Model. Comput. Simulat. **31**(1), 1–32 (2018)
2. Ross, R.: An application of the theory of probabilities to the study of a priori pathometry—Part I. Proc Roy Soc A-Math Phy Eng Sci **92**, 204–230 (1916)
3. Ross, R., Hudson, H.: An application of the theory of probabilities to the study of a priori pathometry—Part II. Proc Roy Soc A-Math Phy Eng Sci **93**, 212–225 (1917)
4. Hethcote, H.W.: Mathematics of infectious diseases. SIAM Rev. **42**, 599–653 (2020)

5. Barman M, Nayak S, Yadav MK, et al. Modeling control, lockdown & exit strategies for COVID-19 pandemic in India. medRxiv, (2020), https://arxiv.org/abs/2007.07988
6. Lin B, Chen Y, Hung C, et al. The data forecast in COVID-19 model with applications to US, South Korea, Brazil, India, Russia and Italy. ArXiv, (2021), https://arxiv.org/abs/2011.04738
7. Sen, D., Sen, D.: Use of a modified SIRD model to analyze COVID-19 data. Ind. Eng. Chem. Res. **60**, 4251–4260 (2021)
8. Davidson, G., et al.: A methodological approach for modeling the spread of disease using geographical discrete-event spatial models. Simulation **100**, 39–70 (2023). https://doi.org/10.1177/00375497231152458
9. Xie, Gang: A novel monte Carlo simulation procedure for modelling COVID-19 spread over time. Scientific Reports **10**(1), 13120 (2020). https://doi.org/10.1038/s41598-020-70091-1
10. Aguilar, J., Bassolas, A., G. Ghoshal, S. Hazarie, A. Kirkley, M. Mazzoli, S. Meloni, S. Mimar, V. Nicosia, J. Ramasco, A. Sadilek, "Impact of urban structure on COVID-19 spread," arXiv: 2007.15367, (2020) https://arxiv.org/abs/2007.15367v1
11. Lorig, F., Johansson, E., Davidsson, P.: Agent-based social simulation of the covid-19 pandemic: a systematic review. J. Artif. Soc. Soc. Simulat. **24**(3), 4601 (2021). https://doi.org/10.18564/jasss.4601
12. Vecherin, S., Chang, D., Wells, E., et al.: Assessment of the COVID-19 infection risk at a workplace through stochastic microexposure modeling. J. Expo. Sci. Environ. Epidemiol. **32**, 712–719 (2022). https://doi.org/10.1038/s41370-022-00411-2
13. Zeigler, B., Praehofer, H., Kim, T.: Theory of Modeling and Simulation, p. 99. Academic Press, Cambridge (2000)
14. Belloli, L., Vicino, D., Ruiz-Martin, C., and Wainer, G.: Building DEVS models with the cadmium tool. In: Proceedings of the Winter Simulation Conference (WSC'19) (2019)
15. Satish, U., Mendell, M.J., Shekhar, K., Hotchi, T., Sullivan, D., Streufert, S., Fisk, W.J.: Is CO2 an indoor pollutant? Direct effects of low-to-moderate CO2 concentrations on human decision-making performance. Environ. Health Perspect. **120**(12), 1671 (2012). https://doi.org/10.1289/ehp.1104789
16. Sidhardhan, S., Das, D.: Indoor Carbon dioxide (CO2) level control using Wearable smart watches over a wireless channel. Int. Conf. Comput. Commun. Inform. (ICCCI) **2021**, 1–4 (2021). https://doi.org/10.1109/ICCCI50826.2021.9402692
17. Hallet,S., Toro, F., Ashurst, J.V.: Physiology, tidal volume. Statpearls Publishing, May, 2021. Accessed on: Sept. 30, 2021. [Online] Available at: https://pubmed.ncbi.nlm.nih.gov/29494108/
18. Priban, I.P.: An analysis of some short-term patterns of breathing in man at rest. J. Physiol. **166**(3), 425–434 (1963). https://doi.org/10.1113/jphysiol.1963.sp007114

Data-Based Model Identification of the Hypothalamus-Pituitary-Thyroid Complex

Clara Horvath, Andreas Körner(✉), and Corinna Modiz

Institute of Analysis and Scientific Computing, TU Wien, Wiedner Hauptstraße 8-10, 1040 Vienna, Austria
{andreas.koerner,corinna.modiz}@tuwien.ac.at

Abstract. The thyroid gland, in conjunction with the pituitary and the hypothalamus, forms a regulated system due to their mutual influence through released hormones. The equilibrium point of this system, commonly referred to as the "set point"', is individually determined. This means that determining the correct amount of medication to be administered to patients with hypothyroidism requires several treatment appointments creating an extended treatment process. Because the dynamics of the system have not yet been fully explored, mathematical models are needed to simulate the mutual influence of the respective hormones as well as their course over time. These models enable a deeper understanding of the functionality in the context of data measurements. Therefore, two time-dependent mathematical models are presented to replicate this overall influence of disparate systems. Both are based on a system of two differential equations modelling the interacting hormones. The parameters of the two models are identified according to different calibration approaches by means of patient data collected in a retrospective study in collaboration with the Medical University of Vienna. The hormonal course in the time domain as well as equilibrium curves including the set-point are then simulated and analyzed with respect to the normalized mean squared error. These calibrated systems allow a more profound insight into the functionality of the HPT complex.

Keywords: simulation of HPT complex · thyroid set-point · system calibration

1 Introduction

A prevalent and progressive problem in the general population are thyroid disorders. According to [1], the degree of endocrine disorders affecting the thyroid function are between 5% and 10% in the general population, therefore research to deepen the understanding of the thyroid gland is becoming increasingly important.

The thyroid gland is an endocrine gland located above the collar bones and below the larynx influencing major physiological functionalities such as metabolism, growth and the cardiovascular system [2]. The functional units of

M. Mujica Mota and P. Scala (Eds.): EUROSIM 2023, CCIS 2032, pp. 119–133, 2024.
https://doi.org/10.1007/978-3-031-68435-7_9

the thyroid gland consists of thyroid follicles responsible for the secretion of the thyroid hormones triiodothyronine (T3), thyroxine (T4), calcitonin and their subsequent release into the bloodstream. The anterior pituitary regulates the secretion of the first two thyroid hormones through the Thyroid-Stimulating Hormone (TSH). The hypothalamus, in turn, regulates the release of TSH by producing the Thyrotropin-Releasing Hormone (TRH). The Hypothalamus-Pituitary-Thyroid (HPT) complex represents a negative-feedback loop since the production of TSH is suppressed if the concentration of thyroid hormones is high [3].

Since the mutual influence of the HPT complex is not yet understood to its full extent, simulation of the course of the included hormones based mathematical models is an emerging research field, see [4–6]. In these models, the hormone TRH is often neglected since accurate data collection is difficult in clinical trials. Thus, hypothalamus and pituitary are combined to form the so-called HP-complex. Mathematical models often focus on the timeless dynamics of the HPT-complex and, in particular, consider only the mutual influence. However, particularly in the context of thyroid disease, time-dependent models are needed to simulate the progression of TSH and FT4 to adjust the corresponding treatment over time. Therefore, two time-dependent mathematical models based on a system of differential equations presented in [3,7] are analyzed in the context of patient data collected in a retrospective study at Medical University of Vienna. The aim of this work is to survey how accurate existing models describe time series of patient measurements and comparing the corresponding results. To achieve this goal, the model parameters are calibrated. Additionally, in [8], the so-called homeostatic set-point theory is established which introduces a patient-specific pair of TSH and FT4 values in their normal range. Thus, the set-point corresponds to the optimal euthyroid state of an individual meaning a healthy condition with respect to the thyroid. As the set-point represents the actual target values for both TSH and FT4 of a patient, gathering more information about the dynamics of the HPT-complex in the time domain and the corresponding equilibrium state could lead to better physiological insight and more advanced treatment methods.

2 Modelling of the HPT System

One possible way to obtain a mathematical model describing the interaction and the corresponding control mechanism between the HP complex and the thyroid is to follow the Michaelis-Menten kinetics, known from system biology. The Michaelis-Menten Equation (MME) forms a commonly used model framework for enzymatic reactions, presented in [9]. The starting point is a system of Ordinary Differential Equations (ODE) representing the dynamics of the enzymatic reaction. A steady-state assumption for the enzyme-substrate complex leads to a simplification of the ODE system and results in the Michaelis-Menten equation of the form

$$\frac{\mathrm{d}[\mathrm{P}]}{\mathrm{d}t} = \frac{v_{\max}[S]}{K_M + [S]}. \tag{1}$$

It denotes the product $[P]$ of the enzyme reaction and the substrate $[S]$, the maximum reaction rate $v_{\max}$ and the so-called Michaelis-Menten constant K_M, which describes the reaction rate of an enzymatic reaction. This basic approach, which recalls transfer functions in behavioural modelling, provides the framework for generating dynamic models in systems biology and is a template for the ODE systems used in this paper.

There are several other models describing the HPT system, which are described in [4,5,10,11]. This work focuses on two mathematical models using differential equations to describe the steady states of TSH and FT4, i.e., the state in which both hormones stabilize and no longer change their behaviour. Both models refer to the hormonal concentration, indicated with [FT4] and [TSH]. The two models are referred to as model A and model B and will be presented in the following sections.

2.1 Fundamentals of Model A

The first model presented in [7] describes the rate of change of the concentrations of [FT4] and [TSH] in the human blood stream through the two differential equations

$$\frac{\mathrm{d}\,[\mathrm{TSH}]}{\mathrm{d}t} = k_1 - \frac{k_1\,[\mathrm{FT4}]}{k_a + [\mathrm{FT4}]} - k_2\,[\mathrm{TSH}]\,, \tag{2}$$

$$\frac{\mathrm{d}\,[\mathrm{FT4}]}{\mathrm{d}t} = \frac{k_3\,[\mathrm{TSH}]}{k_d + [\mathrm{TSH}]} - k_4\,[\mathrm{FT4}]\,, \tag{3}$$

with initial conditions $[\mathrm{TSH}]\,(t_0) = [\mathrm{TSH}]_0$ and $[\mathrm{FT4}]\,(t_0) = [\mathrm{FT4}]_0$ and the parameters $k_1, k_2, k_3, k_4, k_a, k_d \in \mathbb{R}^+$. These parameters will be identified on the basis of patient data in later sections. The right side of both Eqs. (2) and (3) represents the secretion rate minus the excretion rate of the respective hormone and the involved fractional term recall the MM kinetics of Eq. (1) presented in the previous section. The given units for [TSH], [FT4] and time t are mU/L, pg/mL and days, respectively.

Model A is derived from a system of four differential equations presented in [3]. In addition to the concentration of TSH and FT4, labelled below as x and y respectively, the original model includes equations considering the functional size z of the thyroid gland and anti-thyroid peroxidase antibodies concentration w according to

$$\frac{\mathrm{d}x}{\mathrm{d}t} = k_1 - \frac{k_1 y}{k_a + y} - k_2 x, \tag{4}$$

$$\frac{\mathrm{d}y}{\mathrm{d}t} = \frac{(k_3 z)x}{k_d + x} - k_4 y, \tag{5}$$

$$\frac{\mathrm{d}z}{\mathrm{d}t} = k_5\left(\frac{x}{z} - N\right) - k_6 z w, \tag{6}$$

$$\frac{\mathrm{d}w}{\mathrm{d}t} = k_7 z w - k_8 w. \tag{7}$$

The model is represented through the system of differential equations with 11 parameters k_1, k_2, k_3, k_4, k_5, k_6, k_7, k_8, k_a, k_d, N. The differential Eq. (2) describing TSH is directly adopted from Eq. (4). Similar, the differential equation for FT4 is mostly identical, however the first term describing the discretion rate of FT4 is slightly modified as the numerator does seemingly not include the functional size z of the thyroid gland.

In accordance to the normal range of z and the original value of k_3 in (5), the value of k_3 in (3) adjusts and incorporates these two aspects in one variable. The functional size z is described in the original model though a differential equation, however in model A incorporated in the parameter k_3 in (3). Since (6) and (7) no longer influence the differential equations for TSH and FT4, consequently they are not longer included.

2.2 Fundamentals of Model B

The second model describing the mutual influence and resulting dynamics of the HPT-complex over time is presented in [7] and further analyzed in [8,12,13]. The system of differential equations including four parameters k_1, k_2, k_3, k_4 represents the behavior of [TSH] over time in negative exponential dependence on [FT4] and vice versa, according to

$$\begin{aligned} \frac{\mathrm{d}\,[\mathrm{TSH}]}{\mathrm{d}t} &= \frac{k_1}{\exp(k_2\,[\mathrm{FT4}])} - [\mathrm{TSH}]\,, \\ \frac{\mathrm{d}\,[\mathrm{FT4}]}{\mathrm{d}t} &= k_3 - \frac{k_3}{\exp(k_4\,[\mathrm{TSH}])} - [\mathrm{FT4}]\,. \end{aligned} \tag{8}$$

This system holds the equations describing its equilibrium state, which means $\frac{\mathrm{d}X}{\mathrm{d}t} = 0$, with $X \in [\mathrm{TSH}]\,, [\mathrm{FT4}]$, leading to

$$[\mathrm{TSH}] = \frac{k_1}{\exp(k_2\,[\mathrm{FT4}])}, \tag{9}$$

$$[\mathrm{FT4}] = k_3 - \frac{k_3}{\exp(k_4\,[\mathrm{TSH}])}. \tag{10}$$

More specifically, these equations describe the response behavior of the HP-complex and the thyroid, respectively, and can be represented by the corresponding compartments in a control loop, see Fig. 1.

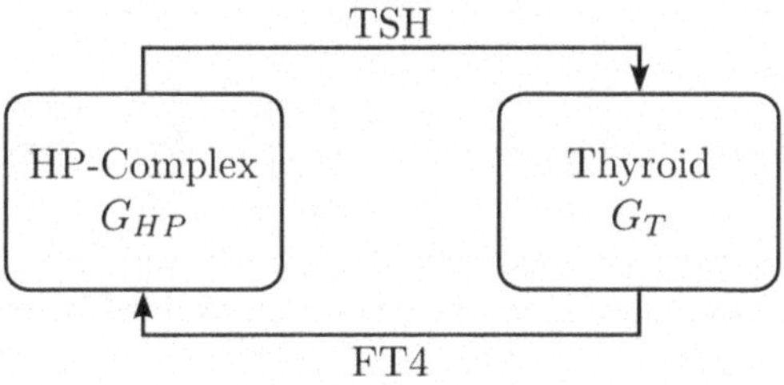

Fig. 1. HPT complex modeled as closed-loop system.

Following the theoretical framework in [8], the equilibrium Eq. (9) of the HP-complex, the so-called HP-function, can be used to compute the set-point. The set-point ($[\mathrm{TSH}]_{sp}$, $[\mathrm{FT4}]_{sp}$) represents the specific values of [TSH] and [FT4] corresponding to a euthyroid state and is patient-specific. It can be expressed explicitly by calculating the point of maximum curvature of the HP-function [12], which is analytically proven in [8] and can be observed in Fig. 2.

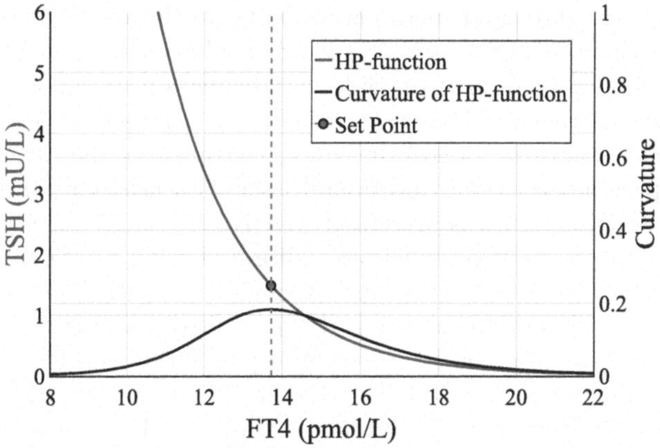

Fig. 2. HP-function with its set-point defined by the corresponding curvature function.

If this theory is further pursued, an explicit representation for the model parameters k_3, k_4 in dependence on the set-point values of [TSH] and [FT4] can be derived. The approach is based on the calculation of the optimal closed-loop gain factor G depending on the individual gain factors of the compartment model shown in Fig. 1.

The optimum of G is found using

$$\frac{\mathrm{d}G}{\mathrm{d}\,[\mathrm{TSH}]} = 0, \tag{11}$$

with

$$G = |G_{HP} \cdot G_T|, \quad G_{HP} = \frac{\mathrm{d}\,[\mathrm{TSH}]}{\mathrm{d}\,[\mathrm{FT4}]}, \quad G_T = \frac{\mathrm{d}\,[\mathrm{FT4}]}{\mathrm{d}\,[\mathrm{TSH}]}, \tag{12}$$

and the equilibrium Eqs. (9) and (10) for [TSH] and [FT4], respectively. As presented in [8] the constants k_3 and k_4 can be specified by

$$k_3 = \frac{[\mathrm{FT4}]_{sp}}{(1-\mathrm{e})^{-1}} \quad \text{and} \quad k_4 = \frac{1}{[\mathrm{TSH}]_{sp}}. \tag{13}$$

Therefore, the parametrization of the differential equation system (8), can be reduced from four to two parameters k_1, k_2. The approach based on the primarily determination of the set-point may yield different parameters than identifying them using the ODE solution in the time domain.

3 Methods

In this section, the data and methods to calibrate the models and simulating the time behaviour of [TSH] and [FT4] are explained in detail. The general framework for the calibration of the differential equation system is equivalent for both models A and B, but differs in the number of parameters to be determined and in whether the approach takes the time-constrained or the timeless model as the basis for parameterization. Additionally, the data used for parameterization of the differential equations, is presented shortly.

Both previously presented models are calibrated using a differential evolution algorithm, which is a population-based stochastic algorithm for solving global numerical optimization problems, see [14]. The control variable that specifies the population size was chosen as 10 to obtain a population consisting of 10 times the number of dimensions, see [15]. To solve the differential equation system, the default explicit Runge-Kutta 45 method is applied. The data points of [TSH] and [FT4] of the respective patients at the first measurement date are chosen as initial condition and the time span is adjusted to represent the entire duration for which measurements of the selected individual are available. Additionally, the Normalized Mean Squared Error (NMSE) for each curve representing one hormone characteristics over time is calculated

$$\text{error} = \mathrm{NMSE}_{[\mathrm{TSH}]} + \mathrm{NMSE}_{[\mathrm{FT4}]}, \tag{14}$$

$$\mathrm{NMSE}_X = \sum_{i=1}^{N} \frac{\left(X(t_i) - \widetilde{X}(t_i)\right)^2}{N(X_{\max} - X_{\min})}. \tag{15}$$

The variable $\widetilde{X}(t_i)$ represents the simulated hormone concentration using described modules and $X(t_i)$ the corresponding measurement point at time t_i

for TSH and FT4, respectively. $X_{\max}$ is the maximum and $X_{\min}$ the minimum of the measured data for each hormone and individual patient. It is important to normalize the respective error in order to take into account the different ranges of hormone values of [TSH] and [FT4].

3.1 Data from Clinical Studies

The data used for the calibration and analysis of both mathematical models was collected in a retrospective study at Vienna General Hospital in cooperation with the Medical University of Vienna. The following calibrations presented in the result section use data from a total of 10 patients, each including three to eight measurements over time. All of the selected individuals are suffering from a type of hypothyroidism (e.g. autoimmunthyreoditis), including patients receiving a certain dosage of artificially produced T4 hormone starting from a specific time during the measurement. The clinical study was originally planned to collect data with respect to thyroid diseases during pregnancy. Therefore, the data includes pregnant patients and measurements on the amount of Human Chorionic Gonadotropin (hCG), see [17]. One patient is selected for an exemplary presentation of the solutions of the respective models. Patient J is pregnant and suffers from hypothyroidism, but is adequately treated with medication throughout her available measurement points.

3.2 Model-Specific Calibration Approaches

Model A contains six parameters, however not the entire set has been taken into consideration for calibration. As mentioned in [3], the parameters k_1, k_2 and k_4 are derived from various literature sources and describe physiological constants, and thus are not optimized during calibration. In accordance with [3], the remaining 3 parameters, namely k_3, k_a and k_d are fitted to [TSH] and [FT4] measurements as these emerge from simulations.

To simulate the course of both hormones over time using model B presented in (8), two approaches are pursued. The first approach B^{sp} focuses on the set point derived from the equilibrium equation for [TSH] (9) to determine the parameters k_1 and k_2. According to the previously explained framework, the set-point values of both hormones correspond to the maximum curvature of the calibrated HP-function. The remaining parameters k_3 and k_4 can then be calculated due to the explicit dependency on the set-point given in Eq. (13). To describe the timeless course of [TSH] as well as [FT4] and their mutual influence, the equilibrium equation of [FT4] (10) is inverted resulting in

$$[\mathrm{TSH}] = -\frac{1}{k_4}\log\left(\frac{k_3 - [\mathrm{FT4}]}{k_3}\right). \tag{16}$$

The system of differential equations is then analyzed using the already calibrated, or calculated, parameters in order to simulate the time-constrained course of both hormones depending on patient-data and to determine the respective NMSE. The second approach B^{ode} directly calibrates the system of differential Eq. (8) in order to fit the two curves to the measurement data in the time domain. Subsequently, the error is directly calculated using the NMSE.

4 Results

This section focuses on establishing a relation between the selected methods and the measurement data. In this context, the determined parameters, the corresponding NMSE and, if applicable, the individual set-point values are presented for each patient. In addition, the resulting hormonal course is shown in the corresponding plots for an exemplary patient described in Sect. 3.1. Although this patient is selected, the hormonal course as a result of both models is similar for all studied patients.

4.1 Findings in the Course of Calibrating Model A

By simulating model A, the behaviour of the hormone concentration of TSH and FT4 over time are described. Depending on the values used in the differential Eqs. (2) and (3) the corresponding solution and steady state differ. The solution settles into a stable state after a few days, therefore patient data with well adjusted medication intake are used to calibrate model A, further elaborated in the discussion section. The biologically significant parameters are fixed at $k_1 = 5000$ mU/(L day), $k_2 = 16.63$ 1/day, $k_4 = 0.099$ 1/day, see [3], while the parameters k_3, k_a, k_d are calibrated and the results for 10 patients are presented in Table 1.

Table 1. Calibrated parameters including NMSE, mean and standard deviation.

id	k_3	k_a	k_d	NMSE
A	1.1947	0.2198	0.0519	0.7744
B	1.1350	0.1033	0.0051	0.4930
C	1.4096	0.0655	0.0373	0.9185
D	0.8905	0.0689	0.0089	0.0754
E	1.0663	0.0603	0.0037	0.5783
F	1.0932	0.0421	0.0007	0.6368
G	1.1262	0.0438	0.0114	0.6004
H	1.1949	0.1001	0.0767	0.3980
I	0.9519	0.0944	0.0979	0.1232
J	1.0981	0.0566	0.0010	0.2752
mean	1.1160	0.0854	0.0294	0.4873
std	0.1414	0.0520	0.0351	0.2722

In comparison with the original model including four differential Eqs. (4), (5), (6), (7), the parameter k_3 changes significantly in the derived model (2), (3). In the original model a possible value for k_3 is given as 86 pg/mL L day. However, the parameter k_3 in the derived model discussed in this work also include the parameter of the functional size z, which has a normal range of $0.005 - 0.125$ L, therefore the resulting values of k_3 is significantly smaller. Nevertheless, the result for k_3 is reasonable, since the functional size z is considered.

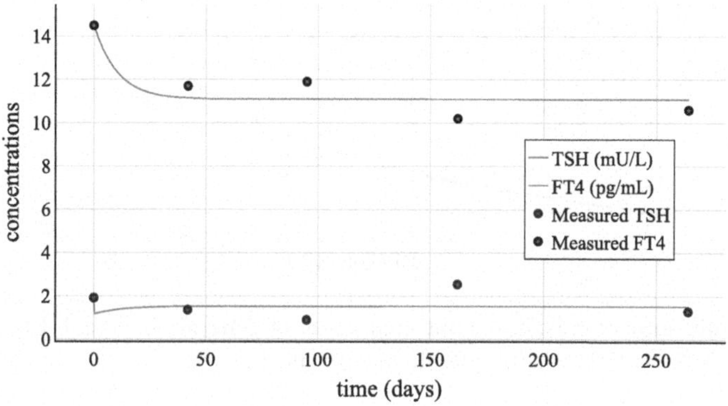

Fig. 3. Calibrated solutions of the Eqs. (2) and (3) for patient J with resulting values $k_3 = 1.0981$, $k_a = 0.0566$, $k_d = 0.001$

Both curves in Fig. 3 describing [TSH] and [FT4] represent the course of the data appropriately and catch the decrease of hormone concentration during the first 50 days. After approximately 60 days, the behaviour of the solution does not change. For both curves, the last measured data point is in close proximity to the solution of model A, which suggests that the medication has been adjusted well to fit the steady state of the differential Eqs. (2) and (3).

4.2 Findings in the Course of Calibrating Model B

For the approach of simulating the course of the hormones of the HPT complex over time using model B, the parameters of the ODE system (8) are determined in two different ways as discussed in the method section. First, following approach B^{sp}, only the parameters k_1 and k_2 are determined and the previously explained framework is applied to calculate the set-point as well as the remaining parameters needed among others for the equilibrium equation of the thyroid compartment (10). In order to plot the equilibrium Eqs. (9), (10) along with the corresponding set-point for each data set of a patient, the inverse thyroid function (16) is selected for the timeless plots. In Fig. 4, the parameterized HP-function is plotted along with the inverse equilibrium function of the thyroid and the data points of corresponding patient J. Both solution graphs intersect at the set-point, which lies within the normal range of FT4 as well as TSH according to [16].

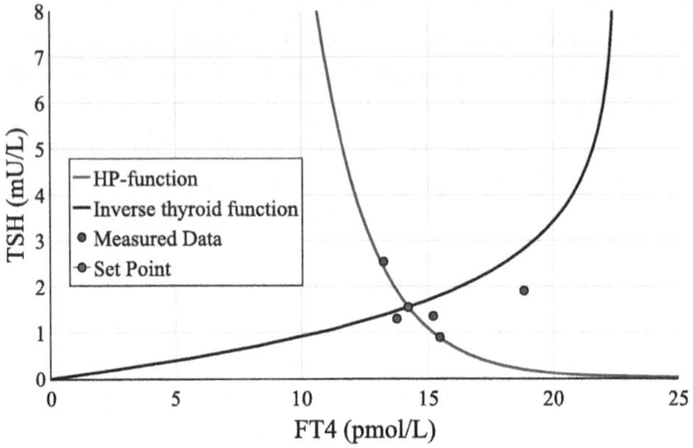

Fig. 4. Equilibrium curves, set-point and data of patient J with $FT4_{sp} = 14.22$, $TSH_{sp} = 1.55$, $k_1 = 1002.24$, $k_2 = 0.46$, $k_3 = 22.49$, $k_4 = 0.64$.

Using the calibrated parameters k_1, k_2, the calculated set-point values and Eq. (13), the course of both hormones over time can be simulated which is presented in Fig. 5.

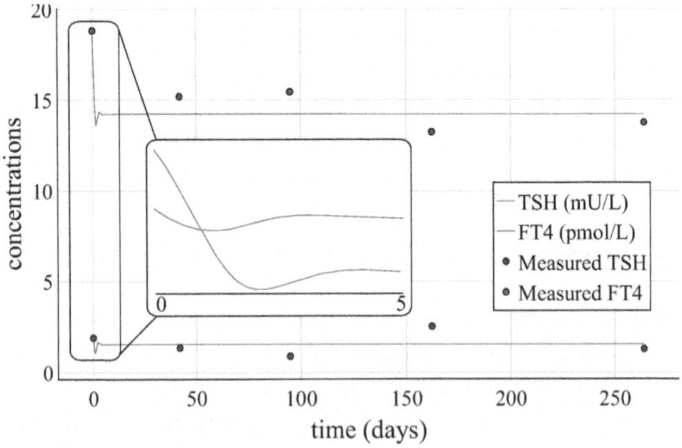

Fig. 5. Solution curves following approach B^{sp} with $FT4_{sp} = 14.22$, $TSH_{sp} = 1.55$, $k_1 = 1002.24$, $k_2 = 0.46$, $k_3 = 22.49$, $k_4 = 0.64$.

Following approach B^{ode}, the course of [TSH] and [FT4] over time resulting from the explicit calibration of both of the solutions of the system (8) to patient-data is shown in Fig. 6 for the selected patient J.

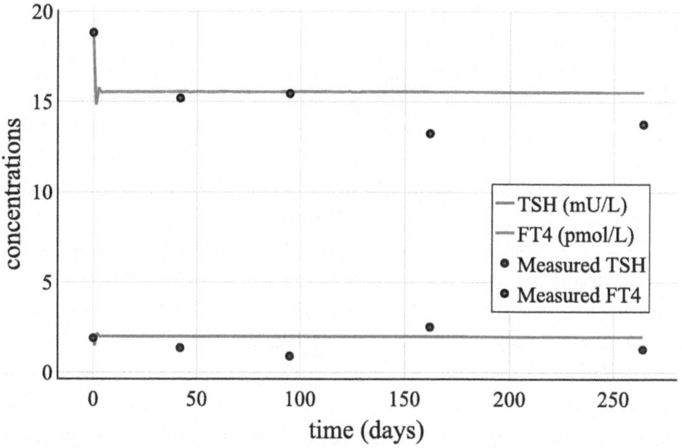

Fig. 6. Solution curves following approach B^{ode} with $k_1^* = 1000.40$, $k_2^* = 0.4$, $k_3^* = 47.03$, $k_4^* = 0.2$.

In contrast to approach B^{sp}, with approach B^{ode} the parameters $k_1^*, k_2^*, k_3^*, k_4^*$ are directly calibrated using the solution in the time domain. The overall results of both approaches are presented in Table 2.

Table 2. Parameters resulting from approach B^{sp} and B^{ode} for 10 patients including set-point, NMSE, mean and standard deviation.

id	$[FT4]_{sp}$	$[TSH]_{sp}$	k_1	k_2	k_3	k_4	NMSE	k_1^*	k_2^*	k_3^*	k_4^*	NMSE*
A	18.23	2.09	1002.83	0.34	28.83	0.48	5.00	1311.66	0.35	21.94	0.23	0.85
B	16.12	1.60	1999.68	0.44	25.50	0.63	1.77	1012.04	0.40	35.38	0.20	0.54
C	14.59	1.60	1006.22	0.44	23.08	0.62	2.40	1390.05	0.38	77.08	0.18	1.15
D	13.12	1.41	1000.00	0.50	20.76	0.71	0.96	1000.27	0.40	50.59	0.11	2.05
E	13.37	1.45	1001.01	0.49	21.15	0.69	0.79	1000.13	0.40	35.88	0.24	1.23
F	13.12	1.41	1000.00	0.50	20.76	0.71	1.11	1003.70	0.40	39.03	0.29	1.39
G	13.72	1.49	1006.84	0.47	21.71	0.67	0.92	1000.88	0.40	58.95	0.15	0.98
H	15.74	1.75	1005.79	0.40	24.91	0.57	0.92	1106.89	0.40	52.05	0.14	0.47
I	13.56	1.47	1000.18	0.48	21.45	0.68	3.40	1000.17	0.40	29.69	0.19	1.57
J	14.22	1.55	1002.24	0.46	22.49	0.64	0.31	1000.40	0.40	47.03	0.20	0.59
mean	14.58	1.58	1102.48	0.45	23.06	0.64	1.76	1082.62	0.39	44.76	0.19	1.08
std	1.65	0.21	315.26	0.05	2.62	0.07	1.46	146.29	0.02	15.94	0.05	0.51

5 Discussion

The course of the hormonal concentration over time of [TSH] and [FT4], respectively, is simulated based on two models and, in the case of model B, on two approaches.

Model A represents an abbreviated model of Eq. (4), (5), (6), (7). However, the model shows the evolution of TSH and FT4 concentration and predicts the equilibrium state of these hormones. This information can be used to reflect on whether the resulting equilibrium state is within the normal range of said hormones. Therefore, if patients are well-adjusted to their medication, the solution describes measurements, even almost resulting in the last data point, see Fig. 3. With regard to Table 1, patient A shows interesting properties, as the parameter k_a is significantly larger than any other calibrated value of k_a. Furthermore, the NMSE is the second highest. A solution to better approximate the measured data using differential equations, could be the original model as patient A suffers from Hashimoto. Therefore Eq. (7) could describe the course of anti-thyroid peroxidase antibodies concentration and a overall lower NMSE could be achieved. Further discussion and simulation are needed in order to investigate the effects of reducing the original system of 4 differential equations to model A, which only consists of two equations. Subsequently, a sensitivity analysis is required to better understand the influence of the parameters in model A. Moreover, a significant amount of solution curves of [FT4] display a sudden bend after a few days. This phenomenon needs to be further investigated. A possible solution could be using a different numerical solver with adaptive step size. Emerging from this, more information on the dynamic of model A could be derived, since the solution do not reflect the physiological influence between [TSH] and [FT4]. Concluding, the model A is limited to these data sets exclusively, as the dynamic does not show any oscillating properties. Model A is not designed to represent these dynamics, but to represent the development using calibrated parameters. A future goal consists of an improved data approximation which includes the dynamics of said measurements and to propose a corresponding drug administration.

Model B assumes an exponential mutual dependence of [TSH] and [FT4] which can be best seen in Fig. 4. This exponential relation differs from the Michaelis-Menten kinetics representing the rational MME, which opens up the potential to investigate this further. The HP-function (9), calibrated within the framework of approach B^{sp}, is derived as equilibrium curve of model B and represents the response behaviour of the HP-complex illustrated in Fig. 1. As shown in Fig. 4, the amount of [TSH] released by the pituitary gland is large when the input signal, i.e. the amount of effective [FT4], is small and vice versa, which is consistent with the expected physiological behaviour of the HPT complex. Also, the response of the thyroid complex, parametrized based on approach B^{sp}, is in line with the underlying physiological principal. A smaller amount of [TSH] as input for the corresponding compartment leads in response to a smaller distribution of [FT4] compared to a larger input signal. The HP function (9) fits the measurement data of the exemplary patient well and crosses two of the data points almost directly. In addition, one point that can be considered an outlier seems to be disregarded for the blue curve, indicating a certain robustness to outliers. One reason for this behaviour could also be the fixed ranges for the calibrated parameters. The set point, also included in Fig. 4, is determined based on the point of maximum curvature, which is reasonable since it represents the point of highest sensitivity of a function and is therefore consistent with the specificity of that point for each individual patient. The exemplary depicted set-point is found in a normal range for [TSH] and [FT4], respectively. This indicates that model B, set up to predict the patient-specific set-point, serves its purpose, but it should be noted that the parameters were set to a specific range during calibration, which then in turn influences the course of both curves as well as the set-point values. In further research, the set-point theory according to [8] can be compared to the model presented in [18], that also includes specific values for the hormonal equilibrium described explicitly by a rational function.

The time-dependent solution curves of the approach B^{sp} in Fig. 5 correspond to the expected physiological behaviour during the first five days, as [FT4] decreases with a time delay in response to decreasing [TSH]. Thereafter, an increase in [FT4] is observed with a short delay associated with the increase in [TSH]. The long-term behaviour shows no dynamics, as the solution curves soon transition to a constant. One reason for this could be that in [7] the time unit is not explicitly specified, so that the dynamic time span could extend if a different unit is determined. Besides, both curves almost meet the last measurement points of patient J. This indicates that the equilibrium of system (8) corresponds with her final hormonal values, which are balanced since she is adequately treated with medication according to her description. Approach B^{ode} results in solution curves exhibiting the same short and long-term behaviour, shown in Fig. 6. In contrast, those curves do not meet the last data points as close as approach B^{sp} and the NMSE is almost twice as high, although the curves are directly calibrated to the curve in approach B^{ode}. As listed in Table 2, in general, the mean NMSE is smaller for approach B^{ode} compared to B^{sp}, which also applies to standard deviation. The maximum NMSE for the approach B^{sp} is found for patient A, but she has a fuzzy anamnesis, e.g. she received medication for hyperthyroidism for five appointments before being described medication for hypothyroidism.

Concluding, the dynamic of the solutions of model A and B results in the corresponding steady state of the related differential equations after a relatively short time span, which lies in accordance with the saturation behaviour of MME. Therefore approximating patients' measurements using the presented model is successful for patients with adequately adjusted dosage of medication. Both models do not reflect any dynamical changes after a certain point in time. However, model B displays the expected physiological behaviour in the time domain as well as in the FT4-TSH plot. There are several possible extensions to better reflect the later dynamics of patients measurements, i.e. including the influence of drug administration, which is not included in any of the discussed models. Additionally, including the hormone hCG [17] as another input parameter could achieve improved results and broaden the field of application possibilities.

Acknowledgements. We would like to express our gratitude to our research partner Prof. Dr. Michael Krebs from Medical University Vienna. He provided us with the necessary data of clinical patients. Furthermore, he continuously offered his insight and detailed knowledge of the functionalities of the thyroid gland.

References

1. Andersen, S., Pedersen, K.M., Bruun, N.H., Laurberg, P.: Narrow individual variations in serum T4 and T3 in normal subjects: a clue to the understanding of subclinical thyroid disease. J. Clin. Endocrinol. Metab. **87**(3), 1068–1072 (2002)
2. Yen, P.M.: Physiological and molecular basis of thyroid hormone action. Physiol. Rev. **81**(3), 1097–1142 (2001)
3. Pandiyan, B., Merrill, S.J., Benvenga, S.: A patient-specific model of the negative-feedback control of the hypothalamus-pituitary-thyroid (HPT) axis in autoimmune (Hashimoto's) thyroiditis. Math. Med. Biol. J. IMA **31**(3), 226–258 (2013)
4. Berberich, J., Dietrich, J.W., Hoermann, R., Müller, M.A.: Mathematical modeling of the pituitary-thyroid feedback loop: role of a TSH-T3-shunt and sensitivity analysis. Front. Endocrinol. **9**, 91 (2018)
5. Mukhopadhyay, B., Bhattacharyya, R.: A mathematical model describing the thyroid-pituitary axis with time delays in hormone transportation. Appl. Math. **51**(6), 549–564 (2006)
6. Leow, M.K.: A mathematical model of pituitary-thyroid interaction to provide an insight into the nature of the thyrotropin-thyroid hormone relationship. J. Theor. Biol. **248**(2), 275–287 (2007)
7. Goede, S.L.: General Review on Mathematical Modeling in the Hypothalamus Pituitary Thyroid System. In: Research Square - Preprint Version 1 (2010). https://doi.org/10.21203/rs.3.rs-1659086/v1
8. Leow, M.K., Goede, S.L.: The homeostatic set point of the hypothalamus-pituitary-thyroid axis-maximum curvature theory for personalized euthyroid targets. Theor. Biol. Med. Model. **11**(1), 35 (2014)
9. Carson, E., Cobelli, C.: Modelling Methodology for Physiology and Medicine, 2nd edn. Elsevier, Amsterdam (2014)
10. Liu, Y., Liu, B., Xie, J., Liu, Y.X.: A new mathematical model of hypothalamo-pituitary-thyroid axis. Math. Comput. Model. **19**(9), 81–90 (1994)
11. Abduvaliev, A., Saydalieva, M., Hidirova, M., Gildieva, M.: Mathematical modeling of the thyroid regulatory mechanisms. Am. J. Med. Sci. Med. **3**(3), 28–32 (2015)
12. Goede, S.L., Leow, M.K., Smit, J.W.A., Dietrich, J.W.: A novel minimal mathematical model of the hypothalamus-pituitary-thyroid axis validated for individualized clinical applications. Math. Biosci. **249**, 1–7 (2014)
13. Goede, S.L., Leow, M.K.: General error analysis in the relationship between free thyroxine and thyrotropin and its clinical relevance. Comput. Math. Methods Med. **2013**, 831275 (2014)
14. Storn, R., Price, K.: Differential evolution - a simple and efficient heuristic for global optimization over continuous spaces. J. Global Optim. **11**(4), 341–359 (1997)
15. Centeno-Telleria, M., Zulueta, E., Fernandez-Gamiz, U., Teso-Fz-Betoño, D., Teso-Fz-Betoño, A.: Differential evolution optimal parameters tuning with artificial neural network. Mathematics **9**(4), 424 (2021)
16. Barth, J., et al.: Comparison of method related reference intervals for thyroid hormones: studies from a prospective reference population and a literature review. Ann. Clin. Biochem. **55**(1), 107–122 (2017)

17. Hershman, J.M.: Physiological and pathological aspects of the effect of human chorionic gonadotropin on the thyroid. Best Pract. Res. Clin. Endocrinol. Metab. **18**(2), 249–265 (2004)
18. Yang, B., Tang, X., Haller, M.J., Schatz, D.A., Rong, L.: A unified mathematical model of thyroid hormone regulation and implication for personalized treatment of thyroid disorders. J. Theor. Biol. **528**, 110853 (2021)

Online Appointment Scheduling of Patients in a Resource Constrained Facility

Simran Lakhani[1(✉)], Ashutosh Mahajan[1], Akshay D. Baheti[2], and Suyash Kulkarni[2]

[1] Indian Institute of Technology Bombay, Mumbai, India
simran@iitb.ac.in
[2] Tata Memorial Hospital Bombay, Mumbai, India

Abstract. An online scheduling problem for a scanning facility in a hospital is described. The facility operates in a resource constrained and high-demand scenario in which machine idleness is undesirable. Other practical constraints dictate that several patients are scheduled in fixed time slots. The objective is to balance the competing goals of minimizing patient waiting time and negligible idle time of the machine. Based on the data available from the facility, patients are divided into three types, and two different policies are proposed for scheduling: one fills all slots one by one, while the other uses the best-fit variant of online bin-packing. To account for variability in the scan times, some slots are overbooked. The proposed appointment policies are simulated and compared. Simulations are based on data received from the MRI department of a cancer-care hospital. The second policy trades off between lesser waiting time for patients and machine idleness.

Keywords: Healthcare · Patient scheduling · Discrete event simulation

1 Introduction

We consider appointment scheduling for a Magnetic Resonance Imaging (MRI) machine in a tertiary care hospital with a very high patient load. The MRI machine serves patients from various departments within the hospital and is capable of performing different types of scans. A stream of patients arrives at the appointment desk, and the secretary allots slots to patients one by one, without having the foresight of the type of patients that may arrive subsequently. Therefore, the slot allocation is an online activity and the patients randomly arrive to book appointments with a request for a particular scan based on a doctor's prescription. The appointment secretary allots a day and a slot to the patient at that instant, and the slot is not changed after allotment. In a recent study [11], the authors addressed the online appointment scheduling problem by performing two separate steps of scheduling and rematching slots to avoid

M. Mujica Mota and P. Scala (Eds.): EUROSIM 2023, CCIS 2032, pp. 134–148, 2024.
https://doi.org/10.1007/978-3-031-68435-7_10

overlap. However, in our setting, such an approach would be difficult to implement as it would require informing patients of their updated slots, which can lead to logistical challenges and patient dissatisfaction. Given the high load and the hospital's objective of serving as many patients at subsidized prices as possible, patients are called early to avoid the machine being idle. As a result, the patients have long waiting times on the day of their scan. The objective of the appointment scheduling problem is to balance the competing goals of minimizing patient waiting time and negligible idle time of the machine.

The day of the scan is usually based on the doctor's recommendation, for example, a cancer patient may require a scan after a prescribed number of weeks for a follow-up visit. The time slot of the scan is selected by the secretary based on how many patients have already been scheduled on that day. Another source of randomness [9] that should be accounted for while making appointments is the variability in the scan times. This randomness is observed during the day of the scan due to the patient's condition, specific recommendations of the doctor, the condition of the machine, and other factors.

2 System Description

Each working day is divided into several time slots, each of duration t units. Patients book appointments for their respective scans well in advance. The secretary allots each patient to one of these slots. A slot may have several patients. Having a slot with several patients naturally results in more waiting as compared to calling each patient at a different time [2,4]. However, the hospital still prefers it because it leads to a lower idle time of the machine, which is important for the limited resources at the hospital. Most of the patients pay only a subsidized fee at the hospital, and they do not have alternate options for getting a scan. Therefore, some amount of waiting is acceptable in this setting.

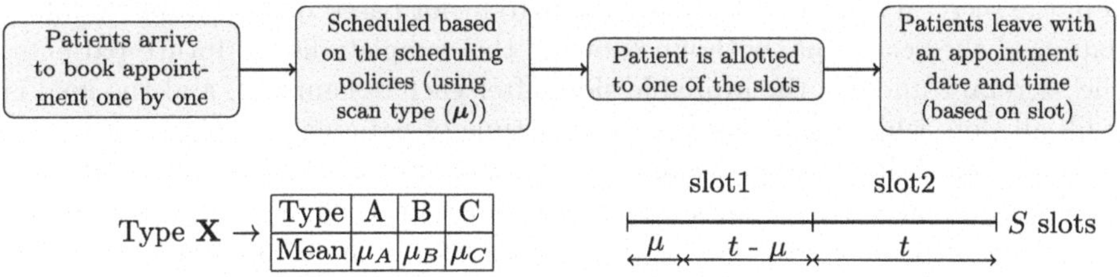

Fig. 1. Flow of patients for booking appointments

There are about 20 different types of scans each taking different amounts of time. Two to three types of scans that were rare with exceptionally long service times were excluded and the remaining types were divided into three categories based on the similarity in their average service times, determined by visual inspection. One of the existing studies [1] concerning the management of a diagnostic facility, assumes a constant scan type for all patients. Even under these simplistic settings, finding an optimal appointment policy is hard. Another study [3] considered the classification of patients based on priority where slots are reserved for specific patient groups and priority-based scheduling is done. We do not prioritize any type of patient and no reservation of slots is done. However, the studies [1,3] did not account for randomness in scan times. In our case, the average scan time for each type of scan, as well as their standard deviations are known and considered in the study. One of the recent studies [12] characterizes an optimal appointment schedule by considering service level constraints but in the offline setting. As, the appointment scheduling problem is inherently complex (see review papers [7,9]), a fixed policy or set of rules may not always be effective [5]. Hence, a dynamic and flexible approach needs to be designed to allocate patients to the pre-designed slots in an online fashion.

3 Model Description

To model the scheduling problem, we consider S identical service slots each of fixed intervals of 't' units of time. The identical-length service slots are common in practice as it ensures easy implementation in the system. The appointments are scheduled for a single day and the available working time of the machine is T mins. Each arriving patient of Type X is mapped to one of the types A, B, and C for which average scan times and the standard deviation are denoted as μ_i, and σ_i respectively with $i \in \{A, B, C\}$ Fig. 1. We denote by p_i with $i \in \{A, B, C\}$, the arrival probabilities of the three types of patients. We assume that the arrivals of the three patient types are independent of each other.

Given the fixed length slots and an online arrival stream of patients to book appointments, the problem can be viewed as an online covering problem to cover all slots with a small amount of overbooking to avoid the machine being idle. In an online covering problem, a set of elements must be assigned to a subset of available slots dynamically and sequentially [6]. Similarly, in our case, at each step, the secretary must allocate a patient to a slot based on the available information of the scan type, without knowing the complete list of future patients. The secretary updates the available slots after each assignment, and the goal is to fill all slots while balancing the waiting time of patients.

While scheduling an equal number of patients in each slot may seem like the simplest approach [7], the variable service times for each type of scan and the unpredictability of patient arrivals (for appointments) in an online setting makes it difficult to determine the exact number of patients that should be scheduled in each slot. If the number of slots is fixed for each type of patient, there is a possibility of idle time for the MRI machine if the expected number of patients of a particular type do not book appointments for a particular day [3]. Therefore, we try to schedule different types of patients in the same slot. This can be done by designing a scheduling algorithm that takes into account the service time variability, patient arrivals, and machine utilization in each slot.

In the case where we have patients of only one type, it is possible to determine the optimal number of patients to schedule in each slot [1,4]. For example, assuming that the service times of the patients X, follow a normal distribution with a mean of 42 mins and a standard deviation of 16 mins, we can calculate the number of patients to be scheduled in each slot such that the probability of the machine being idle is less than a desired limit. Let Y be the total service time for n patients, then scheduling 5 patients in a slot of 120 mins yields the probability of the machine being idle to be 0.005. However, in our case, there are multiple types of scans with varying average and standard deviation of service time.

To prevent idleness, overbooking is necessary. We say that slots are overbooked when the expected total service time of all the scheduled patients in a slot exceeds the slot length t. So, a key decision to be made here is the extent of overbooking in each slot. It is important to decide the appropriate amount of overflow in each slot, such that it does not exceed the day's limit of T mins. One approach is to fix the amount of overbooking factor or buffer to be considered for each slot say, o_s. However, a better way might be to find the optimal combination of the number of slots to overbook and the overbooking factor of each slot. One might consider overbooking the initial k slots by an additional buffer of o and not overbook in the later slots, that is considering $o_s = o$ for $s \leq k$, and 0 otherwise. The total scheduling time can then be written as:

$$T = k(t + o) + (S - k)(t)$$

While scheduling, the secretary may opt to consider only the average scan time of each patient as an easy approach ignoring variance Fig. 1. The overbooking factor acts as a buffer against the machine being idle if some patients finish early. It implicitly accounts for the variance. The various policies to schedule patients need to be compared in terms of machine utilization, patient waiting time, and the number of patients served each day.

4 Appointment Scheduling Policies

We propose two policies and one variant of the second policy in this section to book appointments and later use simulation to compare their performances.

4.1 Fill Slotwise with Overbooking (FAS with OB)

The first policy considered aims to allocate the earliest available time slot to a patient until the slot overflows. The earliest available slot is determined based on the utilization of the slot (u_s) and its position in the list of slots. The patients are scheduled until the total available scheduling time of T mins is reached. This policy allows for overbooking the slots. For example, if two patients have been scheduled in the first slot, and still there is a space of 5 mins, the third patient will be scheduled in the first slot itself, irrespective of its scan type. The policy also considers that the initial k slots are overbooked by o amount, where k and o can be defined accordingly. For these k slots, the slot length is $t + o$. The policy follows a first-come, first-allocated slot for each patient.

Algorithm 1: FAS with OB

```
Input: Patient i with average scan time μ_i, i ∈ {A, B, C}
       Current utilization of all slots u_s, s ∈ {1, 2, ..., S}
Output: Slot s for patient i
Parameter: t, o, k, S, and T
Initialize: s ← 1
while s ≤ S and Σ_{s∈S} u_s < T do
    if u_s < t + o_s then
        Add patient i to slot s
        u_s ← u_s + μ_i
    end
    s = s+1
end
```

The scan type $i \in \{A, B, C\}$ is required to compute the utilization of slots at the beginning of iteration after scheduling patient i in the previous iteration. The utilization of a slot is the expected total service time of the scheduled patients. The output of the algorithm is a schedule of patients slots wise. The policy allows for some inherent overbooking in the remaining slots (other than initial k slots) that ensures high machine utilization.

Consider an example of scheduling patients to $S = 5$ slots with initial $k = 2$ slots of total length 135 mins (additional buffer of $o = 15$ mins), and remaining slots of length $t = 120$ min. The total available time for scheduling is $T = 630$ mins. Suppose the arriving requests are mapped to the three types: Type A, Type B, and Type C with the average scan times as $\mu_A = 45$, $\mu_B = 40$, and $\mu_C = 35$. The patients arrive in an online fashion with one being the first patient to arrive in Table 1 and all others subsequently. The schedule according to this policy is shown in Table 2. Note that the last slot is underutilized and the next patient in the arrival stream cannot be scheduled as the limit of 630 min is reached.

Table 1. Arrival stream of patients

Patient ID	1	2	3	4	5	6	7	8	9	10	11	12	13	14	15	16	17	18	19	20
Patient type	A	C	B	C	C	C	C	C	B	A	B	A	C	C	C	C	A	B	B	C

4.2 Bin Packing All Slots and Then Overbooking (BinPack Then OB)

In this policy, we adapt the concept of the online bin packing problem to schedule patients. Given the patients arriving in an online fashion, the algorithm must decide immediately where to schedule them without any information on future arrivals. We treat appointment slots as bins with fixed capacities, and patients as items to be packed into these bins. Heuristics and approximate algorithms have been proposed in the literature for solving the online bin packing problem,

Table 2. Schedule of patients from FAS with OB

Slots	Patients scheduled	Utilization of the slots (in mins)
Slot 1	1, 2, 3, 4	155
Slot 2	5, 6, 7, 8	140
Slot 3	9, 10, 11	125
Slot 4	12, 13, 14, 15	150
Slot 5	16, 17	80

including First Fit, Next Fit, and Best Fit heuristics [8]. In our policy, a patient is scheduled in a slot if the patient's average service time can be completely accommodated in the slot, akin to packing items in bins. However, if the total service time of all scheduled patients is less than the slot length, then some slots may remain underutilized. To address this, we overbook the initial k slots if the total service time of all scheduled patients, that is utilization of the slot is less than the slot length $t + o$. The first overbooking is done in the slot that has the minimum utilization (u_s), and the process continues until either all k slots are overbooked or the available time limit of T mins is reached. The algorithm is described as follows:

Algorithm 2: BinPack then OB

```
Input: Patient i with average scan time μ_i, i ∈ {A, B, C}
       Current Utilization of all slots u_s, s ∈ {1, 2, ..., S}
Output: Slot s for patient i
Parameter: t, o, k, S, and T
Initialize: allot ← False
for s ∈ {1, 2, ..., S} do
    if u_s + μ_i ≤ t + o_s then
        Add patient i to slot s
        allot ← True
        u_s ← u_s + μ_i
        break
    end
end
if allot = False and Σ_{s∈S} u_s < T then
    a = arg min_{s∈1,2,...,S}(u_s)
    if u_a < t + o_a then
        Add patient i to slot a
        u_a ← u_a + μ_i
    end
end
```

The scheduling depends on the scan type of patient. The output of the algorithm is a schedule of patients slots-wise. Considering the similar arrival stream of patients as in Table 1, the schedule according to this policy is shown in Table 3. Note that the overbooking is done after all the bins are packed. The boldface letters are the patients that are overbooked in the respective slots and utilization is then updated. Further patients cannot be scheduled as there is no slot left for booking as well as the total scheduling time is exceeded.

We can also consider a variant of this policy that involves packing slots without any buffer time added to initial slots, such that the length of each slot is t. The procedure of schedule remains the same as described above except that, $o_s = 0 \quad \forall s \in S$. This policy also considers overbooking the initial k slots when the utilization of the slot is less than the slot length t. The overbooking is done until the last patients scheduled in these slots overflow to the next slot or the scheduling day limit of T mins is reached. We analyze the performance of this variant as well as the two policies described earlier in the next sections. For example, a schedule based on this variant considering the same arrival stream as in Table 1 is presented in Table 4. The boldface letter represents the patients that are overbooked in the respective slots and utilization is then updated.

Table 3. Schedule of patients from BinPack then OB

Slots	Pat scheduled	Utilization(u_s mins)
Slot 1	1, 2, 3, **17**	165
Slot 2	4, 5, 6, **16**	140
Slot 3	7, 8, 9	110
Slot 4	10, 11, 13	120
Slot 5	12, 14, 15	115

Table 4. Schedule of patients from variant of BinPack then OB

Slots	Pat scheduled	Utilization(u_s mins)
Slot 1	1, 2, 3	120
Slot 2	4, 5, 6, **16**	140
Slot 3	7, 8, 9	110
Slot 4	10, 11, 13	120
Slot 5	12, 14, 15	115

5 Simulation Model

Python programming language was used to prepare schedules of patients using three scheduling policies discussed in Sect. 4. A Discrete Event Simulation (DES) model was developed using Anylogic PLE to evaluate the effectiveness of these policies. The model was designed to simulate the schedules generated by the policies for various arrival streams and to account for the variability in service times. Although the variance in scan times was not considered during scheduling, the service time distribution of each type was taken into account during the simulation. The results of the simulation were analyzed to compare the performance of the policies using various metrics defined in the subsequent sections.

5.1 Data Collection and Analysis

Data collection was conducted over a period of one week through observation and extraction from the hospital's information system. The observation involved recording machine start and end times, 8 : 00 AM in the morning till 7 : 00

AM the next morning. Every day is divided into several two-hour slots and appointments are scheduled for these slots based on availability considering the average scan time of each type of patient as 30 min. The IT system was utilized to extract information on the type of scan performed, the scheduled appointment time, the arrival time of patients at the MRI section, and the start and end times of the scans. By calculating the difference between the MRI start time and end time, we were able to derive the scan times for each patient. Figure 2 shows the histogram of the service times recorded over a one-week period, along with the best fit to the distribution. The mean duration is 42 min and the standard deviation of 16 min is significant. This significant variability is due to the broad mix of MRI scans that are done on a single machine [1], ranging from a pelvis scan with a mean of 30 min to cardiac scans with a mean of 65 min. This results in the long waiting time of patients of about five hours because of which the number of patients served in a day was 18 less than the number of patients scheduled.

In our model, the variability in service time was not taken into account while scheduling patients. However, to better reflect the actual scenario on the day of the scan, the distribution of the service times for the three types of scans is considered in the simulation model. For each day in one week's data, the types of scans performed were mapped to three categories mentioned earlier. The service times for each of the three types followed a normal distribution with different parameters Table 6. The hypothesis was verified using the chi-square and KS-test.

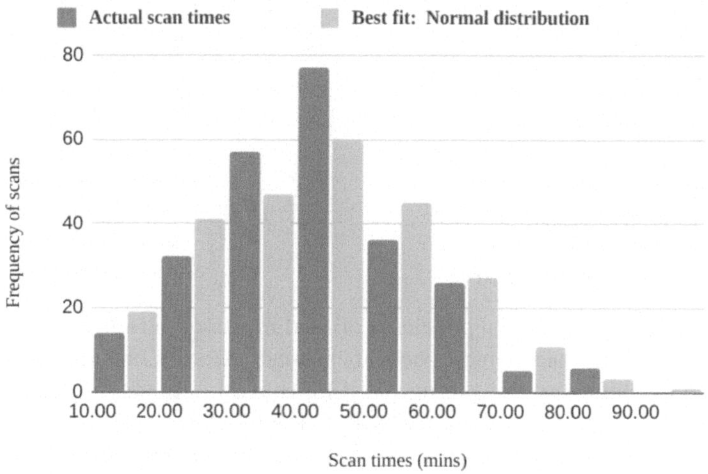

Fig. 2. MRI actual scan times and best fit distribution

5.2 Model Assumptions

The following assumptions were made for scheduling appointments and simulation of the schedules obtained from three policies:

1. Each slot was assumed to be of two hours, with a total of 11 slots in a day, i.e., $S = 11$, and $t = 120$ min.
2. Initial $k = 4$ slots were overbooked by considering the slot length of $t = 135$ mins ($o = 15$ mins).
3. Slots are filled considering the available time of $T = 1380$ mins (except for the variant policy, for which it is 1320 mins only) to account for additional buffer.
4. Machine availability on the day of scan is assumed to be 22 h with some amount of overtime to complete the scheduled patients.
5. The proportions p_i with $i \in \{A, B, C\}$ for the three types of scans were estimated from the one week's data.
6. Patients punctually arrive at the beginning of their scheduled appointment slot.

The values summarized in Tables 5, 6 were used as input parameters in our scheduling and simulation study. While these values may vary across different healthcare systems, they can serve as a general guide for other hospitals facing similar constraints of limited resources and high patient demand.

Table 5. Parameters for scheduling

Parameter	Value
Available time for scheduling T	1380 mins
Slot length t	120 mins
Amt of overbooking o	15 mins
Total number of slots S	11
No of slots overbooked k	4

Table 6. Parameters for simulation

Parameter	Type A	Type B	Type C
μ_i	36.59	44.82	56.80
σ_i	15.60	11.65	18.23
p_i	0.463	0.334	0.202

5.3 Performance Measures

To evaluate the efficacy of the policies proposed in Sect. 4, a range of performance measures have been considered. These measures primarily include the total idle time of the machine and the average waiting time of patients. Furthermore, the extra time required to complete the last patient before the time limit of 22 h is reached, denoted as overtime, has also been taken into account. Additionally, the average number of patients scheduled and served without overtime has been analyzed. These metrics were selected due to their impact on congestion and overcrowding in the system, machine utilization, and patient dissatisfaction caused by long wait times. By analyzing these performance measures, we can compare the effectiveness of the policies and recommend the most suitable approach for different scenarios.

The scheduling process was carried out for 20 days to capture the randomness in the online arrival of patients. The arrival stream was generated based on the proportions of each type estimated from one week's data. The scheduling policies discussed in Sect. 4 were employed to prepare the schedules, and the output consisted of the number and type of patients allocated to each slot Fig. 1. To minimize waiting time for patients scheduled in the same slot, the shortest processing time (SPT) rule was used to sequence the patients within each slot, as proposed in [10]. As the patients were assumed to arrive punctually at the beginning of their respective slots, they were served according to the SPT rule. Following the scheduling and sequencing for 20 different arrival streams/days, each day was simulated and replicated 1000 times to account for the variability in service times Fig. 3.

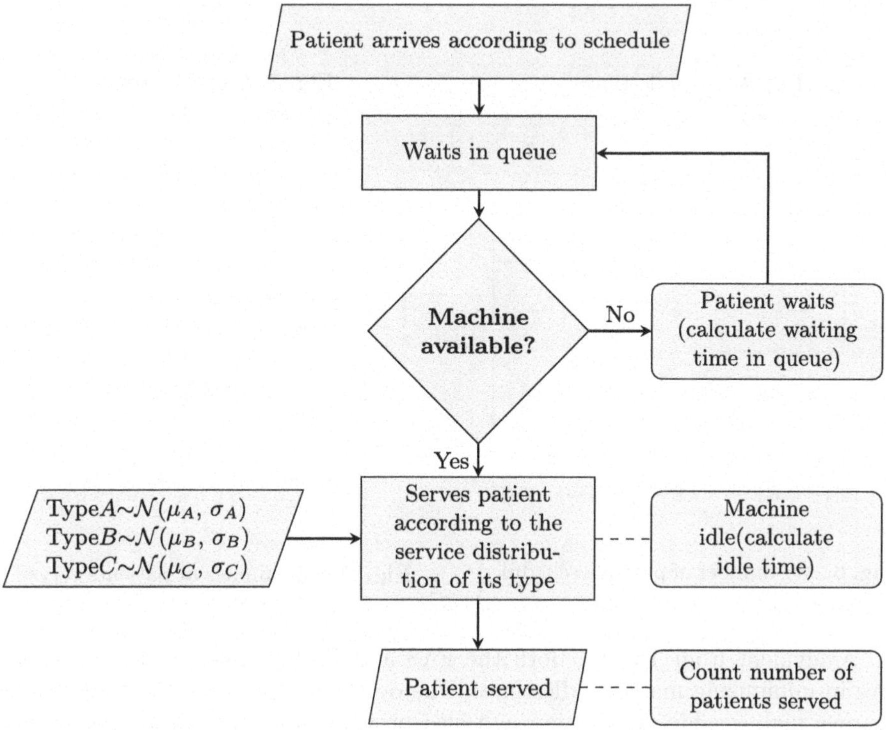

Fig. 3. Flowchart of events in simulation model

6 Results

We denote BinPack135 for the described policy and BinPack120 for the variant of the same. Table 7 shows the various measures (average, standard deviation, minimum and maximum value) of performance metrics for each day replicated 1000 times. Figures 4, 5, 6, 7, and 8 show the averages of metrics across replications corresponding to each day (arrival stream).

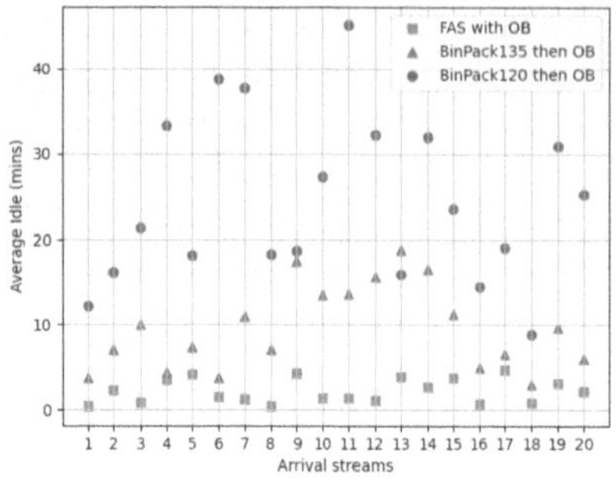

Fig. 4. Avg Idle times

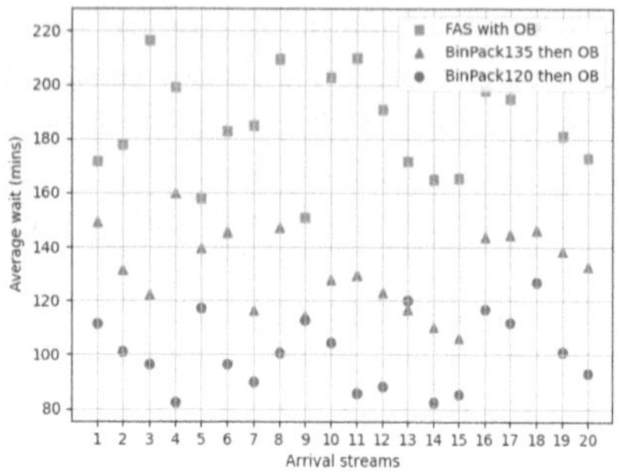

Fig. 5. Avg Wait times

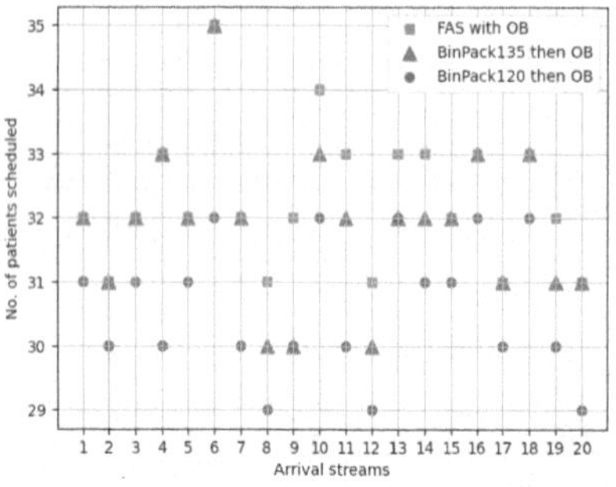

Fig. 6. Avg number of patients scheduled

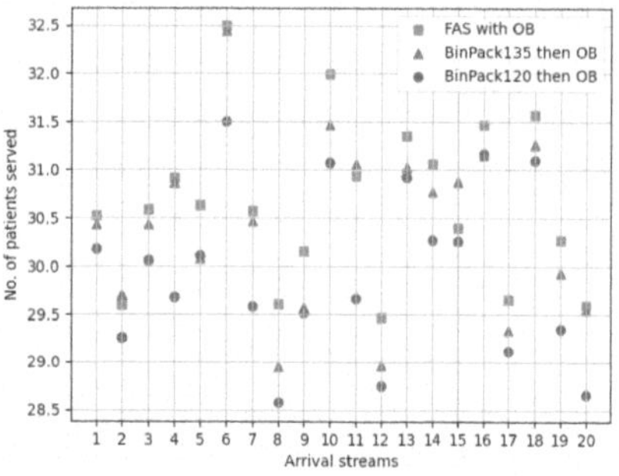

Fig. 7. Avg number of patients served

As evident from Table 7, both the FAS and BinPack135 policies are effective in minimizing machine idle times. The overflowed buffer in the initial slots ensures high machine utilization and also provides some slack time before the machine goes to rest. While BinPack135 has a maximum idle time of 18.82 mins compared to 5 mins of FAS policy, this is still considered acceptable. Additionally, both policies result in a reasonable amount of overtime, a maximum of 25 mins but this is acceptable as the difference between the number of patients scheduled and the number of patients served on average in a day is minimal. This indicates a lower level of patient dissatisfaction compared to the current setting.

BinPack135 policy dominates the FAS policy concerning the average waiting time metric, reducing it by 29%. However, the BinPack135 policy also has a

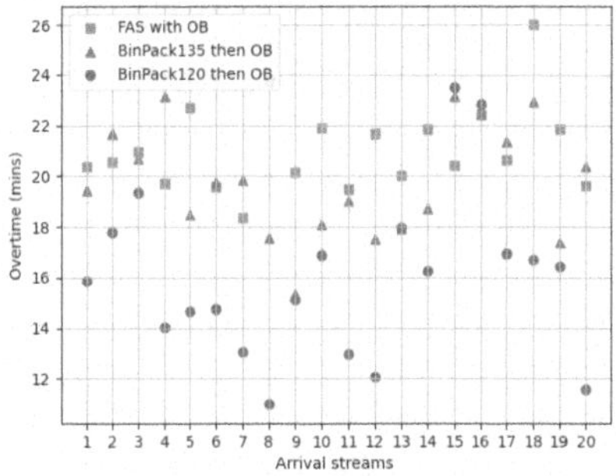

Fig. 8. Average overtime

Table 7. Various performance metrics

Performance measure		FAS	BinPack135	BinPack120
Average idle time (mins)	Avg	2.20	9.56	24.45
	Std dev	1.39	4.83	9.59
	Min	0.36	2.95	8.81
	Max	4.73	18.82	45.12
Average waiting time (mins)	Avg	186.21	132.18	101.09
	Std dev	19.48	14.56	13.23
	Min	150.70	106.11	82.12
	Max	221.67	160.12	126.75
Average overtime in a day (mins)	Avg	21.10	19.74	15.98
	Std dev	1.69	2.15	3.24
	Min	18.37	15.31	10.97
	Max	25.99	23.16	23.49
Number of patients scheduled	Avg	32.3	31.85	30.6
	Std dev	1.05	1.19	1.01
	Min	31	30	29
	Max	35	35	32
Number of patients served	Avg	30.64	30.41	29.93
	Std dev	0.82	0.88	0.85
	Min	29.46	28.95	28.57
	Max	32.49	32.45	31.50

significant waiting time of about 2 h on average. This can be attributed to the overbooking in the initial slots and also the significant variance in scan times for patients. Nevertheless, both policies improve the current waiting times of 5 h on average. The BinPack120 policy outperforms both the FAS and BinPack135 policy in terms of waiting times, achieving a 45% reduction in waiting times

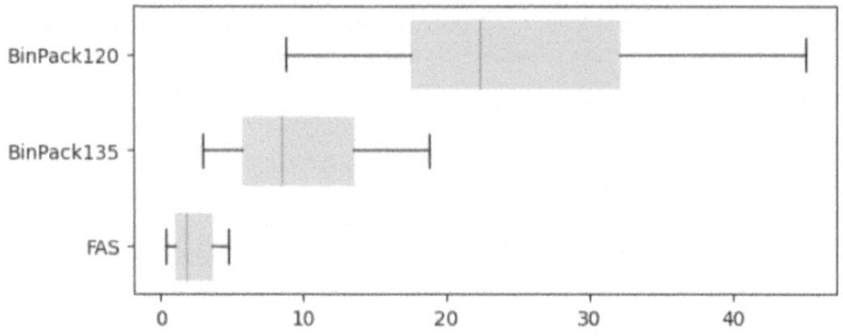

Fig. 9. Avg Idle times across all replications for all streams

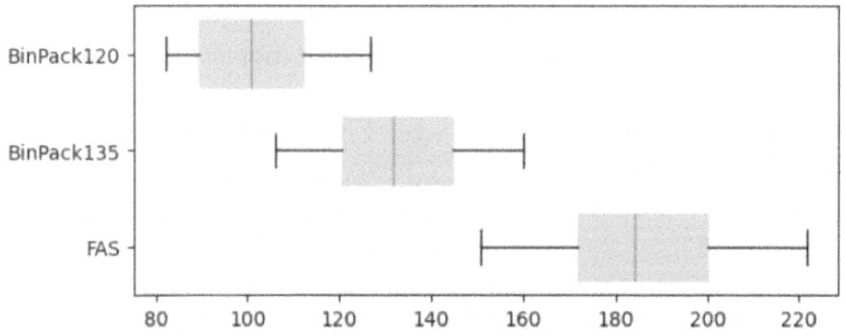

Fig. 10. Avg Waiting times across all replications for all streams

compared to FAS and a 23% reduction compared to BinPack135. As evident from Fig. 10, the minimum value of wait times of FAS is 45% more than that from the minimum value of BinPack120. Additionally, the FAS policy has a larger spread of waiting times, shown in Fig. 10 compared to the more consistent and narrow range of wait times for the other two policies.

However, the lower wait times for BinPack120 come at a cost of increased machine idle time, with an average idle time of about 25 mins and a maximum idle time of 45 mins on some days Fig. 9. This is due to scheduling patients within the range of the slot length and the variability in scan times, resulting in idle time mostly in the later slots. While this policy may not be the preferred choice when aiming for maximum machine utilization, it may work well for systems where idle time in this range is acceptable on some days. At a confidence level of 95% Fig. 8, the simulation model is consistent with the actual results, indicating the reliability and validity of the model in predicting the performance of different policies in the appointment scheduling system.

Table 8. 95% Confidence intervals for various performance metrics

Performance measure	FAS	BinPack135	BinPack120
Average idle time (mins)	9.56 ± 2.117	2.20 ± 0.609	24.45 ± 4.203
Average waiting time (mins)	132.18 ± 6.381	186.21 ± 8.537	101.09 ± 5.798
Average overtime in a day (mins)	19.74 ± 0.942	21.10 ± 0.741	15.98 ± 1.420
Number of patients scheduled	31.85 ± 0.522	32.3 ± 0.460	30.6 ± 0.443
Number of patients served	30.41 ± 0.386	30.64 ± 0.359	29.93 ± 0.373

7 Conclusion

Based on the simulation results, we conclude that the BinPack135 then OB policy offers a trade-off between machine idle times and patient waiting times. On the other hand, BinPack120 provides extremely low waiting times for patients, but it comes at the cost of an average machine idle time of 25 mins, in contrast to the FAS policy, which leads to excessive patient wait times. Therefore, we recommend using BinPack120 in systems where patient service is of utmost importance, as it provides better performance in terms of average waiting times. Although the average idle time is higher than the other policies, it is acceptable given the congestion in an overstressed facility. In contrast, FAS with OB is recommended in systems where machine utilization is of high importance.

Our study highlights the importance of considering an overbooking factor in initial slots to ensure a buffer against variability in service times. The overbooking factor (o) and the slots to be overbooked (k) can be determined based on the specific requirements of the system. The results obtained from the BinPack135 policy are intuitive, as it only allots a patient to a slot if it can be completely accommodated in it. In contrast, the FAS policy allows not only overbooking in initial slots but also inherent overbooking in later slots by allotting patients even if there are only 5 mins of time left in the slot, and scheduling is stopped once the limit is reached. This results in either empty or underutilization of later slots with no patients scheduled in them. This leads to increased patient wait times as patients are called about two slots ahead of when they can be served.

Although implementing bin packing policies in practice may be difficult, it is possible to design IT systems that schedule slots according to bins and monitor slot utilization. The long-term benefits of such policies in terms of improved patient satisfaction and reduced waiting times cannot be overstated. On the other hand, FAS policies may seem easier to implement, but their long waiting times can lead to overcrowding and increased patient stress. However, a potential issue with bin packing policies is that even patients at the end of the queue can get access to the initial available slots, which might create a perception among some patients that they can book appointments later. Despite these limitations, bin-packing policies offer a promising alternative to FAS policies for managing appointments in healthcare systems, and further research is needed to explore their implementation in real-world settings.

8 Future Work

In addition to comparing policies for the current set of arrival schedules, further evaluations can be performed by considering different scenarios. Specifically, different sets of average scan times and variances in their scan times can be tested to evaluate the performance of the designed policies. Another potential comparison that can be made is between online and offline appointment scheduling problems, where the arrival stream is known beforehand. This would provide insights into the effectiveness of policies in real-time scenarios compared to ideal situations. Moreover, an extension of this work could involve analyzing the impact of variable appointment slot lengths on the performance of the policies. This would involve modifying the policies to accommodate for variable appointment lengths and testing their performance on different arrival schedules. In summary, further investigations into the aforementioned aspects can provide a comprehensive understanding of the designed policies and their applicability in different healthcare systems.

References

1. Green, L.V., Savin, S., Wang, B.: Managing patient service in a diagnostic medical facility. Oper. Res. **54**(1), 11–25 (2006)
2. Bailey, N.T.J.: A study of queues and appointment systems in hospital out-patient departments, with special reference to waiting-times. J. Roy. Stat. Soc. Ser. B (Methodol.) **14**(2), 185–199 (1952)
3. Cappanera, P., Visintin, F., Banditori, C., Di Feo, D.: Evaluating the long-term effects of appointment scheduling policies in a magnetic resonance imaging setting. Flex. Serv. Manuf. J. **31**, 212–254 (2019)
4. Soriano, A.: Comparison of two scheduling systems. Oper. Res. **14**(3), 388–397 (1966)
5. Patrick, J., Puterman, M.L., Queyranne, M.: Dynamic multipriority patient scheduling for a diagnostic resource. Oper. Res. **56**(6), 1507–1525 (2008)
6. Alon, N., Awerbuch, B., Azar, Y.: The online set cover problem. In: Proceedings of the Thirty-Fifth Annual ACM Symposium on Theory of Computing, pp. 100–105 (2003)
7. Cayirli, T., Veral, E.: Outpatient scheduling in health care: a review of literature. Prod. Oper. Manag. **12**(4), 519–549 (2003)
8. Gupta, J.N.D., Ho, J.C.: A new heuristic algorithm for the one-dimensional bin-packing problem. Prod. Plan. Control **10**(6), 598–603 (1999)
9. Gupta, D., Denton, B.: Appointment scheduling in health care: challenges and opportunities. IIE Trans. **40**(9), 800–819 (2008)
10. Bhattacharjee, P., Ray, P.K.: Simulation modelling and analysis of appointment system performance for multiple classes of patients in a hospital: a case study. Oper. Res. Health Care **8**, 71–84 (2016)
11. Smedira, D., Shmoys, D.: Scheduling appointments online: the power of deferred decision-making. In: Chalermsook, P., Laekhanukit, B. (eds.) WAOA 2022. LNCS, vol. 13538, pp. 82–115. Springer, Cham (2022). https://doi.org/10.1007/978-3-031-18367-6_5
12. Benjaafar, S., Chen, D., Wang, R., Yan, Z.: Appointment scheduling under a service-level constraint. Manuf. Serv. Oper. Manag. **25**(1), 70–87 (2023)

Identifying Flight Schedule Characteristics Increasing Pilots Absenteeism at an Airline Using a Data Mining and Simulation Approach

Thomas Luke Nibbering and Alejandro Murrieta-Mendoza(✉)

Amsterdam University of Applied Sciences, 1091 GC Amsterdam, The Netherlands
a.murrieta.mendoza@hva.nl

Abstract. Sickness absenteeism among flight crews is a pervasive problem disruptive to operations and costly for the employer. According to literature, exposure to certain schedule attributes has been associated with adverse health issues. However, the relationship between schedule characteristics and sickness absenteeism remains unclear. Therefore, the aim of this study is to identify schedule characteristics increasing the odds of sickness absenteeism based on historical data. Here, data records for each flight crew member were obtained from a Dutch low-cost airline in the period between 1 January 2018 and 24 January 2020. Schedule characteristics with an adverse effect on both the circadian and/or social rhythm, as identified in literature, were extracted from the available data, and included in the model. Exploration on these potential harmful schedule attributes was done using two generalised additive models. After adjusting for the socio-demographic and work-related confounding variables, simulations revealed that employees exposed to night shifts, backward, and forward rotations over a thirty-day period were significantly more likely to report sick. Furthermore, employees who flew four sectors showed higher odds to call in sick compared to employees who flew two sectors. Based on the results, it is recommended to schedule either sufficient rest periods after exposure or limit the occurrence of the identified schedule attributes.

Keywords: Sickness · Absence · Flight Crew · Schedule · Model · Data

1 Introduction

Absenteeism of employees is of interest for organisations due to the disruption of services caused by not having the expected labour in function at a given period. Previous studies have focused on identifying causes of absenteeism in a range of industries. Winkelman [1] developed a Poisson count model to examine the correlation between workplace conditions and salaries for blue collar employees. Although their Markov Chain Monte Carlo analysis found no correlation between salary and absence from work, the study revealed a positive association between firm size and absenteeism. Globerson and Nagarvala [2] used a simulation model to predict the first day of absence along with its duration. They found that workers were more likely to be absent at the start of the work week. Vialatte et al. [3], employed principal component analyses and vector autoregressive

M. Mujica Mota and P. Scala (Eds.): EUROSIM 2023, CCIS 2032, pp. 149–162, 2024.
https://doi.org/10.1007/978-3-031-68435-7_11

models to identify causes of absenteeism among hospital employees. The study found several risk factors that led to higher absenteeism rates, such as having children, working administrative tasks or having a certain age. Goldenberg and Waldman [4] reported that absenteeism was not related to job satisfaction.

An industry that has received limited attention in literature concerning potential causes of absenteeism among its employees is aviation. Here, flight crew members have to endure challenging work related factors, such as extended duty days, early starts, night flights and multiple flight legs per day [5]. These job aspects can cause disruptions of the circadian and social rhythm, due to reduced sleep quality and quantity as well as jet lag symptoms [6]. The resulting fatigue is a common phenomenon that interferes with the pilots' normal social activities and work duties [7] and is also suspected to cause adverse physical [8–10], mental [11–13] and social [11, 14, 15] health problems. In turn, evidence indicates that these type of health issues are associated with increased levels of sickness absence among employees [16, 17].

Besides the individual consequences in terms of affected health, sickness absence is considered a pervasive problem disruptive to operations and costly for the employer [18]. As a result, it is only natural for both the employer and employee to seek solutions to this problem. The airline's possibility to influence certain job aspects when assigning flight crew to scheduled flights might reduce sickness absence rising from fatigue. However, a clear overview of schedule attributes negatively associated with flight crew reporting sick is non-existent.

Some studies found that certain job aspects are related to sickness absence [19, 20], but the associations between these schedule characteristics are not well-established, as other studies found weak or no relationships [21, 22]. A systematic review of Costa et al. [23], identified that schedule characteristics disruptive to the circadian and/or social rhythm pose to be negatively associated with health. Subsequently, this may increase the risk of sickness absenteeism among flight crew employees. These potential harmful schedule characteristics include the type of shift [24, 25], the shift length [26], the rest period between shifts [27], the rotation direction [26], the rotation pace [25], and the work hour distribution [24]. Based on these potential disruptive attributes, this study aims to identify whether these characteristics increase the odds of sickness absenteeism among flight crew. In doing so, this paper sets out to accomplish three main objectives. First, it addresses the scarcity of literature on schedule attributes associated with absenteeism among flight crew. Second, it provides empirical evidence of the detrimental impact of circadian and social rhythm disruptive schedule characteristics on health, as indicated by flight crew reporting sick. Lastly, the results may offer practical insights to help airlines reduce schedule-related sickness absenteeism while serving as inputs in more elaborate simulation models.

This paper is organised as follows: Sect. 2 elaborates on the data cleaning and pre-processing conducted, followed by an explanation of the dependent and independent variables included in the study. Moreover, the section explicates the rationale behind the inclusion of confounders and the use of generalised additive models to examine the association of interest. The results are presented and discussed in Sect. 3, while the conclusion and future work are discussed in Sect. 4.

2 Methodology

2.1 Data Collection

In this study, human resources administration and flight schedule information from a Dutch low-cost airline was obtained for the period between 1 January 2018 and 24 January 2020. The reason for obtaining data from this specific period was primarily due to the lack of historical data recorded in a comparable manner. Additionally, impurities originating from the Covid-19 pandemic [28], with its first European case on 24 January 2020, could potentially distort the results. The obtained data included socio-demographic, work-related, absenteeism and daily schedule information for each flight crew employee. Both datasets were combined and anonymised in order to comply with the General Data Protection Regulation (GDPR) for processing personal data [29]. As a result, the Data Privacy Office (DPO) of the airline exempted this study from further privacy review.

2.2 Data Pre-processing

Inspection of the obtained dataset revealed inconsistencies, such as outliers and erroneous values among the independent variables contract rate and shift duration. In order to overcome these contract rates inconsistencies, inclusion criteria were being in receipt for flight crew with a contract rate between 0.5 and 1. Exclusion criteria existed for contract rates less than 0.5 or more than fulltime, as they were considered flaws in data entry and were corrected manually. Inconsistencies in shift duration were corrected using exclusion criteria. Here, shifts exceeding the maximum extended flight duty period (> 14 h) without a layover at destination were excluded from the data [30]. The reason for this was that legislative limits were otherwise exceeded, making the existence of these types of schedules highly unlikely.

Finally, flight crew with a managerial function or those who performed secondary activities at the airline were excluded from the analysis. The reason for this was that employees with or without managerial or secondary activities were not considered one homogeneous group. Eventually, the data pre-processing procedure resulted in 167,077 observations being available for the analysis, as presented in Fig. 1.

2.3 Outcome Encoding

Sickness absenteeism was defined as absence from work due to health reasons as registered by the occupational health service [31]. In this study, only sickness absence records after publication of the flight schedule were included in the analysis. This was because these employees were more likely to suffer from schedule-related absenteeism [32]. Each sickness absence episode was considered finished when the employee fully returned to work for at least one day. To ease further analysis, the outcome variable was transformed into a binary outcome defined as either presence (value of 1) or absence of an individual from a shift (value of 0).

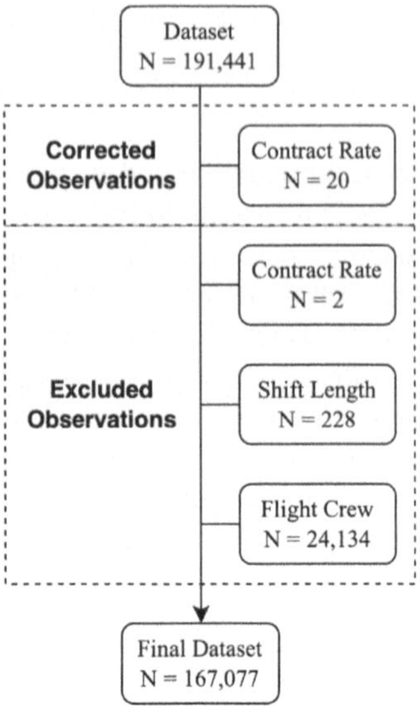

Fig. 1. Schematic overview of the data cleaning procedure.

2.4 Schedule Characteristics Extraction

Independent variables representing schedule characteristics were selected and extracted from the available data. First, the selected independent variables from the airline's content management system included the starting and ending times of a shift, duration of the shift, number of sectors flown, and whether the shift had a layover at the destination. Second, based on these starting and ending times, schedule attributes disruptive to the circadian and/or social rhythm were extracted [23]. These potential harmful schedule characteristics include the type of shift [24, 25], the shift length [26], the rest period between shifts [27], the rotation direction [26], the rotation pace [25], and the work hour distribution [24].

The shift period was defined as either a night or day shift with a night shift encroaching any portion between 01:00 and 05:59, in accordance with collective bargaining agreements [33]. Based on legislative definitions, disruptive shifts were extracted from the start and ending times [30]. Furthermore, shift durations were defined as the period between shift start and shift end (e.g. 07:00–17:00 = a 10 h shift), whereas the rest period was determined based on the previous shift end and the following shift start (e.g. 23:00 – 11:00 = a 11-h rest). Moreover, workdays were considered calendar days with an observation in the dataset, while the absence of an observation within the dataset was considered a rest day. Forward rotating shifts were determined to start later than the previous shift (e.g. previous shift started at 06:00, following started at 07:00). In contrast, backward rotating shifts were considered to start earlier than the previous shift. Furthermore, fast rotating shifts were defined as less than or equal to five similar consecutive shift starts. Finally, several studies emphasised on the adverse health effects when exposed to schedule characteristics for prolonged periods [21, 32]. Therefore, in line with Ropponen et al. [20], cumulative exposure to the type of shift, the rest period, the rotation direction, and the work hour distribution were measured over a rolling period of seven and thirty days and included in the analysis.

2.5 Confounders Determination

To obtain a valid estimate of the association between the identified characteristics and sickness absenteeism, confounding variables, i.e. those variables that influence both the dependent and independent variable, were included in this study [34]. Socio-demographic confounders included age, gender and marital status whereas work-related confounders comprised of crew function, number of employment years and contract rate. The reason for selecting these types of confounders was two-fold. First, although scientific literature proposed additional confounders [21, 32, 35], the limited availability of this type of information stored in the administrative databases of the airline reduced the possibility to do so. Second, literature revealed that the selected confounders were associated with both the independent and dependent variables. Hence, failing to account for these variables might distort the results.

2.6 Generalized Additive Model

The association between schedule characteristics and sickness absenteeism was explored using generalised additive models. Here, data was modelled with a binary outcome. The consideration for this type of data mining model was three-fold. First, this type of model allowed to describe the relationships with a constant effect. Hence, making it possible to distinguish between schedule characteristics that represent a higher or lower odds of sickness absenteeism. Second, the model allowed to estimate the relationship while adjusting for potential confounders [36]. In turn, this reduced the probability that certain schedule attributes were mistakenly considered to influence sickness absence [37]. Third, the model allowed to estimate both linear and non-linear relationships between independent variables and the log odds of the outcome [38], contributing to statistically valid inferences.

Before modelling the data, the independent variables were tested for collinearity ($r > 0.8$) [39]. If collinearity was assumed, the variable with the weakest linear relationship with the outcome was removed [40]. In turn, this excluded the independent variables; employment years, date, shift start and end, rest days last seven and thirty days, the number of hours worked last seven and thirty days, and the number of early disruptive starts last thirty days. Thereafter, independent variables were tested for non-linearity with the log odds of the outcome. This was done using the Box-Tidwell test [41]. Predictor variables with a non-linear relationship included age, the number of early disruptive starts last thirty days, and the number of forward rotations last thirty days. To model the curved relationship of these independent variables, thin plate spline algorithms were used. This technique was used as it automatically detects the optimal parameters during the procedure while estimating the underlying trend with high accuracy [42], as presented in Fig. 2b. Here, an assumed linear relationship would provide incorrect results, as illustrated in Fig. 2a.

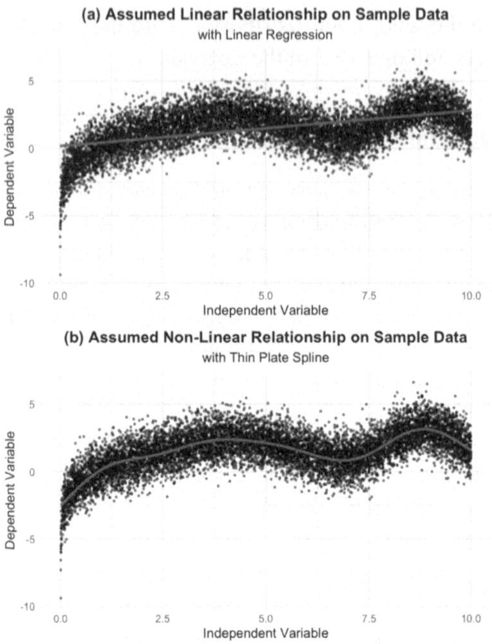

Fig. 2. Linear and non-linear relationship among predictor and outcome sample variables

Next, two generalised additive models were composed. First, the full model consisted of all socio-demographic, work-related, and schedule-related independent variables from the dataset. A second generalised additive model controlled for the potential effect of confounding variables. This adjusted model excluded the variables age, gender, marital status, crew function, and contract rate from the analysis. Independent variables with a significance level of $p < 0.001$ were considered statistically significant. Since this study was interested in finding the true association between schedule characteristics, this rather conservative significance level was chosen in an effort to provide valid results.

3 Results

3.1 Study Population

The final study population consisted of 691 flight crew employees without either a managerial or secondary function The majority of the study population was male (N = 624), had an average age of 40.58 (SD = 10.12) years, was married (N = 227), and worked fulltime (N = 640) as a captain (N = 350). Throughout the study period, 86.25% of the employees reported sick at least once. An overview of the socio-demographic and work-related characteristics of the study population is provided in Table 1.

Table 1. Descriptive statics of socio-demographic and work related characteristics of the study population.

Independent Variable	Frequency	Percentage	Mean	SD[1]	Range[2]
	Study Population (N = 691)				
Age (in Years)	691	–	40.58	10.12	20–60
Gender					
Male	624	90.30%	–	–	–
Female	67	9.70%	–	–	–
Marital Status					
Married	227	32.85%	–	–	–
Cohabit	179	25.90%	–	–	–
Single	179	25.90%	–	–	–
Unknown	54	7.82%	–	–	–
Partnership	38	5.50%	–	–	–
Divorced	10	1.45%	–	–	–
Widowed	4	0.58%	–	–	–
Crew Function					
Captain	350	50.65%	–	–	–
First Officer	341	49.35%	–	–	–
Contract Rate					
1	607	87.85%	–	–	–
0.90	1	0.14%	–	–	–
0.80	1	0.14%	–	–	–
0.75	43	6.23%	–	–	–
0.70	1	0.14%	–	–	–
0.5	38	5.50%	–	–	–
Absence Code					
Absence	596	86.25%	–	–	–
Presence	95	13.75%	–	–	–

[1]SD = standard deviation, [2]Range = minimum and maximum value

3.2 Generalised Additive Model Output

The full and adjusted generalised additive models provided the significance level, estimated odds ratio, and the confidence interval for each independent variable in the model. Here, the significance level indicated the probability that an association between variables could occur by chance [39] whereas odds ratios (OR) represented the strength of the association between the independent and dependent variables [40]. This made it possible to distinguish between schedule characteristics that represented a significant higher (OR > 1) or a lower (OR < 1) risk of sickness absenteeism. Moreover, confidence intervals indicated the level of uncertainty around the odds ratio by representing an upper and lower limit [40].

Simulation results from both the full and adjusted model revealed seven schedule attributes significantly associated with sickness absenteeism. Table 2 indicated that both data mining models provided similar odds ratios and significant relationships between the independent and dependent variables. As a result, this may indicate that the association between predictors and the outcome was only distorted slightly by the included confounders.

Table 2. Statistical significance, odds ratios, and associated confidence intervals in both generalized additive models.

Independent Variable	OR1	99.9% CI[2]	P-value	OR[1]	99.9% CI[2]	P-value
Sectors	*Full Model*			*Adjusted Model*		
0	Ref.	Ref.	Ref.	Ref.	Ref.	Ref.
1	0.71	0.43–1.17	0.026	0.70	0.42–1.15	0.018
2	1.26	1.15–1.36	<0.001	1.26	1.15–1.36	<0.001
3	0.87	0.66–1.15	0.10	0.88	0.67–1.15	0.12
4	1.48	1.26–1.73	<0.001	1.52	1.30–1.79	<0.001
5	4.06	0.10–164	0.20	3.67	0.09–148	0.20
Night Shifts last 30 Days	1.03	1.01–1.06	<0.001	1.04	1.02–1.06	<0.001
Workdays last 30 Days	0.96	0.94–0.98	<0.001	0.94	0.92–0.96	<0.001
Backward Rotation						
No	Ref.	Ref.	Ref.	Ref.	Ref.	Ref.
Yes	0.88	0.78–1.00	<0.001	0.88	0.78–1.00	<0.001
Backward Rotations						
last 30 Days	1.08	1.05–1.11	<0.001	1.09	1.06–1.13	<0.001
Early Disruptive Starts						
last 30 Days[3]	–	–	<0.001	–	–	<0.001
Forward Rotations						
last 30 Days[3]	–	–	<0.001	–	–	<0.001

[1]OR = odds ratio, [2]CI = confidence interval, [3]Non-linear independent variables. Hence, the absence of a constant OR and CI.

In addition, since certain independent variables had a non-linear relationship with the outcome and were modelled using a thin plate spline smoothing algorithm, the constant effects of these predictor variables could vary. Therefore, visual methods were used to estimate the associations. Hence, the absence of ORs for these non-linear independent variables in Table 2. In turn, the simulation results of both models provided four main patterns associated with sickness absence which will be discussed and interpreted next.

First, the full model revealed that employees who flew two (OR = 1.26) or four (OR = 1.48) sectors were more likely to report sick. In other words, employees who flew two sectors were 1.26 times more likely to be absent from work due to sickness compared to flight crew members without a sector flown. In turn, four sectors flown makes employees 1.48 times more likely to call in sick. A possible reason for these findings is provided by Flynn-Evans et al. [45]. This study concluded that fatigue among flight crew is proportional to the number of sectors flown. Since several studies linked fatigue to increased levels of sickness absence [19, 46], an incrementation of sectors flown might be one of the reasons for finding these increased odds ratios. Interestingly, employees who flew three sectors were found to be 0.87 times less likely to report sick. Although this result may appear unexpected, this finding might be attributed to the interaction with an overnight stay at destination, potentially discouraging flight crew members from calling in sick.

Second, the results of both models showed that prolonged exposure to specific schedule characteristics makes employees more likely to be absent from work due to sickness. Specifically, the full model revealed that flight crew members exposed to night shifts (OR = 1.03) or backward rotations (OR = 1.08) over a thirty-day period were more likely to call in sick compared to employees without this exposure. A similar increased odds for sickness absenteeism can be found when employees are exposed to forward rotations over a thirty-day period. Since this latter independent variable had a non-linear relationship with the outcome and was modelled using a thin plate spline algorithm, the OR of this predictor varies. Therefore, Fig. 3 depicts the likelihood of sickness absence when exposed to this schedule attribute for both models. Figure 3 shows that prolonged exposure to forward rotations over a thirty-day period provides increased odds of sickness absenteeism only after more than five rotations, i.e. after five rotations the odds ratio becomes larger than one.

An explanation for this finding is provided by van Drongelen et al. [21]. This study concluded that employees may have been unable to implement these irregular schedules in their social and family life. Hence, the lack of adoption might be one of the reasons for finding increased odds levels for sickness absenteeism, as circadian and/or social rhythms are disrupted without sufficient possibility to recover.

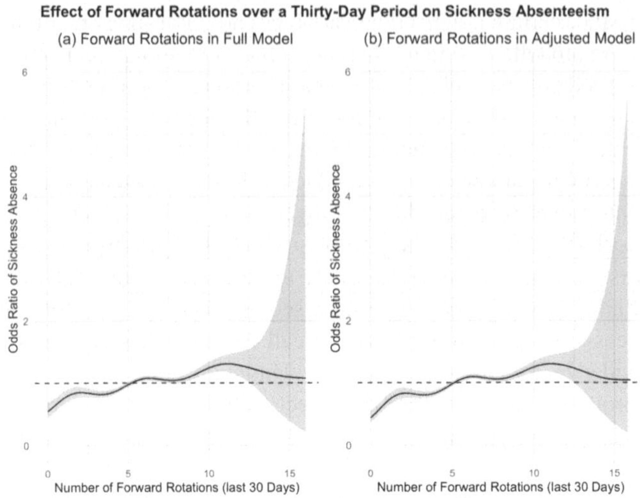

Fig. 3. Effect of the number of forward rotations over a thirty-day period on sickness absenteeism

Third, the results showed that prolonged exposure to specific schedule attributes could also make employees less likely to call in sick. Specifically, the full model revealed that employees exposed to an increased number of workdays over a thirty-day period (OR = 0.96) were less likely to be absent from work due to sickness. In other words, each additional workday over a period of thirty days makes employees 0.96 times less likely to report sick. This result may seem counterintuitive as the possibilities to recover from circadian and/or social rhythm disruptions are limited with an increased number of workdays. However, rest periods at the airline are proportional to the number of days worked. Hence, a compressed number of workdays in a specific month could be compensated for in another month, providing an opportunity for successive rest periods. In addition, it was found that extended exposure to disruptive shift starts over a period of thirty days makes employees less likely to report sick, as depicted in Fig. 4. Again, in contrast to linearly related independent variables, the OR of this non-linear predictor could vary. Since this type of schedule characteristic was linked to increased levels of fatigue [30], this finding may seem contradictory. Specifically, as several studies found associations between fatigue and increased levels of sickness absenteeism [19, 46]. A possible reason for this result is that this attribute had a limited number of data observations. In turn, this may decrease the ability of the model to detect the true effect of this schedule attribute on sickness absenteeism [47].

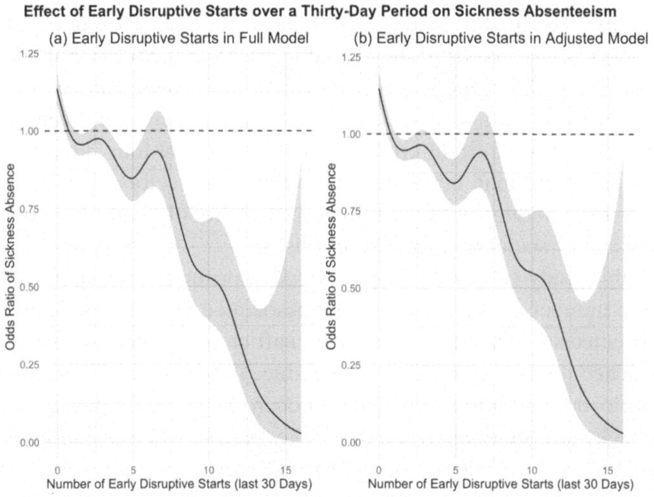

Fig. 4. Effect of the number of early disruptive starts over a thirty-day period on sickness absenteeism

Fourth, both models also revealed that the effect of specific schedule attributes was not consistent for all periods. Specifically, the full model revealed that flight crew members exposed to a backward rotation (OR = 0.88) were less likely to report sick than those without exposure. Interestingly, the same schedule characteristic over a thirty-day period revealed higher odds for sickness absence. One reason for this finding is provided by Dall'Ora et al. [19]. This study concluded that incidental disruptions of the circadian and/or social rhythm are less problematic than extended ones. Hence, this could explain the difference in ORs.

In summary, the results of the full and adjusted models revealed that especially prolonged exposure to night shifts, backward, and forward rotations over a thirty-day period provided flight crew members with an increased odds of reporting sick. Since the ORs of these attributes were larger than one, each unit decrease reduces the odds of sickness absenteeism. As a result, reducing the frequency of these schedule attributes when assigning flight crew to scheduled flights could potentially reduce the odds of sickness absenteeism among flight crew. Moreover, a similar odds reduction could be realised when the number of sectors flown per shift is lowered from four to two.

4 Conclusion

This study aimed to identify schedule characteristics increasing the odds of sickness absenteeism based on modelling and simulating historical data. Human resources administration and flight schedule data provided socio-demographic, work-related, absenteeism, and daily schedule information for each flight crew employee. Literature revealed that employees exposed to schedule characteristics with an adverse effect on both the circadian and/or social rhythm were more likely to suffer from health problems. These schedule attributes included the type of shift, the shift length, the rest period between shifts, the type of shift rotation, and the work hour distribution. In turn, these characteristics were extracted from the available data and included in the final dataset.

The generalised additive models revealed that employees exposed to night shifts, backward, and forward rotations over a thirty-day period were significantly more likely to report sick. Furthermore, flight crew members who flew four sectors were slightly more likely to be absent than employees who flew two sectors. The main theoretical explanation for these findings was the limited possibility of employees to recover from these circadian and/or social rhythm disruptions.

Moreover, this study revealed that prolonged exposure to certain schedule characteristics and the number of sectors flown are associated with a significant higher odds of sickness absenteeism. Therefore, providing sufficient rest periods, when exposed to these types of schedule characteristics, might reduce the odds of sickness absence among flight crew employees. Additionally, each unit decrease in the frequency of night shifts, forward, and backward rotations over a thirty-day period, when assigning flight crew to scheduled flights, could reduce the odds of flight crew reporting sick. A similar odds reduction could be realised when the number of sectors flown per shift is lowered from four to two.

Further research should focus on examining the financial and operational implications of constraining the identified schedule attributes in the airline planning process. This will ensure that recommendations of this study are considered from multiple perspectives. Furthermore, as this study focused more on the association rather than the cause-and-effect relationship, similar research should be performed to validate the findings before implementation. This might be done using a more complex (simulation) model.

References

1. Winkelmann, R.: Markov chain Monte Carlo analysis of underreported count data with an application to worker absenteeism. Empir. Econ. **21**, 575–587 (1996). https://doi.org/10.1007/BF01180702
2. Globerson, S., Nagarvala, P.J.: Unexplained absenteeism. In: Proceedings of the 7th Conference on Winter Simulation-WSC74. ACM Press: New York. pp 55–58 (1974)
3. Vialatte, L., Pereira, B., Guillin, A., et al.: Mathematical modeling of the evolution of absenteeism in a university hospital over 12 years. Int. J. Environ. Res. Public Health **19**, 8236 (2022). https://doi.org/10.3390/ijerph19148236
4. Goldberg, C.B., Waldman, D.A.: Modeling employee absenteeism: testing alternative measures and mediated effects based on job satisfaction. J. Organiz. Behav. Int. J. Ind. Occupat. Organiz. Psychol. Behav. **21**, 665–676 (2000)
5. Powell, D.M.C., Spencer, M.B., Holland, D., et al.: Pilot fatigue in short-haul operations: effects of number of sectors, duty length, and time of day. Aviat. Space Environ. Med. **78**, 698–701 (2007)
6. Caldwell, J.A.: Fatigue in aviation. Travel Med. Infect. Dis. **3**, 85–96 (2005). https://doi.org/10.1016/j.tmaid.2004.07.008
7. Petrie, K.J., Powell, D., Broadbent, E.: Fatigue self-management strategies and reported fatigue in international pilots. Ergonomics **47**, 461–468 (2004). https://doi.org/10.1080/0014013031000085653

8. Vyas, M.V., et al.: Shift work and vascular events: systematic review and meta-analysis. BMJ. **345**, e4800–e4800 (2012). https://doi.org/10.1136/bmj.e4800
9. van Drongelen, A., Boot, C., Merkus, S., et al.: The effects of shift work on body weight change—a systematic review of longitudinal studies. Scand. J. Work Environ. Health **37**, 263–275 (2011). https://doi.org/10.5271/sjweh.3143
10. Lin, X., Chen, W., Wei, F., et al.: Night-shift work increases morbidity of breast cancer and all-cause mortality: a meta-analysis of 16 prospective cohort studies. Sleep Med. **16**, 1381–1387 (2015). https://doi.org/10.1016/j.sleep.2015.02.543
11. Moreno, C.R.C., Marqueze, E.C., Sargent, C., et al.: Working Time Society consensus statements: evidence-based effects of shift work on physical and mental health. Ind. Health **57**, 139–157 (2019). https://doi.org/10.2486/indhealth.SW-1
12. Angerer, P., Schmook, R., Elfantel, I., Li, J.: Night work and the risk of depression. Dtsch. Arztebl. Int. **114**, 404 (2017). https://doi.org/10.3238/arztebl.2017.0404
13. Zhao, Y., Richardson, A., Poyser, C., et al.: Shift work and mental health: a systematic review and meta-analysis. Int. Arch. Occup. Environ. Health **92**, 763–793 (2019). https://doi.org/10.1007/s00420-019-01434-3
14. Albertsen, K., Rafnsdottir, G.L., Grimsmo, A., et al.: Workhours and worklife balance. Scand. J. Work Environ. Health **34**, 14–21 (2008)
15. Greubel, J., Arlinghaus, A., Nachreiner, F., Lombardi, D.A.: Higher risks when working unusual times? A cross-validation of the effects on safety, health, and work–life balance. Int. Arch. Occup. Environ. Health **89**, 1205–1214 (2016). https://doi.org/10.1007/s00420-016-1157-z
16. Booker, L.A., Sletten, T.L., Alvaro, P.K., et al.: Exploring the associations between shift work disorder, depression, anxiety and sick leave taken amongst nurses. J. Sleep Res. **29**, 1205 (2020). https://doi.org/10.1111/jsr.12872
17. Marmot, M., Feeney, A., Shipley, M., et al.: (1995) Sickness absence as a measure of health status and functioning: from the UK Whitehall II study. J. Epidemiol. Community Health **49**, 124–130 (1978). https://doi.org/10.1136/jech.49.2.124
18. de Jong, P.: Sickness, disability and work: breaking the barriers. OECD (2010)
19. Dallora, C., Ball, J., Redfern, O., et al.: Are long nursing shifts on hospital wards associated with sickness absence? A longitudinal retrospective observational study. J. Nurs. Manag. **27**, 19–26 (2019). https://doi.org/10.1111/jonm.12643
20. Ropponen, A., Koskinen, A., Puttonen, S., Härmä, M.: Exposure to working-hour characteristics and short sickness absence in hospital workers: a case-crossover study using objective data. Int. J. Nurs. Stud. **91**, 14–21 (2019). https://doi.org/10.1016/j.ijnurstu.2018.11.002
21. van Drongelen, A., Boot, C.R.L., Hlobil, H., et al.: Cumulative exposure to shift work and sickness absence: associations in a five-year historic cohort. BMC Public Health **17**, 67 (2017). https://doi.org/10.1186/s12889-016-3906-z
22. Bernstrøm, V.H., Houkes, I.: Shift work and sickness absence at a Norwegian hospital: a longitudinal multilevel study. Occup. Environ. Med. **77**, 555–563 (2020). https://doi.org/10.1136/oemed-2019-106240
23. Costa, G., Tommasi, E., Giovannini, L., Mucci N.: Shiftwork Organization. In: Textbook of Patient Safety and Clinical Risk Management. Springer: Cham, pp 403–412 (2021). https://doi.org/10.1007/978-3-030-59403-9_29
24. Åkerstedt, T.: Shift work and disturbed sleep/wakefulness. Occup Med (Chic Ill) **53**, 89–94 (2003). https://doi.org/10.1093/occmed/kqg046
25. Arlinghaus, A., Bohle, P., Iskra-Golec, I., et al.: Working Time Society consensus statements: evidence-based effects of shift work and non-standard working hours on workers, family and community. Ind. Health **57**, 184–200 (2019). https://doi.org/10.2486/indhealth.SW-4
26. Tucker P, Folkard S (2012) Working time, health and safety: a research synthesis paper

27. Åkerstedt, T., Kecklund, G.: What work schedule characteristics constitute a problem to the individual? A representative study of Swedish shift workers. Appl. Ergon. **59**, 320–325 (2017). https://doi.org/10.1016/j.apergo.2016.09.007
28. ECDC (2021) Timeline of ECDC's response to COVID-19. https://www.ecdc.europa.eu/en/covid-19/timeline-ecdc-response. Accessed 24 Jun 2022
29. European Parliament and Council: Regulation 2016/679. In: EUR-Lex (2016). https://eur-lex.europa.eu/eli/reg/2016/679/oj. Accessed 24 Jun 2022
30. EASA (2022) Easy Access Rules for Air Operations (Regulation (EU) No 965/2012). https://www.easa.europa.eu/document-library/easy-access-rules/easy-access- rules-air-operations-regulation-eu-no-9652012. Accessed 24 Jun 2022
31. Boot, C.R.L., van Drongelen, A., Wolbers, I., et al.: Prediction of long-term and frequent sickness absence using company data. Occup Med (Chic Ill) **67**, 176–181 (2017). https://doi.org/10.1093/occmed/kqx014
32. van Drongelen, A., van der Beek, A.J., Penders, G.B.S., et al.: Sickness absence and flight type exposure in flight crew members. Occup Med (Chic Ill) **65**, 61–66 (2015). https://doi.org/10.1093/occmed/kqu169
33. Transavia Airlines (2021) CAO Transavia Vliegers. https://loonwijzer.nl/arbeidsvoorwaarden/caowijzer/cao/1373/transavia- vliegers. Accessed 25 Jun 2022
34. Baños, J.H.: Confounding variables. In: Kreutzer J, DeLuca J, Caplan B. (eds) Encyclopedia of Clinical Neuropsychology. Springer International Publishing, p 7 (2018)
35. Vedaa, Ø., Pallesen, S., Waage, S., et al.: Short rest between shift intervals increases the risk of sick leave: a prospective registry study. Occup. Environ. Med. **74**, 496–501 (2017). https://doi.org/10.1136/oemed-2016-103920
36. Ranganathan, P., Pramesh, C., Aggerwal, R.: Common pitfalls in statistical analysis: logistic regression. Perspect. Clin. Res. **8**, 148–151 (2017)
37. Shao, Y., Goštautaitė, B., Wang, M., Ng, T.W.H.: Age and sickness absence: testing physical health issues and work engagement as countervailing mechanisms in a cross-national context. Pers. Psychol. (2022). https://doi.org/10.1111/peps.12498
38. Hastie, T., Tibshirani, R.: Generalized additive models. Stat. Sci. **1**, 101 (1986). https://doi.org/10.1214/ss/1177013604
39. Mansfield, E.R., Helms, B.P.: Detecting multicollinearity. Am. Stat. **36**, 158–160 (1982). https://doi.org/10.1080/00031305.1982.10482818
40. Vatcheva, K., Lee, M.: Multicollinearity in regression analyses conducted in epidemiologic studies. Epidemiology **6**, 1–20 (2016). https://doi.org/10.4172/2161-1165.1000227
41. Box, G., Tidwell, P.: Transformation of the independent variables. Technometrics **4**, 531–550 (1962)
42. Wood, S.N.: Thin plate regression splines. J. R. Stat. Soc. Ser. B Stat. Methodol. **65**, 95–114 (2003). https://doi.org/10.1111/1467-9868.00374
43. Cicchetti, D.: Statistical Significance. In: Volkmar, F. (ed) Encyclopedia of Autism Spectrum Disorders. Springer Publishing, p 2995 (2013)
44. Szumilas, M.: Explaining odds ratios. J. Canad. Acad. Child Adolesc. Psychiatry **19**, 227–229 (2010)
45. Flynn-Evans, E.E., Arsintescu, L., Gregory, K., et al.: Sleep and neurobehavioral performance vary by work start time during non-traditional day shifts. Sleep Health **4**, 476–484 (2018). https://doi.org/10.1016/j.sleh.2018.08.002
46. Sagherian, K., Unick, G.J., Zhu, S., et al.: Acute fatigue predicts sickness absence in the workplace: a 1-year retrospective cohort study in paediatric nurses. J. Adv. Nurs. **73**, 2933–2941 (2017). https://doi.org/10.1111/jan.13357
47. Dufek, J.S., Bates, B.T., Davis, H.P.: The effect of trial size and variability on statistical power. Med. Sci. Sports Exerc. **27**, 288–295 (1995). https://doi.org/10.1249/00005768-199502000-00021

Medical Device Safety Notification Process Analysis Using Queuing Theory and Simulation

J. Dorbecker[1], J. Meraz[1](✉), J. P. Cisneros[1,2], J. Luna[1,2], S. Lara[1,2], L. Mata[1,2], A. Almeida[1,2], and I. Flores de la Mota[1,2]

[1] Siemens Healthineers, 350 National Army Avenue 350 Floor 3, 11560 Mexico City, Mexico
julianmerazrdz@gmail.com

[2] Faculty of Engineering UNAM, University City, 04510 Mexico City, Mexico

Abstract. This paper presents the application of Queuing Theory and Simulation in a case study of a company dedicated to the manufacture of medical equipment for the health care sector, which has the need to execute the communication process to its customers with the intention of reducing the risk of incidents and thus improve the safety and performance of a medical device daily. The main objective is to demonstrate the benefits of analysis and the improvements that can be made in a system to increase its efficiency through these two tools. In addition, a system was selected in which data could be collected to perform the diagnosis, analysis and conclude with the proposal of actions to improve the service capacity of the system. Finally, simulations of different scenarios were carried out to confirm the operation of the system with the improvement proposals made.

Keywords: Queue · processes · mathematical and simulation model · medical device

1 Introduction

1.1 A Subsection Sample

A Field Safety Corrective Action (FSCA) is an action taken by a manufacturer to reduce a risk of death or serious deterioration in health status associated with the use of a medical device already on the market. Such actions, whether associated with direct or indirect harm, must be reported, and notified through a company-wide process of communicating these actions to authorities, customers, hospitals, and clinics.

Field safety actions for medical devices may include:

- Return to supplier.
- Inspection/Examination.
- Modification.
- Exchange.
- Destruction.
- Upgrade by purchaser or manufacturer's design change.

M. Mujica Mota and P. Scala (Eds.): EUROSIM 2023, CCIS 2032, pp. 163–173, 2024.
https://doi.org/10.1007/978-3-031-68435-7_12

- Manufacturer's advice related to use and/or related to patient, user, or other monitoring.

A field safety report should include the following information:

- Manufacturer's name
- Name of the product, product code and lot number affected.
- If the notification relates only to certain lots, an explanation of why the other lots are not affected should be included.
- List of all affected countries
- Background and reason(s) for issuing the safety action.
- Description of the deficiency or malfunction of the medical device, clarification of the potential hazard associated with continued use of the device and the associated risk to the patient, the user or any other person, and any potential risk to patients associated with previous use of the device.
- Relevant elements of the risk analysis.
- Description and justification of corrective and/or preventative action
- Advice on actions to be taken by the distributor and the user, for example:

    o Identify and quarantine the device.
    o Method of recovery, elimination, or modification of the device.
    o Recommended patient follow-up
    o A request to send the attached field safety notice (FSN) to anyone who should know about it.

In Mexico, the applicable regulatory framework on the subject is comprised in the Mexican Official Standard issued by the Ministry of Health, NOM-SSA-240-2012 entitled "Installation and operation of Postmarket Surveillance" published in the Official Gazette of the Federation on October 30, 2012.

The objective of the company is to analyze the application of queuing models directly on the process of field safety actions. They are justified in the significance that all notifications are communicated efficiently, helping the organization to build confidence in customers, patients, government agency, investors, and the community in general since the company will be able to fulfill its obligations and will be prepared to act with the agility expected from such an important process, such as spreading the safety corrective actions.

Some questions of interest to the company are as follows:

- What opportunities can be detected from the analysis that will allow generating action plans to improve the performance of the process?
- Is the number of personnel available to perform the activity sufficient?

To answer these questions, queuing theory we will use queueing theory and simulation. In particular, the M/M/c model in different scenarios, in which the number of servers will be the independent variable.

2 State of the Art

The mathematical models of queuing theory have been of great value for systems analysis, Portilla, Arias and Fernandez (2010) conducted an analysis and simulation exercise in a fictitious scenario of type M/M/1. They applied the theory and simulation; however, the study system was proposed by the authors for the understanding of the models.

Muñoz-Vergara (2019) elaborated an analysis of the waiting lines of the library service of an educational organization in Cartagena, Colombia. He analyzed three types of independent queues that are generated in this service, collected the data, and proceeded to perform the statistical analysis and calculations of the performance indicators of the system, concluding in his work with the state in which the system was and provided recommendations to improve the efficiency of the service.

Within the hospital branch, the use of queuing theory and simulation has been applied for the analysis of the emergency service, the authors Rodríguez, Gonzáles and Hernández (2017) related existing models of waiting lines to calculate the minimum number of physicians needed to meet the current and future demand for service.

Systems of great interest for the application of queuing theory and simulation have been airports. Jhala and Bhathawala (2017) used modeling and simulation for the analysis and improvement of problems that commonly occur in airport systems. Like Portilla et al. (2010) they assumed an M/M/1 system and performed the mathematical analysis without modeling and only characterizing the study system.

3 Theoretical Framework

3.1 Queuing Theory

Queuing theory is the set of mathematical models that describe waiting line systems, its objective is to find the steady state of the systems analyzed and determine their service capacity that guarantees a balance between the quantitative factor and the qualitative factor, the quantitative factor referring to the costs of the system and the qualitative factor to the satisfaction of the customers receiving the service (Giraldo, 2006).

The main participants in the analysis are the clients and the servers. Customers can be a person, a piece of equipment, a vehicle, among other entities that require a service. On the other hand, the servers are those who have the capacity to perform the service activity on the customer, they can be a cashier, a machine, a gasoline dispenser, among others. While queuing and simulation study waiting lines, they are not optimization techniques; rather they determine performance measures of waiting lines, such as average waiting time in the queue, average waiting time for service, and the use of service facilities. Queuing models use probabilistic and stochastic models to analyze queuing lines, and simulation estimates performance measures by mimicking the behavior of the real system (Taha, 2012). The key components for analysis are as follows (Winston, 2005):

1. Arrival of customers.
2. Queue size.
3. Queue discipline.
4. Service times.

5. Number of servers.
6. System stages.

The notation used to indicate the type of system being analyzed is Kendall's notation and is expressed as A/B/C, where:

A. Is the probability distribution followed by the arrivals to the system.
B. Is the probability distribution that follows the service time.
C. Is the number of servers in the system.

The distribution of arrival times A and B are detailed in the following table (Table 1).

Table 1. Notation of the probabilistic functions (Portilla et al., 2010).

Heading level	Example
M	Exponential
D	Deterministic
Ek	Type k Erlang
G	General

Having established the above, to have an integral knowledge of the system, it is necessary to define the key components of the analysis (Portilla et al., 2010).

- **Inter-arrival time**: This concept analyzes how the queuing system is fed and determines how much time elapses between two successive arrivals to the system. It is necessary to analyze the probability distribution presented by this variable, its associated parameter is lambda (λ).
- **Queue size**: It can be finite or, for all practical purposes, infinite. It is important to know the queue capacity since there may be cases where it is limited.
- **Queue discipline**: Represents the order in which customers are selected in a queue, the most common discipline is first-come, first-served; other disciplines include last-come, first-served and random order service. Customers may also be selected based on some previously established order of priority.
- **Service time**: There may be only one server or several servers in the system, this varies from system to system. The expected service time depends on the average service rate which is associated with the parameter mu (μ).
- **Number of Servers**: It is important to know and identify how many servers are available to serve the clients that enter the system. Different queuing system structures may be present. The most common are two, the first is the single queue model feeding multiple servers and the second model is a parallel system with a queue for each server (Eppen, Gould & Schmidt, 1999).

The equations used for the case study are as follows (Table 2):

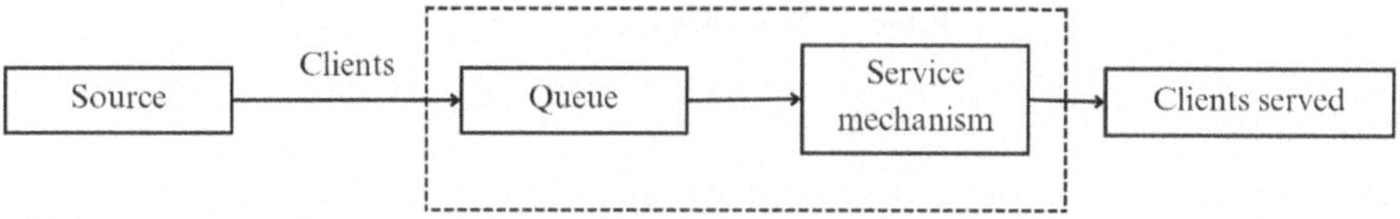

Fig. 1. Basic operation of a queuing system (Hillier and Liebermann, 2010).

Table 2. Formulas used for the queuing system (Hillier and Liebermann, 2010).

Measure	Formula	
Stability	�< c�	(1)
Utilization factor	$\rho =$ � / (� c)	(2)
Average queue length	$L =$ (�) �� / � ρ	(3)
Average queue waiting time	$W_q =$ L_q / �	(4)

4 Case Study

4.1 Data Collection and Model Selection

As an initial stage of the present work, the data provided by the area in charge of handling the requests was collected and prepared. The data table contains information on the stages of the process between October 2021 and November 2022, with a total of 47 records.

The table is made up of 16 variables, of which only 8 were selected to analyze information corresponding to waiting times in the attention of system updates. The relevant variables are described as follows (Table 3).

With the relevant variables identified and the times calculated, the excel data analysis tool was used to make a histogram of the inter-arrival time variable (λ) and the service time variable (μ) to visually identify the type of probability distribution.

Based on what was observed in the resulting graphs, it is assumed that the behavior of the inter-arrival time (λ) and service time (μ) follow an exponential distribution. To verify, R software was used to perform a goodness-of-fit test of the data. Although in the hypothesis test the fit with an exponential distribution is rejected, when comparing with other types of distributions, the Akaike Information Criterion (AIC) and Bayesian Information Criterion (BIC) tell us that the exponential model gives the best fit (Table 4). Therefore, it is considered that the appropriate model to solve the queuing problem is the M/M/c type.

Table 3. Variables and meaning

Variable name	Description
Date Issued	Date of issuance of the notification by the manufac- turer
Interarrival time	Variable generated from 'Date Issued', resulting from subtracting date "n" minus date "n-1"
Status	Status of the update at the time of the data table download query cut-off open closed
Date Customers notified	Date on which the customer is notified that the request has been initiated
Service start date	Variable generated with information from the corresponding area indicating the start of the service request
Waiting time	Variable generated from the difference between 'date issued' and 'service start date'. The resultant corre- sponds to the number of days it took to start the service (time in queue)
Date Closed	Date on which the request was closed because the request was completed
Service time	Variable generated from the difference of 'date closed' and 'service start date'. The result corresponds to the number of days it took for the request to be notified and receive the acknowledgement (Service Time)

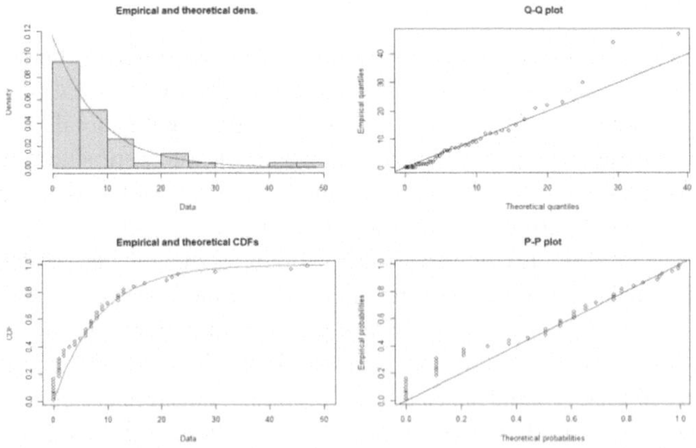

Fig. 2. Distribution analysis for inter-arrival time

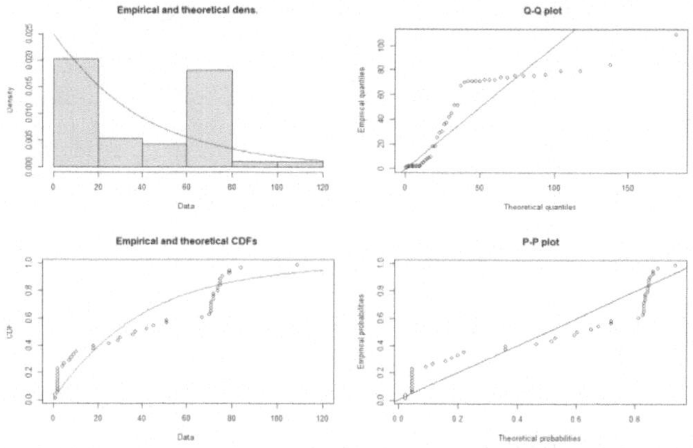

Fig. 3. Distribution analysis for service time

Using the average of the 47 records of the variable 'inter-arrival time', it is obtained that on average every 8.53 days a new request is added to the queue.

Likewise, using the average of the 'service time' variable shows that the average service time for each request is 40 days (Table 4).

Table 4. Statistical analysis

Interarrival time

Family	logL	AIC	BIC
Normal	-177.09	358.17	361.87
Cauchy	-170.82	345.64	349.34
Logistic	-171.97	347.93	351.63
Exponential	-147.76	297.52	299.37

Service time

Family	logL	AIC	BIC
Uniform	NULL	NULL	NULL
Normal	-230.39	464.77	468.47
Cauchy	-247.05	498.1	501.8
Logistic	-233.19	470.38	474.08
Exponential	-220.43	442.85	444.7

Counting with the average inter-arrival time and the average service time with their respective parameters $\lambda = 1/8.53 = 0.1172$ notifications per day and $\mu = 1/40.04 = 0.025$ notifications per day, we proceed to calculate the service utilization factor given by $\rho = \lambda/s\mu$. Knowing that the objective of the work is to diagnose and give a proposal for improvement, the utilization factor was calculated varying from 1 to 6 servers, to identify the number of servers needed to obtain a stable system (Table 5).

Table 5. Utilization times with different servers

Servers	ρ	P0
1	4.69	
2	2.34	
3	1.56	
4	1.17	
5	0.94	0.39
6	0.78	0.46

4.2 Simulation and Validation

To experiment with the data collected, a simulation was developed using the discrete simulation software FlexSim. We tried to experiment by modifying the server parameter, which was initially thought to vary from 3 to 6, but since analytically a non-stable state had been obtained with 3 and 4, the simulation would not yield a definitive result.

To validate the simulation, the stability of the system was tested with four servers since, if with this number the queue only increased, then when experimenting with three servers the same unstable state would have to be obtained.

Therefore, we proceeded to create the simulation with 4 servers. The input parameters for the simulation are only two: the inter-arrival rate (λ) which was fed as 8.5319 days/notification and the mean service time (μ), which was entered with a value of 40.0426 days/notification; this time is fed to each of the servers. Finally, the connections between the source of arrivals, the queue, the servers, and the output of the system are generated.

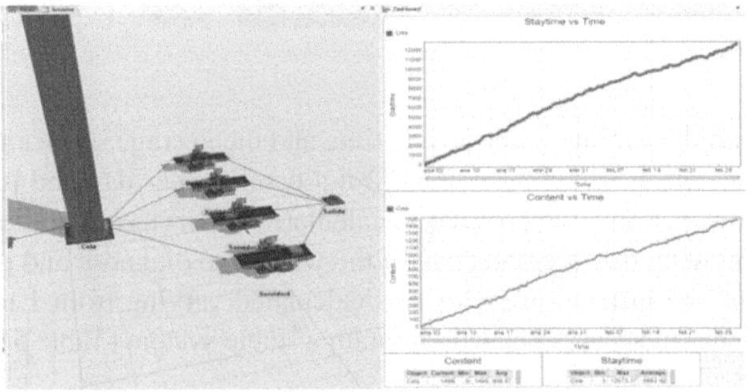

Fig. 4. Simulation with 4 servers

Figure 4, a) shows the 3D visualization produced by the software after running the scenario with 4 servers. In this case each box represents an arrival, it can be seen in the queue that these arrivals have been stacked since the system is unstable, as it had been obtained analytically since $\rho > 1$. In Fig. 4, b) you can see graphically the average time in queue and in Fig. 4, c) the average content in the queue. In both graphs there is an indefinite growth, thus, the model is validated showing that in the scenario with both 4 and 3 servers is insufficient for the demand.

Therefore, we proceeded to simulate with 5 and 6 servers, which have a ρ value of 0.94 and 0.78, respectively. The simulations can be seen in Figs. 5 and 6. With these initial values we know that a steady state will be achieved by having $\rho < 1$. To create the simulation additional servers were added and connected to the system as shown in Figs. 2 and 3. Thus also, it can be seen from the figures that we no longer have indefinite growth from the graphs. Although there are fluctuations because it is a stochastic process, it always returns to stability. With these two simulations we can arrive at the analytical results described below.

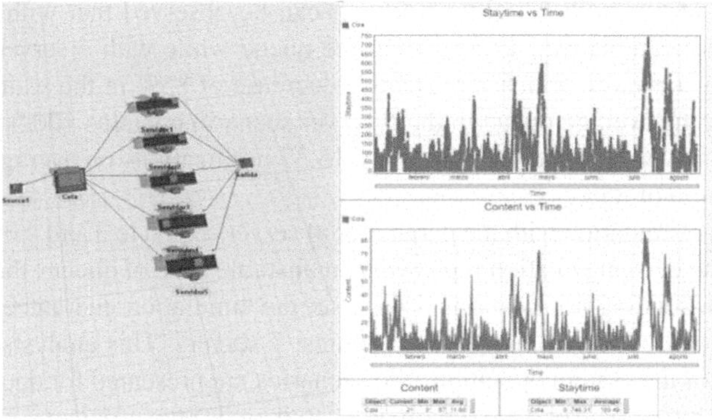

Fig. 5. Simulation with 5 servers

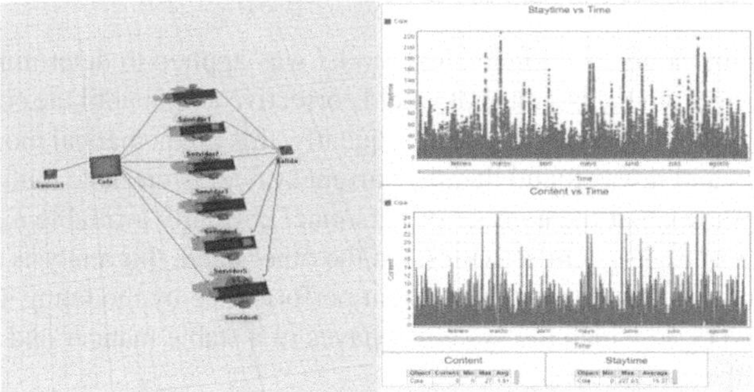

Fig. 6. Simulation with 6 servers

4.3 Results and Discussion

From the simulation, the indicators of average time in the queue (Wq), as well as the average content in the queue (Lq) were obtained. The results can be seen in the following table(Table 6):

Table 6. Results for Lq and Wq with 3 to 6 servers.

Servers	Lq (notifications)	Wq (days)
3	N/A	N/A
4	N/A	N/A
5	11.80	100.5
6	1.81	15.37

As previously mentioned, for the scenarios with three and four servers, it is not possible to obtain Lq and Wq data, since a steady state is not reached. Thus, we proceed to analyze the results with 5 and 6 servers. It can be observed that with 5 servers on average the notifications wait 11.8 days in the queue while with 6 servers the wait is reduced to only 1.8 days, which represents a decrease of 85% in the waiting time. On the other hand, the average queue in the 5-server scenario contains 100.5 notifications, while in the 6-server scenario there are only 15.37 notifications on average. Again, we obtain a reduction of 85%.

Through the simulation with the parameter of servers equal to 3 and 4 it was possible to verify that the current situation represents an unstable state of queues that is reflected in the delays in sending notifications. Likewise, the simulation guarantees that having 6 servers results in an 85% improvement against 5 servers. This analysis enriches the understanding of the system operation, and scenarios are presented for decision making oriented to the hiring of 2 or 3 servers in order to increase the efficiency of the system.

5 Conclusion

Based on queuing theory, a mathematical model was applied to determine the average time in the system from the issuance of a field corrective action until the communication process to the recipient is completed. Subsequently, this mathematical model was simulated in FlexSim and it was found that the current system is unstable with less than five servers, which shows that the number of personnel currently available is not sufficient to perform the required system activities. On the other hand, this analysis also allows to start drawing action plans to improve system performance by modeling scenarios with five and six servers, in which the system behaves in a stable manner and with a higher efficiency than the current state.

Naturally, another way to make the system stable would be to reduce the service time; however, given the complexity that would entail, the approach used instead was to increase the number of servers.

References

1. Banks, J., Carson, J.S., Nelson, B.L., Nicol, D.M.: Discrete-event system simulation. Prentice Hall, USA (2005)

2. Díaz-Martínez, M.A., Zárate-Cruz, R., Román-Salinas, R.V.: Simulación FlexSim, una nueva alternativa para la ingeniería hacia la toma de decisiones en la operación de un sistema de múltiples estaciones de prueba. Científica **22**(2), 97–104 (2018)
3. Dunna, E.G., Reyes, H.G., & Barrón, L.E.C.: Simulación y análisis de sistemas con ProModel. Pearson Education (2006)
4. Eppen, G., Gould, F., Schmidt, C.: Investigación de Operaciones en Ciencia Administrativa. Prentice Hall, New Jersey (1999)
5. Flores de la Mota, I. Introducción a la Teoría de Espera. CDMX: UNAM (2022)
6. Fuentes Zenón, A.F.Z.: El Enfoque de Sistemas en la Solución de Problemas: La elaboración del modelo conceptual. México: Facultad de Ingeniería, UNAM (1990)
7. Giraldo, N.: Procesos Estocásticos. Universidad Nacional de Colombia, Medellín (2006)
8. Hiller, F., Lieberman, G.: Introducción a la Investigación de Operaciones. McGraw Hill, México (2010)
9. Jhala, N., Bhathawala, P.: Application of queueing theory to airport related problems. Global J. Pure Appl. Math. **13**, 3863–3868 (2017)
10. Muñoz-Vergara, A.: Aplicación de la Teoría de Líneas de Espera en el servicio de biblioteca de una organización educativa en cartagena colombia. Saber, Ciencia y Libertad **14**, 171–178 (2019)
11. Portilla, L., Arias, L., Fernández, S.: Análisis de Líneas de Espera a Través de Teoría de Colas y Simulación, pp. 56–61. Scientia et Technica, XVII (2010)
12. Rodríguez, G., González, A., Hernández, S., & Hernández, M.: Análisis del Servicio de Urgencias aplicando Teoría de Líneas de Espera. Contaduría y Administración UNAM, 719732 (2017)
13. Taha, H.A.: Investigación de Operaciones. México: Pearson Educación. (2012)
14. Winston, W.: Investigación de Operaciones: Aplicaciones y Algoritmos. Thomson, New York (2005)

Analysis of Covid-19 Transmission Using Complex Networks

Sashiko Shirai Reyna[1], Oroselfia Sánchez[2,3](✉), Carmen A. García-Cerrud[3], and Idalia Flores-De la Mota[3]

[1] Instituto de Investigaciones en Matemáticas Aplicadas y Sistemas, Universidad Nacional Autónoma de México, 04510 Mexico City, Mexico

[2] Department of Chemical, Industrial and Food Engineering, Universidad Iberoamericana, 01219 Mexico City, Mexico

oroselfia.sanchez@ibero.mx

[3] Facultad de Ingeniería, Universidad Nacional Autónoma de México, 04510 Mexico City, Mexico

Abstract. The global public health crisis caused by the SARS-CoV-2-19 (COVID-19) pandemic has highlighted the need for research into contagion complexity. This challenge necessitates the development and testing of various approaches to manage rapidly changing information with high impact. In this paper, we employ time series analysis and complex networks analysis to compare the evolution, spread, and containment of COVID-19 pandemics in eleven countries and globally. Our analysis enables us to observe the dynamics of spread and the impact of different strategies employed by each country in increasing and decreasing cases through complex network techniques. Additionally, we explore the transformation of data behavior over time as our understanding of the virus improves. Our findings provide important insights into the limitations of using statistical models and suggest that simulation of new cases of COVID-19 data can be modeled using complex networks. The complex network model provides a general description of contagion dynamics in the 11 countries and worldwide situation. This paper contributes by highlighting the limitations of using statistical models to infer and study early time series data and proposing the use of a complex network approach to study contagion dynamics.

Keywords: Pandemic · SARS-CoV-2-19 (COVID-19-9) · Visibility algorithm · Time series · Complex networks · Modeling

1 Introduction

In December 2019, a worldwide crisis was unleashed by a new coronavirus (SARS-CoV-2-19), first discovered in China, that quickly spread throughout the world, with millions of confirmed cases [1, 8]. A behavior analysis on the spread of new cases using time series analysis enables us to observe patterns of behavior as regards trends, seasonality, and randomness. The different methods used for modeling aspects of the COVID-19 pandemic include mathematical models for studying its spread and complex networks

M. Mujica Mota and P. Scala (Eds.): EUROSIM 2023, CCIS 2032, pp. 174–189, 2024.
https://doi.org/10.1007/978-3-031-68435-7_13

for analyzing outbreaks like Wuhan [4, 15–17]. Information about the dynamics of virus spread enables us to study the effects of the actions taken by the public health authorities and be able to estimate the numbers of the population that are infected [2, 6, 7, 10, 11]. Thus, we can make reliable predictions about its future evolution in a specific timeframe and consequently be able to prepare the health system to deal with it [3, 5, 9].

Using control approaches gives us quick treatments for a small part of the population but their effects help to fight the ravages of infectious diseases on large populations by administering the smallest possible amount of medicine, thus enabling its greater availability. Statistical methods for estimating representative distributions for certain variables of COVID-19 such as hospitalization, medical appointments when the symptoms appear, and other specific clinical cases have also been used to describe the dynamics of disease transmission [12]. These analyses support the decision-makers reports and the implementation of policies, such as closing borders and international flights, enacting national lockdowns, closing parts of a city or the economy, relaxing said restrictions, and deploying equipment and resources for field epidemiology [13].

The policies applied throughout the world for the mitigation of COVID-19 include [5, 10, 14] restrictions such as mobility, socio-economic, physical distancing, hygiene measures and changes in public communication about the situation. In this paper, we highlight the use of algorithms that, in combination with network models, contribute to modeling the efficiency of drugs that were not initially de-signed for viral treatments but that have been used to treat COVID-19 [18]. This comparison is based on time series characterization, a visualization algorithm to convert all data into networks, which will be analyzed using complex network metrics.

The paper is organized as follows: Sect. 2 presents the analysis of COVID-19 new cases data in many countries at the same time. Aiming to conserve and describe the infection behavior with population response to sanitary measures. All countries decided to implement different strategies to avoid the increase in COVID-19 new cases, the reason why every country has a particular infection behavior. Finally, the results are presented in Sect. 3, preceding the conclusion and discussion.

2 Case Study Analysis

These countries were selected due to the data with the greatest spread of the epidemic and considering that all continents are considered.

The countries selected in this paper are: United States (US), Mexico (MX), Japan (JP), India (IN), Russia (RU), South Africa (ZA), Italy (IL), Germany (DE), Morocco (MA), Netherlands (NTH), Brazil (BZ) and we included the global information as a frame for the worldwide behavior of the COVID-19 pandemic.

The data on the new Covid-19 cases were obtained from the World Health Organization, defining the confirmed cases listed on its website, in daily reports as an analysis unit [19]. Figure 1, shows the comparison between the calculated percentage ranking with the population density of each country versus the new cases published by the WHO ranking. Graph A and graph B contain the same new cases data but the first one presents the ranking of countries with more population contagion. Meanwhile, graph B shows the COVID-19 new cases reported ranking. For example, the US keeps the largest number of new cases reported and is the country with the most infected population. Besides, public portals report India as the second country with more cases in the world, but with a large demography population can be placed in the tenth place of the eleven countries analyzed in this work. Different from Italy and Netherlands ranked in lower order but both are countries with the largest population infected. Owing to the number of graphics generated for said analysis, we have used the ones corresponding to Mexico as an example. Graphics for the other countries are given in the Appendix A (Appendixes can be seen on: https://github.com/SashikoSR/COVID-19).

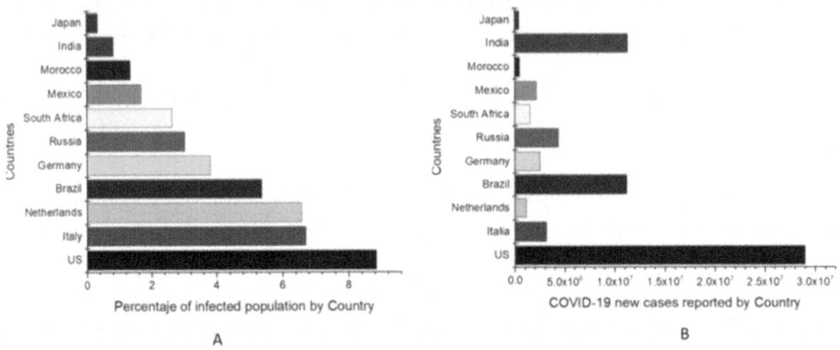

Fig. 1. Comparison between population density ranking and the World ranking of daily new cases published by the WHO.

2.1 Data Path: From the Repository to the Complex Network Model

The complex network model is built using the networks of new COVID-19 infection cases from each of the 11 countries. To determine the countries' analysis feasibility, the new cases data published in public and open repositories were compared. For example, in the portal of each country, the total new cases recorded were reviewed and compared on three additional databases (WHO, Hopkins, and the country's database). For some countries, there was a huge difference between the published records, both in the total number of new cases and in each day's records. These inconsistencies are a natural result of the methodological adjustments made by each country, but this does not conserve the propagation dynamics, it only accumulates the data to reach a total of new cases that is closer to reality. The difference between the databases and the data corresponding to each country is wide and there is a lot of inconsistency between the data, which has an impact on the feasibility of using this database in the complex network model owing to the impossibility of propagating data inconsistencies in a way that represents the dynamics of propagation.

For the model to describe the individual characteristics of the countries, a careful analysis of the published public data was made.

Figure 2 available in GitHub, shows the data comparison, transformation, and preparation states for the complex network model.

The records of new cases are reported per day in the data sources. The data reports depend on factors such as the working day, the distance traveled to collect the data, to name but a few, and this generates noise in the databases. Grouping the data every week preserves the behavior while at the same time eliminating the noise that the daily period presents.

It is important to mention that the aim is to analyze whether the data present a seasonality or a specific trend. However, this prediction is not yet possible due to the state of the pandemic worldwide. In the future, with more collected and recorded data, trend and seasonality analysis can be applied to determine the stability of the time series. Once the data has been selected, pooled, and analyzed, the visibility algorithm is applied for its analysis. This algorithm connects the weeks that are visible to each other according to the newly reported cases. If there is a data segment that can be plotted in a polygonal shape, then all the nodes will be connected. Finally, the network is built from the newly reported cases for each week of the pandemic.

2.2 Hot Issues of Data in Complex Networks

New cases of COVID-19 reports reflect the different updates of methodologies for collecting the transmission. Over time, all the countries changed the criteria for the new cases records. In the data treatment framework, the natural evolution of the COVID-19 pandemic content the accumulation of the new cases total, but the time series do not show the propagation dynamics. It is common in complex networks to deal with different types of data, and always an opportunity to demonstrate the characteristics of such data with classical tests.

Figure 45 available in GitHub, shows the classical treatments that demonstrate how the behavior of complex network data differs from others. It would be suitable to model the data under the same conditions for all the periods. Table 1 contains the results of the different tests applied to new case data. Table 2 shows the normality test results of the time series for every country analyzed in the complex networks model.

Kolmogorov-Smirnov demonstrates if the samples are different enough between them to infer future behaviors. The model variation represented by Lilliefors is for review of the normal distribution adjustment. On the other hand, chi-squares relate two variables and show if the samples are associated or independent. And the run test allows testing if there is randomness in the subsequent order of the data. The statistical models were applied for the eleven selected countries according to the flow chart (see Fig. 44 available in GitHub). For each time series, data were prepared to be tested in the metamodel using RStudio packages.

Through this additional data analysis, we conclude that it is not an option to transform data into other distributions despite inconsistencies in all the time series. The data used in the model represents a particular moment of COVID-19 transmission. Therefore, as mentioned before this type of data should not be transformed in other distributions.

Table 1 demonstrates that the data we collected cannot be used directly for statistical purposes or as input without prior characterization. Table 2, on the other hand, illustrates that it is not appropriate to transform the time series into other distributions due to two main reasons. First, complex network models commonly utilize the raw time series data to avoid losing important information during transformation. Additionally, time series analysis allows us to identify important features and patterns in the data that may be lost during the transformation process. Second, we propose characterizing the time series through time series analysis, which involves modeling the underlying trends and patterns of the time series. If this method is unable to infer information, a complex network model may be applied. This is particularly relevant in the case of epidemics, such as COVID-19, where the simulation of new cases can offer valuable insights about the available data.

Table 1. Statistical tests for new cases of COVID-19.

Data Analysis by week						
Countries	KS (Two samples)	Conclusion	Chi square	Conclusion	Box-Ljung	Conclusion
Mexico-Italy	8.43E-03	Without similarities	2.12E-02	Without similarities	2.30E-14	Dependency
Mexico-South Africa	1.54E-04		1.10E-06		2.30E-14	
South Africa-Russia	4.95E-08		3.82E-04		9.66E-15	
Russia - Morocco	2.20E-16		3.82E-04		9.66E-15	
Morocco - India	5.77E-15		3.61E-03		7.77E-16	
India-Brazil	4.14E-02		3.37E-03		7.77E-16	
Brazil-Netherlands	1.33E-15		9.01E-07		6.66E-16	
Netherlands-Germany	4.14E-02		2.12E-02		2.00E-15	
Germany-Italy	7.17E-01	Similarity	1.80E-02		2.78E-15	
Italy-Japan	1.39E-05	Without similarities	8.47E-02	Similarity	2.11E-15	
Japan-US	< 2.2e-16		8.13E-02		1.27E-14	

Table 2. Normality Test.

Normality test KS (Lilliefors)			
Country	Value	Country	Value
Mexico	7.47E-02	Netherlands	6.56E-10
South Africa	1.23E-10	Germany	1.69E-09
Russia	2.13E-02	Italy	8.24E-09
Morocco	6.59E-09	Japan	1.42E-08
India	8.45E-06	US	1.71E-06
Brazil	3.45E-03	Global	1.64E-02

2.3 Time Series Analysis

Time series decomposition is a mathematical procedure that splits the series into its three components: Seasonality Patterns that repeat with a fixed periodicity. Trend underlying trend of the metrics, ascendant or descendant. And Random or noise: the remainder of the original time series after the seasonal and trend series are removed. ACF (Auto-Correlation Function) was used to determine the autocorrelation values for any series with its lagged values. It describes how well the present value of the series relates to its past values. ACF considers all the series components while finding correlations. The use of PACF (Partial Auto-Correlation Function) enables the determination of residual correlation by removing the effects of the anterior lag(s) and the posterior lag(s). An accurate correlation is obtained if there is any hidden data displayed by the following lag. The subsequent lag is taken as an element while modeling, but the elimination is preferable of correlated highlights that create multicollinearity issues [20]. The ACF and PACF plots are more commonly used to obtain the values of p and q to feed into ARIMA model frequently used for forecasts.

In this work we used time series as a based for the visibility algorithm to transform time series to complex network and then analyze the complex network topology. It transforms the time series into a complex network through a visibility graph that provides a complete analysis while showing the properties of the time series in the network. Section 2.5 explains the visibility algorithm and a small instance is also set to define how the algorithm sets the data as a network.

2.4 Time Series Analysis for the Eleven Countries Comparison

Time series metrics were used to analyze reported data for new cases of COVID-19. The data used correspond to daily new cases of COVID-19 in 66 weeks since January 2020 for the eleven countries and the global data reported. The data was analyzed using MicroStrategy, Excel, and R.

Figure 3 available in GitHub, shows the graphs obtained in this analysis. The stage of the pandemic in each country (first wave and second wave), was used for its division. Graphs A and D correspond to the countries in the first wave, while graphs B and E represent countries in the second wave of infections. The behaviors analyzed were the seasonal variation index and non-seasonal data for each country vs. global behavior. Moreover, graphs C and F show Mexico's behavior in respect of the global behavior for both metrics.

These findings are only comparisons for the eleven studied countries, and the total and partial autocorrelation functions obtained for each country make it impossible to get accurate predictions for an overall view of the pandemic period of newly registered cases. However, when there is a higher number of new daily cases records, the correlational estimates between the values and the delays (Xt and Xt-2) shall be used to compare autoregressive models (Fig. 3, available in GitHub). It is important to observe that the non-seasonal data for Mexico (graph F) shows an acceptable control of the pandemic, as the initial peaks were much lower in respect of the behavior during the first wave (graphs A and D) in the other countries. Nevertheless, the data recorded by the countries have changed over time, which can influence the observed dynamics. As we can note in Fig. 3, available in GitHub, no countries present seasonally or a clear tendency in COVID-19 new cases data. In other words, the lack of the two aspects mentioned before in the time series cannot allow (during the period under review) to know the future behavior dynamics. As a tool, time series analysis is useful for understanding behaviors. This analysis can be used with another data period to study if the behavior presents predictive characteristics. Meanwhile, complex networks can complete the methodology for describing the governmental strategies and the effects of society's response to the COVID-19 crisis dynamics.

The analysis of the time series conducted in this study did not reveal any clear seasonality or trend in accordance with the models mentioned before. However, upon examining the data, we found that organizing it on a weekly basis was necessary to address noise, despite the absence of any discernible pattern. Our analysis also revealed that peaks in the data that deviated from the norm corresponded with holidays, weekends, or non-working days, though it was difficult to quantify the overall impact of these events on the time series. Furthermore, the practice of governments adjusting the data by adding the total number of deaths or non-deaths on a specific day made it impossible to maintain or project any seasonality or trend over the entire time series. Instead, this practice only reflects a more accurate representation of the total number of new cases, which, in turn, affected the overall behavior of the time series. These findings can contribute to the preparation of data for visibility algorithms to be used in complex networks. The data analysis was performed using RStudio and Microsoft Excel.

2.5 Visibility Algorithm

The main goal of the visibility algorithm is to map a time series into a complex network. To study and analyze it with all the techniques and properties of network theory, resulting in a more complete analysis and not just the time-series study. Another important thing is that this network inherits several properties of the time series. The algorithm criterion sets up two arbitrary data values (ta, ya) and (tb, yb) that will have visibility, and consequently will become two connected nodes of the associated graph if any other data (tc, yc) placed between them satisfies:

$$y_c < y_b + (y_a - y_b)\frac{tb - tc}{tb - ta} \tag{1}$$

Figure 4 available in GitHub, gives an example of a time series of 20 data values plotted as a periodic series by using vertical bars (the data values are displayed above the plot). Considering this as a landscape, we link every bar (every point on the time series could be days, weeks, months, years, etc.) with all those that can be seen from the top of the one being considered (gray lines), thus obtaining the associated network (shown in the lower part of the figure). In this network, every node corresponds in the same order to series data and two nodes are connected when there is visibility between the corresponding data and if there is a straight line connecting the series data, provided this "visibility line" does not intersect any intermediate data height. Using the algorithm, the associated network extracted from a time series is always Connected meaning that each node can "see" at least its nearest neighbors (left and right), the first and the last one at least sees one. And undirected, where there is no direct definition in the links. Also, invariant under affine transformations of the series data, the visibility criterion is invariant under rescaling of both horizontal and vertical axes and horizontal and vertical translations. Figure 4b available in GitHub, shows the visibility algorithm generated for the Mexico case.

The visibility network of a time series remains invariant under several transformations of the time series as shown in Fig. 3 available in GitHub (a) Original time series with visibility links. (b) Translation of the data. (c) Vertical rescaling. (d) Horizontal rescaling. (e) Addition of a linear trend to the data. As can be seen in the bottom diagram, in all these cases the visibility network remains invariant. The key question is to know whether the associated network inherits some structure of the time series, and consequently whether the process that generated the time series may be characterized by using network theory. In the first step, we will consider periodic series. The example plotted in Fig. 4a available in GitHub is nothing but a periodic series with period 4. The associated visibility network is regular, as long as it is constructed by periodic repetition of a pattern. The degree distribution of this network is formed by a finite number of peaks related to the series period, much in the vein of the Fourier power spectrum of a time series. Generically speaking, all periodic time series are mapped into regular networks, the discrete degree distribution being the fingerprint of the time series periods. In the case of periodic time series, its regularity seems, therefore, to be conserved or inherited structurally in the network through the visibility map.

2.6 Complex Network Analysis

The complex network model is made up of 11 networks that represent each of the chosen countries, in addition to the global behavior of new cases. Each network is made up of nodes that represent every week of the pandemic, containing the accumulated cases for 7 days. In particular, the networks consist of 66 nodes each covering a reporting period for new cases from January 2020 to August 2021. The links are connected through the visibility algorithm, which determines all the weeks (nodes) that are seen from each point, meaning that this algorithm makes it possible to relate to each other those numbers of new cases for each week that have the same maximum number of closest cases, both: backward and forwards.

After the creation of the network, network metrics such as clustering, closeness, betweenness, assortativity, degree distribution, among others, are obtained to perform the complex network analysis. This provides an accurate network behavior approach. Figure 5 shows the network created for Mexico's case.

When analyzing the metrics of the network associated with each country, the proportion of infected cases and the degree of control of the spread of the disease are compared.

2.7 The Analysis of Complex Networks Metrics for the Eleven Countries

The comparison of the metrics obtained by the complex network analysis allows the characterization of the spread and disease spread control degree to determine how several variables such as the implemented policies, the population density, the degree policies fulfillment by the population among others change the behavior dynamics.

Concepts such as degree (i.e., the number of links per node) or clustering (the number of neighbors that are also connected) are truly local quantities, depending on the state of a single node and its neighbors. Measurements of centrality, such as betweenness, depend on the state of the entire system and will be defined below as global. In between these two scales, we have the study of communities that range from a few nodes to the entire network [21] (Fig. 2).

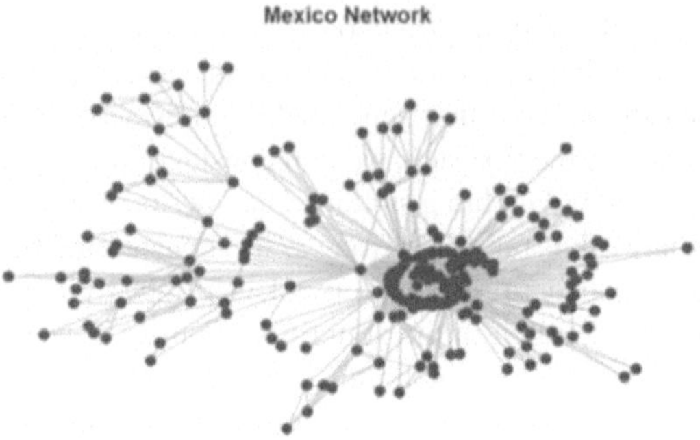

Fig. 2. Visibility algorithm for Mexico.

3 Results

The time series analysis was carried out in Excel and Origin. The visibility algorithm and the complex network were solved in RStudio. Table 3 presents the results obtained by complex networks metrics calculations.

These results mark the contribution of the use of complex networks. This tool helps to achieve a general description of the 11 countries without missing the specific behavior of each country. For instance, the form in which COVID-19 new cases are increasing because of the health measures implemented during the lockdown.

Table 3 shows the metrics obtained by using complex network analysis, and their interpretation is presented below:

Diameter: The diameter D of a network is defined as the longest distance you can find between two nodes in the network. Some other definitions (e.g., average distance) are possible [21]. In this complex network model, the diameter indicates how much time there is between weeks with a higher or lower number of infections, in other words, the cycles of recorded new cases. The results of the complex network analysis (Table 3) show that Japan and Italy presented the smaller diameter, while Brazil, Morocco, Germany, Russia, and Mexico presented the bigger diameter. It is possible to determine the control of the spread of the virus using the strategies applied if population density is considered.

Mean distance: This metric gives the distance in a more general way since it tells us the average of the distances between every pair of nodes in the network. In our case, the mean distance represents the variation of new cases between the weeks before and after each node. If the diameter is small, then the weekly rise in new cases will be similar from one week to the next. For high diameter values, the new cases between one week and the weeks before and after will change without distinction, representing increasing uncertainty. In this case, for all countries, the mean distance is very similar with a very low variation, meaning that all countries have almost the same distance between nodes.

According to the mean distance results (Table 3), the average number of weeks needed to observe a change in the behavior of the propagation dynamics that occurs between 2 or 3 weeks.

Cliques: One of the first papers on community structure, published in 1949, defined a community as a group of individuals whose members all know each other. In terms of graph theory, this means that a community is a complete subgraph or a clique. A clique is a connected subgraph with maximal link density. Two k-cliques are considered adjacent if they share k – 1 nodes [22]. In our problem, cliques indicate the time between the changes (adjacency), in other words, the behavior of the virus spread and the pandemic control period. The clique value translates into two behaviors: the number of weeks when the virus is spreading or the number of weeks when it is under control. A high clique value could mean more stability from one node to another (valley or peaks), while a smaller value represents a virus spread variability. The dynamics of propagation in the US are characterized by being very unstable and consistent with the value obtained from the complex network analysis (10 cliques). India, South Africa, and Italy have the highest number of cliques in the entire network (17 cliques). These values show minor variation in new cases from the previous week to the next one. The differences between the clique values for each country show that a higher clique value represents the countries that have contained the pandemic for extended periods, or when the increase in new cases takes more time. India and South Africa have the same clique value. However, for South Africa, it represents the number of weeks that the pandemic has been under control (valley), and for India, it indicates the number of weeks the new cases have increased (peak).

Density: The density of a network is defined as the ratio between <k>, the average value of the degree (the degree of a node), and N, which is the order (number of links) of the network: $\rho = \langle k \rangle /N\text{-}1$, with $\rho \leq 1$ [21]. For this complex network model, this metric establishes whether each country's dynamics are scattered or dense, in other words, the connectivity degree between one week and the following ones or the impact of one on the other. India, South Africa, and Italy (0.32, 0.30, 0.29 respectively) present the highest density values, while Brazil, Mexico, and the US have the lowest values (see Table 3). This metric indicates new cases recorded over the weeks (66). For example, South Africa shows that new cases are mostly together in the peaks of the propagation dynamic. In contrast to the situations in Brazil, Mexico, and the US, where all new cases are recorded over 60 or 50 weeks which causes a propagation dynamic more scatter.

Assortativity: We can identify the presence of a community from nodes with equivalent properties. The level of 'similarity' of two nodes is usually computed using a mathematical quantity called correlation. Usually, since the most immediate property of a node is the degree, we look for the presence of a correlation between nodes with a similar degree. There is no reason, in principle, to expect a particular correlation. Actually, in some situations, there is a tendency for high-degree nodes to be connected to other high-degree nodes. In this case, the network displays what is called assortative mixing or assortativity [21]. In our networks assortativity measures two connected nodes that have a similar increase (or decrease) in the number of new cases, being positive when these changes vary slightly from week to week, and negative when this behavior is not alike. Japan is the only country with negative assortativity (−0.03; see Table 3), which describes the occurrence of three gradual peaks and some valleys. As a metric, this behavior implies that there are several nodes of different degrees, meaning that it has many weeks in valleys or peaks. India has the highest assortativity value because these propagation dynamics (data for 66 weeks) show a noticeable increase and a long duration of a high number of cases, which implies that the degrees of the nodes are remarkably similar to each other throughout the entire time series. Another relevant behavior is the one presented by South Africa (0.149, see Table 3), where the new cases present two marked increments, which determines a valley that remains for several weeks. Despite that, it is an assortative network like India, the results of the model show the difference between the countries.

Clustering: The clustering coefficient Ci of node i is a measurement of the number of links 'around' vertex i. Ci is given by the average fraction of pairs of neighbors (of the same node) that are also neighbors of each other. In general, we can write the clustering coefficient as the fraction of actual links over the possible ones between vertices i, j, k. [21]. Global Clustering refers to the probability of forming triangles between the nodes. It measures the mean proportion of neighbors a node has and presents two variations: global clustering refers to the entire network and local clustering that is calculated for each of the nodes, providing the mean of all the nodes. In this model, global clustering shows the influence of the government measures adopted by each country to contain the COVID-19 contagion rate. Global clustering for the eleven countries is between 0.58 and 0.81 (Table 3), meaning that many groups have formed during the 66 weeks. On the other hand, the local clustering represents the increase of recorded new cases. For example, South Africa and Japan have high increments (Table 3). The eleven countries display similar values, but their behavior differs. If local clustering is higher, then there is also a higher probability of transitivity per node.

Betweenness centrality: Sensitive measurement of centrality is given by the number of times we cross one node k in going from one node I to another j following the path of minimal length (distance d (I, j)). This number is called site-betweenness b(i) [21]. Another definition is a link or link betweenness, defining xij as the number of shorter paths that go through the link (I, j). Links connecting different communities are expected to have large xij while links within a community have small xij [22]. Betweenness centrality indicates the proportion of increased infections in respect of the previous and subsequent weeks. As a complex metric, this can be interpreted as the propagation period, including the rise in new cases, peak width, and decrease, representing how many weeks is COVID-19 active in the whole period. According to Table 3, there is not a wide difference in the countries' values. The betweenness centrality values for Germany and the US are the same, corresponding to 0.33 both. This metric analysis and interpretation conclude that the COVID-19 pandemic is still propagating.

Closeness: Centrality closeness is aimed at measuring how close a node is to other nodes in the network. This is done in terms of communication distance, as measured by the number of links between two nodes if connected by the shortest path, as the closeness metric is the mean shortest path to all other nodes [23]. Closeness represents the time when the new cases remain low or high for each country. In complex networks, a high closeness considers the existence of a few nodes with an extremely high degree. These nodes are called hubs. Table 3 shows that Russia (0.28) and Brazil (0.33) present the lowest closeness value, meaning these countries have the highest variation between subsequent weeks. Japan (0.61) and the Netherlands (0.54) present several similar nodes, which means that the number of new cases is similar for a longer period, or it is remaining in the valley.

Degree centrality: The degree centrality of a node refers to the number of links attached to the node. The standardized score is found by dividing each score by n − 1 (n = the number of nodes). As for the centrality degree, the pattern of behavior regarding closeness is accentuated for the following countries: Japan (0.52), Germany (.42), Netherlands (0.41), and Italy (0.35). Brazil (0.18), Mexico (0.19), India (0.19), and Russia (0.19) have similar values and the lowest values of the eleven countries. However, the dynamics between them are different, for Brazil and Mexico, this value means that the centrality degree is low because of the variability in the number of infections that arise on consecutive weeks. Whereas, in the case of Russia and India, the similarity between the nodes represents the accumulation or drop in newly recorded cases. The highest centrality degree corresponds to Japan (0.52), which represents an accelerated rise in the number of COVID-19 infections in a few weeks, implying closed borders.

Degree distribution: Degree distributions can be calculated by the ratio between the number of nodes in network G of degree k and N denotes the size of G (number of nodes). The equation p(k) = |dk(n)|/ N is exactly the proportion of nodes in G having degree k. Degree ki of node I is the number of links connected with node i. From this, it follows that this equation also has the meaning that a randomly chosen node in the network has, with probability p(k), k links.

It is an interesting and important fact that many real-world networks like the World Wide Web (www), the Internet, social networks, citation networks, or food webs are not Poisson distributed but follow a power law: $p(k) = k - \gamma, \gamma > 1$ [23]. According to the complex network analysis, the eleven countries are free-scale networks. Free-scale networks consider that few nodes have a lot of connections while several nodes have few connections. It means that there are many weeks with records of COVID-19 new cases and that nodes with high values are connected to the rest of the nodes, which in complex network metrics are known as hubs. Figure 6F available in GitHub, shows the distribution degree for Mexico, and the global distribution degree (Fig. 6A available in GitHub), which presents a p-law distribution, even the tail is longer. Graphs E and F show the evolution of the pandemic over time. Comparing plot E which contains new cases reported until September 2020, and plot F that corresponds to January 2020 until March 2021, the degree distribution shows the differences between nodes' degrees.

Other information obtained from the degree distribution metric is that countries with more infected populations behave similarly to graphs B, D, or F, such as the US, that reported 8.84% of its population infected. Countries with a lower number of infected populations follow a pattern like graphs A, C, or E, such as Japan, which reported 0.35% of its population infected, both these findings are based on data up to March 2021. All of this means that with the new cases recorded and the increased number of weeks elapsed, the distribution degree tends to function as a homogenous network.

The advantage of using a complex network technique is to describe the entire dynamic behavior of the selected countries. The methodology proposed in this research does not use only time series analysis, due to the dynamic behavior of the data that is not reflected if this method is used alone. On the other hand, statistical methods only describe proportions and parts of the total behavior, while modeling a system by complex networks shows the variations and dynamic nature of data.

When the complex networks metrics are understood in a social behavior context, the inference about the relation of data behavior with sanitary measures can be done.

Table 3. Complex networks results.

Global Ranking		1	2	3	6	8	9	15	20	21	36	41
Country	Global	US	India	Brazil	Russia	Italy	Germany	Mexico	South Africa	Netherlands	Japan	Morocco
Max degree	33	28	34	24	28	42	43	25	36	43	48	37
Min degree	2	1	1	1	1	1	1	1	5	1	4	2
Mean degree	9.7	12.7	21.2	11.9	15.2	19.1	15.6	12.2	20.1	16.1	13.9	16.4
Diameter	5	5	5	6	6	4	6	6	5	5	4	6
Mean distance	2.57	2.30	2.34	2.93	3.09	2.29	2.37	2.68	2.20	2.22	2.08	2.23
Cliques	11	10	17	13	14	17	16	14	17	15	13	12
Density	0.147	0.195	0.326	0.183	0.233	0.294	0.240	0.188	0.309	0.248	0.214	0.252
Assortativity	0.084	0.087	0.712	0.675	0.610	0.579	0.217	0.609	0.149	0.176	-0.04	0.05
Global Clustering	0.589	0.563	0.816	0.755	0.707	0.777	0.713	0.762	0.807	0.671	0.659	0.607
Mean Local Clustering	0.792	0.688	0.799	0.769	0.748	0.795	0.821	0.790	0.845	0.763	0.822	0.738
Closeness Centrality	0.476	0.367	0.375	0.331	0.286	0.432	0.457	0.374	0.412	0.545	0.611	0.351
Degree Centrality	0.353	0.236	0.197	0.186	0.198	0.352	0.422	0.197	0.245	0.414	0.524	0.316
Betweenness Centrality	0.514	0.331	0.513	0.490	0.376	0.410	0.332	0.395	0.485	0.496	0.476	0.418
Eigencentrality	0.702	0.645	0.559	0.685	0.639	0.554	0.623	0.676	0.564	0.625	0.675	0.616

4 Conclusions

Time series analysis was used to organize each country's behavior, which made it possible to determine that it is too early for there to be enough data (owing to the time the pandemic has lasted) to be able to establish a seasonality or trend of the dynamics of virus spread. The techniques used in this methodology can conserve particular and individual dynamics for each country, thus providing a specific contribution using complex network modeling in pandemic data while performing a whole analysis of the behavior of the eleven countries and overall behavior of covid across the world. We can discuss here the correlation between the metrics and the countries listed, as well as the exceptional cases where it can be assumed that this is because of how the policies for the contention of the pandemic were implemented in each country.

A ratio has already been mentioned between the network's diameter and population density and to illustrate this we are giving the following tables that present the countries with smaller and larger diameters and their respective population densities. In the case of these countries, it can be expected that for a high population density, there is a shorter time between weeks for peaks or valleys and the dynamics of the pandemic are faster. Cliques refer to a higher number of weeks of contention or new cases, while density refers to newly recorded cases, which is where the two metrics coincide. On the other hand, the density metric is low for Brazil, Mexico, and the US, which means that the spread is low over several weeks, which coincides with their low number of cliques. However, there are countries, such as Morocco, that have a low number of cliques, but higher density, so we would have to examine, among other factors, the country 's contention policies once again. For the results to be better understood, we should clarify that the density of the network is equivalent to the mean degree divided by the number of nodes minus 1. So, for example, in the case of Mexico has a density of 0.1878, this is equivalent to the mean degree, which is 12.212 between the 66 nodes minus 1, in other words, 65 nodes, resulting in 0.1878. Global Clustering and Betweenness centrality: In these two metrics, we can observe that South Africa and India have a high probability of the nodes (weeks) forming triangles (global clustering) while having at the same time a high (Table 4) Betweenness metric as well as more cliques and higher density. This is important because, at the same time as triangles are formed in India and South Africa, they have a higher number of shorter paths that involve a node; in this case the weeks that have reported more cases.

Table 4. Comparison between clustering and betweenness.

Country	Clustering	Betweenness
India	0.815658	0.512773
South Africa	0.806672	0.484567

In a unique way from the papers found during the literary review, our paper seeks to describe, through complex network metrics, how government policies and the social response to them, affect the propagation dynamics in each country. But with the use of techniques, the components can be analyzed and described noticing that, from an appropriate association through modeling, we get results that contribute to an understanding of the evolution of the pandemic as a function of each country's singularities such as population density, economic situation, type of government, or geography. The contribution of this work is given by using complex networks with their metrics and how each country applied the restrictions and the influence on the global behavior is confirmed. Furthermore, we can analyze cases over the world, analyze the impact of the vaccination and other analysis like mortality or analyze the different stages or waves from the pandemic.

The main contributions of this work are that we analyzed the different time series for COVID-19, we applied the visibility algorithm to transform the time series into complex networks, analyze and compare the complex network metrics. As part of the future work, we can extend this work for the new cases, analyze the different stages of the pandemic, analyze the influence of the vaccination in different countries and apply the methodology to other countries and make a comparison among them. Simulation of complex networks offers several advantages over traditional statistical approaches, as demonstrated in our study. By using simulation, we were able to analyze the time series data from eleven different countries simultaneously, while preserving each country's unique characteristics. This approach would not have been possible using classical statistical methods that rely on analyzing new cases of COVID-19 individually. The networks we constructed for each country and their global behavior enabled us not only to examine the data as a network and observe its behavior but also to compare the behavior between countries and the global data. This comparison allowed us to gain insights into the similarities and differences in the spread of COVID-19 among the different countries and how it impacted the global community.

References

1. WHO. World Health Organization (2020). https://covid19.who.int/table
2. Snacken, R.: Pandemic planning. Vaccine **20**, 88–90 (2002)
3. Waterer, G.W., Hui, D.S., Jenkin, C.: Public health management of pandemic (H1N1) 2009 infection in Australia: a failure! Official J. Asian Pac. Soc. Respirology **15**, 51–56 (2010)
4. Liu, C., Wu, X., Niu, R., Wu, X.: A new SAIR model on complex networks for analyzing the 2019 novel coronavirus (COVID-19). Nonlinear Dyn. 1–11 (2020)
5. Emanuele, J., Koetter, L.: Workflow opportunities and challenges in healthcare. In: BPM & Workflow Handbook, pp. 2–11 (2007)
6. Liu, D., et al.: A machine learning methodology for real-time forecasting of the 2019–2020 COVID-19 outbreak using Internet searches, news alerts, and estimates from mechanistic models. e-prints posted on arXiv (2020)
7. Zhang, J., et al.: Changes in contact patterns shape of dynamics of the COVID-19 outbreak in China. Science **368**(6498), 1481–1486 (2020). https://doi.org/10.1126/science.abb8001
8. Davis, J.T., et al.: Cryptic transmission of SARS-COV-2 and the first COVID-19 wave. Nature **600**, 127–132 (2021). https://doi.org/10.1038/s41586-021-04130-w
9. Díaz-Pinzón, J.E.: Precisión del pronóstico de la propagación del COVID-19 en Colombia. Revista Repertorio De Medicina Y Cirugía (2020)
10. Scabini, L., Ribas, L., Neiva, M., Junior, A., Farfán, A., Bruno, O.: Social interaction layers in complex networks for the dynamical epidemic modeling of COVID-19 in Brazil. Physica A Stat. Mech. Appl. **564**, 125498 (2020). https://doi.org/10.1016/j.physa.2020.125498
11. Alanis, A.Y., Hernández-Vargas, E.A., Ramírez, N.F., Ríos-Rivera, D.: Neural control for epidemic model of covid-19 with a complex network approach. IEEE Lat. Am. Trans. **19**(6), 866–873 (2020)
12. Zhang, J., Litvinova, M., Wang, W., Wang, Y., Deng, X., Chen, X.: Evolving epidemiology and transmission dynamics of coronavirus disease 2019 outside Hubei province, China: a descriptive and modeling study. Lancet Infect. Dis. **20**(7), 793–802 (2020)
13. Villamizar, E., et al.: Políticas públicas, grandes datos, teoría de redes y COVID-19. Desafíos 1–19 (2020)
14. Bruinen de Bruin, Y., et al.: Initial impacts of global risk mitigation measures taken during the combatting of the COVID-19 pandemic. Saf. Sci. **128** (2020)
15. Stella, L., Pinel, A., Bausso, D., Colaneri, P.: The role of asymptomatic individuals in the COVID-19 pandemic via complex networks. SSRN Electron. J. (2020)
16. Montes-Orozco, E.E., et al.: Identification of COVID-9 Spreaders using Multiplex Networks Approach (2020)
17. Vespignani, A., et al.: Modeling COVID-19. Nat. Rev. Phys. **2**, 279–281 (2020)

18. Morselli Gysi, D., et al.: Network medicine framework for identifying drug-repurposing opportunities for COVID-19. Proc. Natl. Acad. Sci. USA **118**(19) (2021)
19. WHO. World Health Organization (2020). https://www.who.int/emergencies/diseases/novel-coronavirus-2019/situation-reports
20. Salvi, J.: Towards Data Science - Significance of ACF and PACF Plots In Time Series Analysis (2020). https://towardsdatascience.com/significance-of-acf-and-pacf-plots-in-time-series-analysis-2fa11a5d10a8. Accessed 2020
21. Caldarelli, G.: Scale-Free Networks. Complex Webs in Nature and Technology. Oxford University Press, Oxford (2013)
22. Barabási, A.: Network Science. Cambridge University Press, Cambridge (2016)
23. Dehmer, M.: Structural Analysis of Complex Networks. Birkhäuser-Springer, Boston (2011). https://doi.org/10.1007/978-0-8176-4789-6

Contributions of Process Mining and Simulation to the Process Management Lifecycle Using the Example of Healthcare Case Studies

Marianne Schmolke(✉)

Wismar - University of Applied Sciences Technology, Business and Design, Philipp-Müller-Str. 14, 23966 Wismar, Germany
marianne.schmolke@hs-wismar.de

Abstract. Process mining and simulation show considerable benefits taken for themselves. Process mining aims at building a process model to describe the behavior contained in event logs of information systems and is concerned with the past and the present, whereas simulation is concerned with future scenarios. A synergetic effect can be derived from combining these two approaches to conduct qualified Process Management. This paper contains a structuring analysis of relevant literature on process mining and simulation. As a large number of studies found was from the healthcare domain the focus was set on healthcare applications. The results are mapped to the (Business) Process Management Lifecycle by discovering patterns in the topic categories addressed by the different studies. The main takeaway is the comparison of application fields of process mining and simulation between healthcare and non-healthcare studies. These results can be used to develop case study designs dealing with core issues in healthcare management systems and processes to obtain more transparency and a more reliable basis for decision-making.

Keywords: Workflows · Healthcare · Process Models · Conformance

1 Introduction

The digitization of society – considering the globally distributed value chains and fast pace of trends, especially in Information Technology can distort the view of individual possibilities and the consequences of one's own actions. Business Informatics but particularly Process Management is a knowledge offer to grasp the given environment and to use it for solution approaches [1].

Process mining and simulation approaches offer a considerable range of benefit potentials [2]. As Wagner et al. [3] found, simulation modeling helps understand the complexity of business processes – the main challenge being the creation of realistic models. Mesabbah et al. [4] argue: Complex processes, uncertain parameters, time-critical activities, budget and resource constraints typify an environment suitable for

M. Mujica Mota and P. Scala (Eds.): EUROSIM 2023, CCIS 2032, pp. 190–204, 2024.
https://doi.org/10.1007/978-3-031-68435-7_14

scenario testing. As Norambuena [5] states: The main difference between the two disciplines is that process mining is used to understand a process's current or past behavior and simulation can be applied to predict future behavior.

Process mining allows us to analyze the system's actual behavior [6] in contrast to the behavior, e.g., a workflow designer assumes it would have. This helps to calibrate the system. Simulation helps to predict future behaviors, which is crucial for decision-making [7].

The Business Process Management Lifecycle – which can also be applied as less specialized "Process Management Lifecycle" - offers a helpful structure to define the different benefit potentials of process mining combined with simulation. The process identification leads to the construction of a process architecture which embraces the different abstraction levels of the process categories. Process discovery leads to an As-Is process model which is the basis for further analyses. If insights on weaknesses and their impact arise, the process can be redesigned and a To-Be process results. This process can be implemented, ideally including an executable process model. Process monitoring takes place and conformance as well as performance insights might lead to a restart of the cycle [8].

2 Related Work and Search Criteria

The body of literature for this paper was found according to these search criteria:

Reviewed English articles (more than 4 pages) containing "process mining" AND "simulation" in title from "2007 on" from the following databases: ACM Digital Library (dl.acm.org), IEEE Xplore (ieeexplore.ieee.org), Proquest (proquest.com), Science Direct (sciencedirect.com) and Springer (link.springer.com). The result was a set of 26 papers considered for the analysis at hand.

The motivation was to find literature with a strong focus on process mining and simulation in the sense that both approaches appear in the document title. This search strategy showed that a huge amount of the identified papers took a healthcare perspective (10 of 26). This domain is characterized by considerable impact. Therefore, a whole Sect. 3.2 is dedicated to these approaches. Furthermore, healthcare should be the main application field of the findings of this article.

Considerable research has been conducted in the area of process mining simulation. The comparison to the existing works can be summarized as follows:

Whereas van der Aalst [2] defines the formal prerequisites for the match of process mining and simulation, this paper focuses on practical insights along the Process Management Lifecycle as designed by Dumas et al. [8].

Norambuena [5] and Martin et al. [9] conduct systematic (literature) reviews on the integration of process mining and (Business Process) simulation. While Norambuena [5] and Martin et al. [9] focus more on challenges or open issues, this paper rather focuses on the enhancement of the Process Management Lifecycle with findings mainly from the healthcare domain – including achievements. This is carried out by mapping the results of the literature analysis conducted with the search criteria mentioned above to the different phases, adapted to the work of Aguirre et al. [10]. This work focuses on Business Process Reengineering. Additionally, a core insight should be how process mining and simulation are mainly applied in practical healthcare studies compared to studies with different focus.

3 Literature Analysis

3.1 Non-health Approaches Categorized by Methodology

Conceptual Work/Exploratory Studies. Conceptual modeling is the non-software-specific description of a simulation model from the real-life component of a process it represents [6]. The models are mainly based on insights from information sources such as documentation, interviews, and direct process observations [5].

Exploratory Studies do not necessarily validate an existing concept but open up new approaches for following validating studies.

Guiseppe et al. [11] experience good mapping results by the insertion of a new sequence of elements "critical node" and "conformance controller" into their process models which they simulate and test on errors. It is proposed to assign to each node of the map

- a risk value equal to the product of the deviation's probability
- calculated on the basis of time series obtained from log files of similar processes and similar nodes,
- and a measurement of the effects on the organization of a deviation from the specific node, and related to potential fines, recovery costs which affect the whole process.

Jadrić et al. [12] demonstrate and assess the potential of using process mining results as an input for Discrete-Event Simulation (DES is a method to reproduce time series of events to investigate their mechanism) modeling. It is an exploratory study that aims to address particular scientific challenges noted by the process mining manifesto [13]. They apply two datasets in two different domains, one dataset is considered data-poor (smart parking) and the other one data-rich (learning management system) in to evaluate the process mining procedure. A core insight is that they can confirm: event data incompleteness is one of the important critical future challenges.

Liu et al. [14] present a generic approach of Business Process Simulation for operational decision support, including Business Process Modeling and workflow simulation. Their design consists of constructing process models of operational decision support based on historical data and simulation of processes for typical and joint problems using event graphs. They analyze the generated simulation models and compare event graph modeling to Petri Net modeling. Their result of comparing the event graph to Petri nets is:

- Simplicity of event graphs is better using event graph modeling,
- expressibility of event graphs: Instead of one focus they are integrating various aspects of processes,
- event graphs are less complex,
- event graphs are more flexible,
- it is easy to handle changes with event graphs.

Martin et al. [15] conduct the extraction of building blocks in simulation and model components of process mining. Critical challenges regarding entity modeling according to them include (1) identification of relevant entity attributes at an appropriate level of abstraction, (2) detecting entity types that influence process execution and can be profiled using entity attributes, and (3) retrieving the entity arrival rate when queues are formed.

Nedopetalski & Freitas [16] conduct a method comparison with the result that new types of resources can be "tested to reach a new level of business performance".

Concept Application/Case Studies. Wynn et al. [7] demonstrate that the same toolset for analyzing simulation logs and the actual workflow logs can be used. Therefore, the view of the natural process execution log ("up to now") can be combined with the simulation log ("from now on").

Pileggi et al. [17] aim to find a way to obtain alternative models of a system of processes using the context (sensor) attributes as indicators of actual system activity. They use a real-world example with smartphone data of social interaction collected by a software suite, called nodobo and the process mining α-algorithm to mine for the actual process activity. They find "useful models that may offer lower complexity at a cost of degraded accuracy, should they be used in some model-driven solution, such as context inference", i.e. "the process of making context information explicitly available from other context sources" [18].

Pourbafrani & van der Aalst [19] demonstrate the capability of generating process models using system dynamics and process mining. According to [20], they describe system dynamics techniques as being employed in order to model complex systems and the relationships between system variables and their environment. It was possible to uncover hidden relations and construct valid simulation models.

Septiyanto et al. [21] analyze the possibilities of process mining discovering bottlenecks with the Discrete-Event Simulation hierarchical process model. They find that the applied method of bottleneck identification in the hierarchical process model can show the location of the problem in outline or specifically.

Dorrer &Dorrer [22] focus on Agent simulation models (agents represent students here), which are generated using process mining. They implement the automation of e-learning processes in a discrete math course and analyze it with ProM Heuristic Miner. This method provides an accurate process model and the possibility of conversion to the EPC model. They find that conversions worked regarding fitness requirements, whereas regarding the noise level several iterations were necessary.

Uerhara et al. [23] test an application of process mining to model simulation logs by use of ProM (Process Mining Framework) with Petri Nets with an airplane passenger scenario illustration. Looking at the validity of process models they investigate the applicability of process mining to the DES context. They identify process mining as a valid instrument.

Zakarija et al. [24] propose model checking techniques for the automated analysis of their discovered process models using actual data gathered from a hotel data management system. By conducting process synthesis, analysis and repair with the so-called Spin model checker they identify benefit potentials of the approach: "Spin model checker can be successfully used to automatically simulate and verify discovered process models." Verification was performed under six different specifications, five using

LTL formulae and one using an assert statement. In addition to the verification of process model soundness, they state that experts can detect irregular traces (false positive examples), generate negative examples, check if the sequence of events is satisfied in process model executions, as well as other specifications.

3.2 Healthcare Applications

The works found about process mining simulation in healthcare can all be attributed to the "Concept Application/Case Study"- category.

Augusto et al. [25] define a new methodology to automatically build a simulation model of the patients' clinical pathways from a raw national hospital database using process mining techniques. They analyze clinical pathways from event logs in care processes to extract knowledge about the impact of medical decisions on the relapse rate, death rate and cost. The process mining approach creates a causal net which is converted into a formal state chart used to model the health state development of a patient and the associated care pathways. Numerical results validate their modeling approach and the balance mechanism between costs and care quality positively.

Abo-Hamad et al. [26] pursue the integration of process mining techniques in the modeling phase to support developing simulation models. They analyze sample data from an emergency department (ED) to evaluate the quality of control flow including bottlenecks etc. They find 'practical evidence', i.e. results show that unblocking of ED outflows have a significant impact on patients' length of stay (over 80% improvement) rather than increasing the physical ED capacity.

Johnson et al. [27] extend the process mining project methodology (Planning, Extraction, Data Processing, Mining Analysis, Evaluation, Process Improvement and Support) from [28] with a process simulation approach for the "what-if"- options for improvement. They apply the framework to case studies. Nevertheless, the Electronic Health Record (EHR) data was to unreliable for process mining analyses. They then used audited EHR data: clinical letters and discharge notes, i.e. using all available resources to measure the extent to which process mining of routine data was possible.

Mesabbah et al. [4] propose the ASMB-H (Auto Simulation Model Builder for Hospital Patients Logs Data) framework. Hybridization is applied by combining paradigms from different categories of techniques. The key approach of the ASMB was the reliance on the system events logs to extract the underlying process structure along with the statistical analysis for patterns of occurrences for the different events and their processing times. The ASMB-H framework includes new and modified components alongside few unmodified components from the ASMB. The extended framework was proposed to overcome the problem of resource planning. (i.e., hospital staff) with various availability for different shifts. The framework introduced several critical improvements for simulating the agents arriving at the system during simulation time.

Phan et al. [29] analyze the clinical pathway and associated costs of patients having an incisional hernia using a medico-administrative database. With certain limitations to the medical insights, they conclude that it is difficult to generalize to other data sources or case studies because of the unique features related to each (medical) case.

Tabmuris & Esposito [30] figure out a DES model using process mining techniques specific to hospital-related processes. They adapt the process mining tool, ProM6, to the commercial Simulation tool, Simul8. They conduct a critical evaluation of both tools and aim at identifying the degree to which issues related to the following aspects could be overcome [31]:

- treatment of the data provided by the organization,
- building time of a DES tool, gaining knowledge of the meaningful process attributes,
- and validation of the resulting model.

They compare: Simul8 is a commercial tool that does not allow any system changes, but the user can import, export, connect with Excel, save the work, to name some of other features. ProM6 is a platform that makes it possible to build different projects related to the PM techniques in Java source; this allows in many cases to export, and to import the standardized 'XES' format files'.

Franck et al. [32] design a generic simulation model in order to analyze all possible patient pathways from the Emergency Department to hospital discharge. They use a global framework and a framework for studied patients and analyze case studies of pathways of stroke patients. By analyzing medical unit capacity variations, they define measures to improve a number of optimal pathways. They find out that the genericity of the model allows a large extension of the study.

Ferronato & Scalabrin [33] construct PM2Sim, a simulation framework to automate the creation of a simulation model from process mining. It integrates different facilities to extract from event logs, historical data, the current state, organizational structure, and discovery of processes to create scenarios. PM2Sim allows the extraction of parameters from event logs of the current state of instances of processes to automatize the support of operational systems in the process environment. The procedure starts with the standardized XES-format files, continues with several process mining and statistical steps, and moves on to the Multi-Criteria Decision Analysis that is supported by the enhanced value benefit analysis approach "Promethee". Finally, the visualization is conducted.

Halawa et al. [34] propose a three-phase framework to leverage hospital tracking data of patient visits while designing healthcare layouts with pod structures ("data structure that is only changed by external threads or processes" ([35])). They showed how historical Electronic Health Record data could be used to extract the most significant workflows so that patient visits could be clustered. This way "more realistic simulation models" could be built "to better align demand and capacity".

Le Lay et al. [36] pursue the development of a framework to automate the generation of clinical patient pathways from the data preprocessing stage to the use in simulations. They look at the generated pathways and LoS (length of stay) as well as effects such as bed shortage. They then analyze pathway outputs and changes. They conclude that the model's ability to accurately replay pathways and to queue patients makes it a reliable tool for strategic decision making.

4 Improving the Process Management Lifecycle

The work adapts a Business Process Reeingeering approach of [10] to BPM lifecycle.

The following work is considered: van der Aalst et al. [37] investigate the representational bias in process discovery with process mining. They name the quality dimensions fitness ("able to replay event log"), simplicity ("Occam's razor"), generalization ("not overfitting the log"), and precision ("not underfitting the log"). Problems regarding process mining event logs can be categorized by the data quality dimensions: "missing", "incorrect", "imprecise" and "irrelevant". In addition to the process mining perspective, Norambuena [5] summarizes common pitfalls in simulation modeling. These need to be considered when investigating the real-world settings represented by the process mining and simulation models (Table 1).

Table 1. Contributions of simulation and process mining to Process Management (adapted, based on [10])

Phase	Contribution of process mining (PM)	Contribution of simulation
Process Identification	Conformance checks can open up new perspectives and lead to the need for model enhancement with bias possibilities in mind (see [37])	Simulation can be evaluated with modeling pitfalls in mind [5]
Process Investigation or Discovery	Conformance Checks can be investigated on and the results can be re-evaluated	Simulation can help to test the assumptions
Process Analysis	The results of the preceding step can be applied to conduct a thorough analysis	Simulation can help to test the analysis
Process Improvement	The actual process model, which PM discovers, is an essential input for process design and redesign	Through simulation, it is possible to perform "what if" analyses of the different process design or redesign options
Process Implementation	In this phase, PM is used to verify that the process complies with business policies and rules. It is also possible to predict the remaining execution of a running case	Business processes are implemented in this phase after being tested and improved through simulation
Process Monitoring	In this phase, PM is used for identifying loops and bottlenecks and further checking for conformance with business rules	Simulation allows business process analysis to eliminate bottlenecks to improve throughput times

Van der Aalst et al. [37] add dimensions to the above-mentioned quality dimensions such as correctness and expressiveness. Key takeaways are: The representational bias may exclude desirable models. At least the basic workflow patterns must be supported, such as sequence, parallel split, etc. Additionally, as particularly process trees are analyzed, just randomly searching for a process tree is not a good idea when mining a process model.

To assess the quality of a process model, Schmolke [38] provides a structured framework. The example of software engineering is used. Still, it can be applied a to other domains as well.

Process Management Lifecycle (PML)→	**Process Mining Contribution** *(italic)* **to PML elements (bold):**							
	Identify		**Investigate**	**Analyze**	**Improve**		**Implement**	**Monitor**
	Conformance Checks	*Avoid bias*	*Reevaluation*		*Business Policies*	*Predict*		
Core findings related to contributions								
[11] New BPMN element improving quality of mapping	X		X	X	X	X	X	X
[12] Deficits of method application on poor-data sets		X	X	X				
[14] Comparison of event graph modeling to Petri Net modeling					X	X	X	
[15] Core research challenges		X	X		X	X		X
[16] Validation of mining of real time scheduling workflow	X						X	
[17] Construct alternative context perspectives for future automization	X		X	X				
[19] Identify bottlenecks in hierachical process model in outline/ Specifically								X
[21] Model complex systems and relationships between variables and their environment		X						X
[22] Generate (e-learning) agent simulation models	X	X					X	
[23] Airplaine passenger application of Process Mining to model simulation logs	X	X			X			
[24] Approach for process model repair	X		X		X			X

Fig. 1. Mapping process mining study outputs to Process Management Lifecycle phases – non-healthcare

Process Management Lifecycle (PML)→	**Process Mining Contribution** *(italic)* **to PML elements (bold):**							
	Identify		**Investigate**	**Analyze**	**Improve**		**Implement**	**Monitor**
	Confor-mance Checks	*Avoid bias*	*Reeva-luation*		*(Business) Policies*	*Predict*		
Core findings related to contributions								
[25] Causal net to model the health state evolution of a patient and the associated care pathways		X			X	X		X
[26] Large number of uncovered control-flows		X			X			X
[27] Application of framework to health case studies	X	X			X	X	X	X
[4] Framework to overcome resource planning problems					X	X	X	X
[29] Clinical pathway analysis				X		X		X
[30] Adapt ProM by commercial tool Simul8	X			X	X			
[32] Simulation model for patient pathways						X		X
[33] Automation of creation of PM simulation model	X		X	X	X	X	X	
[34] Optimization of hospital data tracking for healthcare design	X		X	X	X		X	X
[36] Automation to mimick patient pathways	X		X	X	X	X		X

Fig. 2. Mapping process mining study outputs to Process Management Lifecycle phases –health-care

It can be stated as demonstrated in Figs. 1, 2, 3 and 4 that the non-healthcare works do not show a clear pattern regarding their contributions to the Process Management Lifecycle categories. Instead the healthcare applications can be characterized by a clear focus (huge number of what if scenarios, mainly eliminating bottlenecks) and the goal to optimize ("improve" and "monitor" are frequently addressed). A category included seldomly is "implement" Surprisingly even process mining combined with simulation does not often imply the implementation of the mining results in Workflow Management Systems in the approaches analyzed here. The reason for the "what if" category not applying in all cases is that also alternative approaches can be used aside of process mining.

Process Management Lifecycle (PML)→	**Contribution of Simulation** (italic) **to PML elements (bold):**					
	Identify	**Investigate**	**Analyze**	**Improve**	**Implement**	**Monitor**
	Avoid Modeling Bias	*Test assump-tions*	*Test analysis*	*„what if“*		*Eliminate bottle-necks*
Core findings related to contributions						
[11] New BPMN element improving quality of mapping				X		
[12] Deficits of method application on poor-data sets	X	X				
[14] Comparison of event graph modeling to Petri Net modeling		X		X		
[15] Core research challenges	X	X		X		
[16] Validation of mining of real time scheduling workflow		X	X			
[17] Construct alternative context perspectives for future automization				X		
[19] Identify bottle-necks in hierachical process model in outline/ Specifically				X		
[21] Model complex systems and relation-ships between variables and their environment						X
[22] Generate (e-learning) agent simulation models	X					
[23] Airplaine passenger application of Process Mining to model simulation logs		X	X	X		
[24] Approach for process model repair		X	X	X		

Fig. 3. Mapping simulation study outputs to Process Management Lifecycle phases – non-healthcare

Process Management Lifecycle (PML)→	**Contribution of Simulation** *(italic)* **to PML elements (bold):**					
	Identify	**Investigate**	**Analyze**	**Improve**	**Implement**	**Monitor**
	Avoid Modeling Bias	*Test assumptions*	*Test analysis*	*„what if"*		*Eliminate bottlenecks*
Core findings related to contributions						
[25] Causal net to model the health state evolution of a patient and the associated care pathways				X		
[26] Large number of uncovered control-flows	X	X		X		X
[27] Application of framework to health case studies	X	X	X	X		X
[4] Framework to overcome resource planning problems			X	X		X
[29] Clinical pathway analysis				X		X
[30] Adapt ProM by commercial tool Simul8				X		
[32] Simulation model for patient pathways				X		X
[33] Automation of creation of PM simulation model				X		
[34] Optimization of hospital data tracking for healthcare design				X		X
[36] Automation to mimick patient pathways				X		X

Fig. 4. Mapping simulation study outputs to Process Management Lifecycle phases –healthcare

5 Conclusions

Summing up, process mining and simulation were evaluated as valid instruments in the conducted studies. Potentials discovered are:

- The possibility to uncover hidden relationships and construct valid simulation models [17] – which can be more realistic to align demand and capacity [33] for strategic decision making [36].
- It is possible to ensure a genericity that helps extend the models, e.g. in [32].

Weaknesses of the studies were:

- Lack of reliability [27]/difficulties to generalize to other data sources or studies because of the unique features related to each (medical) case [29]. This group of issues can be treated by using audited Electronic Health Record data: clinical letters, discharge notes. i.e. using all available resources to measure the extent to which process mining of routine data is possible [27].
- "Incompleteness" [12]. This issue can be treated by focusing on the detection of irregular traces (false positive examples), generation of negative examples, check if the sequence of events is satisfied in process model executions, as well as other specifications [24].

The analysis showed that main application fields of process mining and simulation in the healthcare domain are classical simulation, i.e. to create what-if-scenarios and the elimination of bottlenecks. Additionally, the optimization focus of processes is strongly present. The implementation category is of minor relevance as process mining and simulation are not automatically connected to a Workflow Management logic.

It is evident that very few papers even include the term 'bias' or 'pitfalls' in their explanations. This does not mean that the subject is completely ignored. Still, it is recommendable for future works to actively embrace this quality criterion.

The categories identified here can guide the discussion about relevant future applications in Process Management and Business Process Reengineering that use the basis of state-of-the-art methods such as process mining and simulation.

References

1. Benker, T., Jürck, C., Wolf, M.: Geschäftsprozessorientierte Systementwicklung: Von der Unternehmensarchitektur zum IT-System. Springer, Wiesbaden (2016). https://doi.org/10.1007/978-3-658-14826-3
2. van der Aalst, W.M.P.: Process mining and simulation: a match made in heaven! Summer Simulation Multiconference, Bordeaux, France (2018). https://doi.org/10.22360/summersim.2018.scsc.005
3. Wagner, G., Seck, M., McKenzie, F.D.: Process modeling for simulation: observations and open issues. In: Online World Conference on Soft Computing in Industrial Applications, Washington, DC, USA, pp. 1072–1083 (2016). https://doi.org/10.1109/wsc.2016.7822166
4. Mesabbah, M., Abo-Hamad, W., McKeever, S.: A hybrid process mining framework for automated simulation modelling for healthcare. In: Online World Conference on Soft Computing in Industrial Applications, National Harbor, MD, USA, pp. 1094–1102 (2019). https://doi.org/10.1109/wsc40007.2019.9004800
5. Keith Norambuena, B.: Integration of process mining and simulation: a survey of applications and current research. In: Rocha, Á., Adeli, H., Reis, L.P., Costanzo, S. (eds.) WorldCIST 2018, pp. 287–294. Springer, Cham (2018). https://doi.org/10.1007/978-3-319-77703-0_29
6. Weijters, A., Aalst, W.M.P., van der, Alves De Medeiros, A.K.: Process mining with the HeuristicsMiner algorithm. BETA publicatie: working papers. Technische Universiteit Eindhoven, Eindhoven (2006)

7. Wynn, M., Rozinat, A., van der Aalst, W., ter Hofstede, A., Fidge, C.: Process mining and simulation. In: Hofstede, A., Aalst, W., Adams, M., Russell, N. (eds.) Modern Business Process Automation: YAWL and Its Support Environment, pp. 437–457. Springer, Heidelberg (2009). https://doi.org/10.1007/978-3-642-03121-2_17
8. Dumas, M., La Rosa, M., Mendling, J., Reijers, H.A.: Introduction to business process management. In: Dumas, M., La Rosa, M., Mendling, J., Reijers, H.A. (eds.) Fundamentals of Business Process Management, pp. 1–33. Springer, Heidelberg (2018). https://doi.org/10.1007/978-3-662-56509-4_1
9. Martin, N., Depaire, B., Caris, A.: The use of process mining in a business process simulation context: overview and challenges. In: IEEE Symposium on Computational Intelligence and Data Mining, Orlando, FL, USA, pp. 381–388 (2014). https://doi.org/10.1109/cidm.2014.7008693
10. Aguirre, S., Parra, C., Alvarado, J.: Combination of process mining and simulation techniques for business process redesign: a methodological approach. In: Cudre-Mauroux, P., Ceravolo, P., Gašević, D. (eds.) SIMPDA 2012, pp. 24–43. Springer, Heidelberg (2012). https://doi.org/10.1007/978-3-642-40919-6_2
11. Giuseppe, C., Valerio, M., Teresa, M., Carmela, S.L.: A simulation approach in process mining conformance analysis. The introduction of a brand new BPMN element. IERI Procedia **6**, 45–51 (2014). https://doi.org/10.1016/j.ieri.2014.03.008
12. Jadrić, M., Pašalić, I.N., Ćukušić, M.: Process mining contributions to discrete-event simulation modelling. Bus. Syst. Res. **11**(2), 51–72 (2020). https://doi.org/10.2478/bsrj-2020-0015
13. Process mining Manifesto - IEEE Task Force on process mining. https://www.tf-pm.org/resources/manifesto. Accessed 14 Feb 2023
14. Liu, Y., Zhang, H., Li, C., Jiao, R.J.: Workflow simulation for operational decision support using event graph through process mining. Decis. Support. Syst. **52**(3), 685–697 (2012). https://doi.org/10.1016/j.dss.2011.11.003
15. Martin, N., Depaire, B., Caris, A.: The use of process mining in business process simulation model construction - structuring the field. Bus. Inf. Syst. Eng. **58**(1), 73–87 (2016). https://doi.org/10.1007/s12599-015-0410-4
16. Nedopetalski, F., de Freitas, J.C.J.: Process mining and simulation for a p-time petri net model with hybrid resources. In: Systems and Information Engineering Design Symposium Charlottesville, VA, USA (2021). https://doi.org/10.1109/sieds52267.2021.9483768
17. Pileggi, P., Rivero-Rodriguez, A., Nykänen, O.: Towards traditional simulation models of context using process mining. In: International Conference on Computational Intelligence, Communication Systems and Networks, 1730 Massachusetts Ave., NW Washington, DC, United States, pp. 70–75 (2015). https://doi.org/10.1109/cicsyn.2015.23
18. Roussaki, I., Kalatzis, N., Liampotis, N., Frank, K., Sykas, E.D., Anagnostou, M.: Developing Context-Aware Personal Smart Spaces. Handbook of Research on Mobile Software Engineering: Design, Implementation, and Emergent Applications (2012). https://www.igi-global.com/chapter/developing-context-aware-personal-smart/www.igi-global.com/chapter/developing-context-aware-personal-smart/66491. Accessed 14 Feb 2023
19. Pourbafrani, M., van der Aalst, W.M.: Discovering system dynamics simulation models using process mining. IEEE Access **10**, 78527–78547 (2022). https://doi.org/10.1109/access.2022.3193507
20. Sterman, J.: Business Dynamics. Irwin McGraw-Hill, Boston (2000)

21. Septiyanto, A.F., Sarno, R., Sungkono, K.R.: Identifying bottlenecks of hierarchical process model by using discrete event simulation and process mining. In: International Conference Computing Engineering and Design Sukabumi, Indonesia (2022). https://doi.org/10.1109/icced56140.2022.10010450
22. Dorrer, M., Dorrer, A.: Generation of agent simulation models by using process mining methods on the example of E-learning process. J. Phys. Conf. Ser. **1399**(3), 33077 (2019). https://doi.org/10.1088/1742-6596/1399/3/033077
23. Uehara, K., Kunihiko, H.: Process mining approach for the conformance checking of discrete-event simulation model. In: Annual Conference of the Society of Instrument and Control Engineers of Japan, vol. 118, no. 296, pp. 121–126 (2018). https://doi.org/10.23919/sice.2019.8859905
24. Zakarija, I., Skopljanac-Macina, F., Blašković, B.: Automated simulation and verification of process models discovered by process mining. arXiv: Artificial Intelligence (2020). https://doi.org/10.1080/00051144.2020.1734716
25. Augusto, V., Xie, X., Prodel, M., Jouaneto, B., Lamarsalle, L.: Evaluation of discovered clinical pathways using process mining and joint agent-based discrete-event simulation. In: Online World Conference on Soft Computing in Industrial Applications, pp. 2135–2146 (2016). https://doi.org/10.1109/wsc.2016.7822256
26. Abo-Hamad, W., Ramy, A., Arisha, A.: A hybrid process-mining approach for simulation modeling. In: Online World Conference on Soft Computing in Industrial Applications, pp. 1527–1538 (2017). https://doi.org/10.1109/wsc.2017.8247894
27. Johnson, O., Dhafari, T.B., Kurniati, A.P., Fox, F., Rojas, E.: The ClearPath method for care pathway process mining and simulation. In: Daniel, F., Sheng, Q., Motahari, H. (eds.) BPM 2018, pp. 239–250. Springer, Cham (2018). https://doi.org/10.1007/978-3-030-11641-5_19
28. van Eck, M.L., Lu, X., Leemans, S.J.J., van der Aalst, W.M.P.: PMA process mining project methodology. In: Zdravkovic, J., Kirikova, M., Johannesson, P. (eds.) CAiSE 2015, pp. 297–313. Springer, Cham (2015). https://doi.org/10.1007/978-3-319-19069-3_19
29. Phan, R., Augusto, V., Martin, D., Sarazin, M.: Clinical pathway analysis using process mining and discrete-event simulation: an application to incisional hernia. In: Online World Conference on Soft Computing in Industrial Applications, pp. 1172–1183 (2019). https://doi.org/10.1109/wsc40007.2019.9004944
30. Tamburis, O., Esposito, C.: Process mining as support to simulation modeling: a hospital-based case study. Simul. Model. Pract. Theory **104**, 102149 (2020). https://doi.org/10.1016/j.simpat.2020.102149
31. Tamburis, O.: Bridging the gap between process mining and DES modeling in the healthcare domain. In: E-Health and Bioengineering Conference Iasi, Romania (2019). https://doi.org/10.1109/ehb47216.2019.8969912
32. Franck, T., Bercelli, P., Aloui, S., Augusto, V.: A generic framework to analyze and improve patient pathways within a healthcare network using process mining and discrete-event simulation. In: Online World Conference on Soft Computing in Industrial Applications, pp. 968–979 (2020). https://doi.org/10.1109/wsc48552.2020.9384021
33. Ferronato, J.J., Scalabrin, E.E.: PM2Sim: the automated creation of a simulation model from process mining. In: IEEE International Conference on Systems, Man and Cybernetics Melbourne, Australia (2021). https://doi.org/10.1109/smc52423.2021.9659211
34. Halawa, F., Madathil, S.C., Khasawneh, M.T.: Integrated framework of process mining and simulation–optimization for pod structured clinical layout design. Expert Syst. Appl. **185**, 115696 (2021). https://doi.org/10.1016/j.eswa.2021.115696
35. Passive data structure. https://xlinux.nist.gov/dads/HTML/passiveDataStruc.html. Accessed 14 Feb 2023

36. Le Lay, J., Neveu, J., Dalmas, B., Augusto, V.: Automated generation of patient population for discrete-event simulation using process mining. In: Annual Modeling and Simulation Conference San Diego, USA (2022). https://doi.org/10.23919/annsim55834.2022.9859406
37. van der Aalst, W.W., Buijs, J.J., van Dongen, B.F.: Towards improving the representational bias of process mining. In: Aberer, K., Damiani, E., Dillon, T. (eds.) SIMPDA 2011, pp. 39–54. Springer, Heidelberg (2011). https://doi.org/10.1007/978-3-642-34044-4_3
38. Schmolke, M.: Process mining model quality in software development case studies: an analysis. In: Tagungsband zur 35. Jahrestagung des Arbeitskreises Wirtschaftsinformatik an Hochschulen für Angewandte Wissenschaften im deutschsprachigen Raum (AKWI) vom 11.09. bis 13.09.2022, Berlin (2022)

Permutation Entropy as a Conceptual Model to Analyse Brain Activity in Sleep

Alexander Edthofer[1], Iris Feldhammer[1], Thomas Fenzl[2], Andreas Körner[1(✉)], and Matthias Kreuzer[2]

[1] Institute of Analysis and Scientific Computing, TU Wien, Wiedner Hauptstraße 8, 1040 Vienna, Austria
andreas.koerner@tuwien.ac.at

[2] Department of Anaesthesiology and Intensive Care Medicine, University Hospital Rechts der Isar, Technical University of Munich, Ismaninger Straße 22, 81675 Munich, Germany

Abstract. Sleep stage classification is a widely discussed topic, due to its importance in the diagnosis of sleep disorders, e.g. insomnia. Analysis of the brain activity during sleep is necessary to gain further insight into the processing that occurs in our brains. We want to use permutation entropy as a model for this analysis. Therefore, the signal processing in terms of electroencephalography is described. This results in a time discrete signal, that can be further processed by applying the method of permutation entropy, which is a modification of the Shannon entropy as a measure of information processing. The method is applied to 18 data sets, nine electroencephalography measurements of patients suffering from insomnia and nine of people without a sleep disorder. A strong correlation between the permutation entropy value and the sleep stages was found during the simulation runs. The results are analysed and presented using boxplot diagrams of the permutation entropy over the sleep stages. Furthermore, it is investigated that there is a steady decrease in the value when the patient is in a deeper sleep. This suggests that the method is a good parameter for sleep stage classification. Finally, we propose an extension of the conceptual model to other pathological conditions and also to the analysis of brain activity during surgery.

Keywords: permutation entropy · sleep stage classification · EEG monitoring

1 Introduction

Every day humans have to process a lot of information. This results in brain activity, which can be measured by electroencephalography (EEG). Plenty of techniques and methods for analysing the complex processes of our brains already exist. One of them is the permutation entropy (PE), which was introduced in [1]. It is a modification of the Shannon entropy, which is a specific measure of information processing, as it describes the rate at which information is produced

M. Mujica Mota and P. Scala (Eds.): EUROSIM 2023, CCIS 2032, pp. 205–218, 2024.
https://doi.org/10.1007/978-3-031-68435-7_15

when discrete information is modelled as a Markov process, see [2]. Based on this, the PE is a complexity measure for time discrete signals as it maps tuples of a time series to a set of predefined patterns. The probability distribution of these is then used for the calculation of the measure.

The aim of this work is to apply the concept of the PE to sleep data and extract information about the sleep states of patients. The strong correlation between the entropy measure and the sleep stages was the fundamental motivation. On the one hand, high values of PE indicate noise-like behaviour and, therefore, wakefulness, where as on the other hand, low values indicate low activity and sleep. This allows the analysis of sleep behaviour and performance of sleep stage classification. The activity patterns are very differentiated during different sleep stages. Some areas are still very active, information is still being processed. Due to the great importance of sleep, the topic is widely discussed using a variety of methods [3–7]. The long-term aim of this work is to have an open source parameter using PE that allows anaesthetists to monitor the level of a patient's coma during surgery, as was already suggested in [8].

Before applying the method, we first introduce the signal processing of brain activity and how to obtain a time discrete signal [9,10]. It is described how different technical subsystems interfere during the measurement and how this is smoothed again by filtering. Afterwards, the Shannon entropy is explained [2] and, as a modification, the PE is introduced and discussed mathematically in detail in order to have a robust model, following [1,11,12].

Next, in Sect. 5, we consider 18 data sets of the Cyclic Alternating Pattern (CAP) sleep data base [13]. As open source data on this topic is already available, we will determine the PE with the data and simulations to test our conceptual model. Reproducibility is given as well. In [4] the method is used as well, but only performed on three patients, so we want to take a look at a larger sample size and also consider patients suffering from a sleep disorder. The processing of the EEG signals of the patients is described. Afterwards, the results of the application of our method are presented. The correlation between the PE and the sleep stages is calculated and the sleep stage classifications are presented in boxplot diagrams.

2 Signal Processing of Continuous Brain Activity into a Time Discrete Signal

The human brain responds with an electrical impulse to every stimulus it perceives, whether by sight, touch, taste, smell or hearing. This signal is transmitted from the peripheral nervous system to the brain, where the information is processed. This processing is done by changing the electrical membrane properties, i.e. the ionic conductivity, see [14]. The measurement of this brain activity is of great importance to neurologists in the diagnosis of diseases, but also for anaesthetists in monitoring during surgery. This measurement is called EEG.

Several things are needed to measure an EEG signal, as it is stated in [9]. First, there are electrodes, which are the connection between the tissue and the technical system. Usually at least three electrodes are used for an EEG during a surgery, when it is done for diagnostic reasons there are more in use. In order to compare results and have reproducibility, the placement of the electrodes is standardised with the ten-twenty system. It is not rigid but dynamic with specifications given in percentages, so that the application also works for smaller heads, such as infants or children have. The reference for the voltage is either the average of the value of all electrodes used, if it is a multichannel EEG, or there is a reference and a ground electrode in addition to the electrode for the measurement.

With the measurement done by the electrodes one can proceed with signal processing. An amplifier increases the strength of the signal, since the deviating value of the reference is usually small. Further signal processing consists of the detection of artefacts. There are biological ones produced by the patient, such as movement of the patient, sweating or twitching of muscles. Other artefacts can be caused by the technical equipment or instruments, e.g. the electrodes, the amplifier and all other used components. Hence, it is necessary to apply highpass and lowpass filtering to the signal. On the one side, the highpass filter eliminates the low-frequent disturbances, which usually consist of the biological artefacts, on the other side, the lowpass filter neglects the high-frequent noise that usually occurs when using technical instruments or electromyography (EMG), which measures muscle activity.

With the signal processing also a discretisation goes hand in hand. This is done after filtering by a multichannel analogue-to-digital converter (ADC). To determine the correct sampling frequency, it is necessary to take a look at the frequency of human brain waves. They are classified in four categories, according to their frequency range: β (13–30 Hz), α (8–13 Hz), θ (4–8 Hz), δ (0.5–4 Hz). Each of them appears at different stages of age or consciousness. Due to the Nyquist theorem [15], which states that a sampling rate of two times the maximal frequency is required, a sampling rate of at least 200 samples/s should suffice [10]. After the discretisation, further filtering can be done.

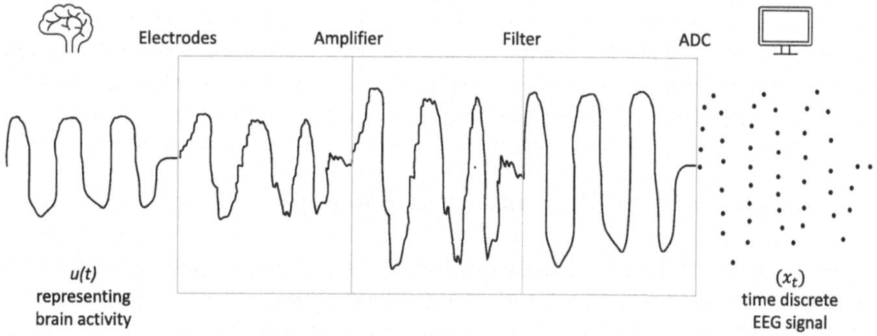

Fig. 1. Schematic diagram of the signal processing of continous brain activity to get a time discrete signal.

In Fig. 1 a schematic diagram of the whole process is shown. We now have a time discrete signal that represents our brain activity and can be further processed.

3 Entropy as a Measure of Information

After recording an EEG, we want to analyse it. Therefore, we want to calculate the PE value of the time discrete signal and see if it tells us anything about the brain activity. First, a few words about entropy in general. It has various applications but its origin is in physics, according to [16]. There it is an important term in the field of thermodynamics. From there, entropy has spread to other areas of physics, from mechanics to astronomy. It has also found its way into chemistry as well as social sciences. The different fields of application interpret the term entropy in their own way, as [17] states. We will now look at its application in information theory, as Claude Shannon described it in his 1948 article "A Mathematical Theory of Communication" [2].

The so-called Shannon entropy should describe the rate at which information is produced if discrete information is modelled as a Markov process. This means that we have n possible events with certain probabilities $p_1, ..., p_n$. The Shannon entropy $H = H(p_1, ..., p_n)$ is "a measure of how much 'choice' is involved in the selection of the event or of how uncertain we are of the outcome?", see [2], p. 10. It is derived from the assumptions:

1. H is continuous in its arguments p_j.
2. H is monotonously increasing in n if $p_j = \frac{1}{n}$ for all $j = 1, ..., n$.
3. If there are three possible outcomes, there is a decomposition from three to two by adding another level of choice.

With these assumptions and $K \in \mathbb{R}^+$, Shannon showed in [2]

$$H(p_1, ..., p_n) = -K \sum_{j=1}^{n} p_j \log p_j. \tag{1}$$

A modification of the Shannon entropy, the PE, was introduced by Christoph Bandt and Bernd Pompe in 2002 [1], which will be the focus of this work. In the next section, the PE will be defined and discussed in detail.

4 The Method of Permutation Entropy

In this section, the PE, as Christoph Bandt and Bernd Pompe introduced in 2002 [1], is discussed in more detail. It is a modification of the Shannon entropy as defined in Eq. (1). It is a statistical measure and describes the complexity of a time discrete series. We orient the notation and definition according to [11,12]. For a given time series $(x_t)_{t=1,...,n}$ of length n we can choose an order m and a time delay τ. The order specifies how long the investigated tuples of the time series will be and later we will see that it has a huge impact on the accuracy of the method. The time delay tells us the index distance between two values in a tuple. For $\tau = 1$ we only have neighbouring values of the time series, for $\tau > 1$ this is no longer the case. If two values of the time sequence are equal $x_i = x_j$, for $i \neq j$, we neglect the value or add some random noise. Since we want to calculate the permutation of the time sequence, we divide in $k := n - (m-1)\tau$ tuples of length m. The mapping rule is

$$(x_1, ..., x_n) \mapsto ((x_1, x_{1+\tau}, x_{1+2\tau}, ..., x_{1+(m-1)\tau}),$$
$$(x_2, x_{2+\tau}, x_{2+2\tau}, ..., x_{2+(m-1)\tau}), ..., (x_k, x_{k+\tau}, x_{k+2\tau}, ..., x_{k+(m-1)\tau})).$$

Next, we take one tuple starting with the value x_j and order the values from smallest to largest value. For numbers $r_1, r_2, ..., r_m \in \{1, 2, ..., m\}$, where r_1 is associated with the first value of the tuple, which is x_j, r_2 is associated with the second value $x_{j+\tau}$, and so on, such that it holds

$$x_{j+(r_s-1)\tau} < ... < x_{j+(r_1-1)\tau} < ... < x_{j+(r_{\tilde{s}}-1)\tau}.$$

Hence, r_s is the number, such that $x_{j+(r_s-1)\tau}$ is the lowest value of the whole tuple and $r_{\tilde{s}}$ such that $x_{j+(r_{\tilde{s}}-1)\tau}$ is the highest. The pattern $(r_1)...(r_m)$ is a permutation ω_j of the pattern $(1)...(m)$. The set of all patterns is defined as $\Omega_m := \{\omega | \omega \text{ is a permutation of} (1)(2)...(m)\}$. The number of possible permutations of natural positive numbers up to m is $m!$. Consequently, the cardinality of the set Ω_m is $|\Omega_m| = m!$. As an example, $\omega_1 = (1)(2)...(m-1)(m)$ and $\omega_{m!} = (m)(m-1)...(2)(1)$. By using the notation of brackets ( ) around each digit, one can also use patterns of order greater than 9.

Now the idea of the PE is to determine the permutation type ω_j of each tuple, so we get k patterns, as it is done in Fig. 2 for $m = 3$ and $\tau = 1$. The probability distribution is then calculated and the probability of each type is defined as $p_j := P(\omega_j)$, where

$$P(\omega_j) = \frac{\#\{i | 1 \leq i \leq k, (x_i, ..., x_{i+(m-1)\tau}) \text{ has type } \omega_j\}}{k}.$$

According to the Shannon entropy, the PE is defined as

$$H(m) := H(p_1, ..., p_{m!}) = -\sum_{j=1}^{m!} p_j \log p_j.$$

The base of the logarithm is 2. If the probability p_j were equal to 0, the summand would not be defined. Therefore, we will take the limit of such a value towards 0 and, with the rule of de l'Hospital, see that

$$\lim_{p_j \to 0} p_j \log p_j = 0.$$

Hence, we can neglect such a summand in the overall sum. According to [2] the range of H is $[0, \log(m!)]$. The minimal value 0 is obtained, if there is a $\tilde{j}$

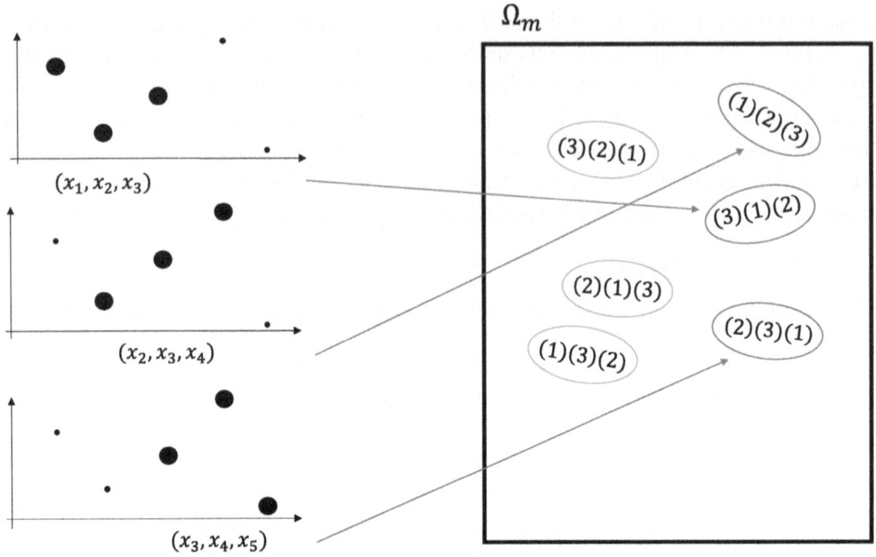

Fig. 2. Schematic diagram of the correspondence between the tuples of the time discrete signal and its permutation type in Ω_m for $m = 3$ and $\tau = 1$.

such that $p_{\tilde{j}} = 1$ and correspondingly $p_j = 0$ for all $j \neq \tilde{j}$. Then

$$H(m) = -\sum_{j=1}^{m!} p_j \log p_j = -p_{\tilde{j}} \log p_{\tilde{j}} = -1 \cdot 0 = 0.$$

The maximal value $\log(m!)$ is obtained if there is an equal distribution over all patterns such that $p_j = \frac{1}{m!}$ for all $j \in 1, ..., m!$.

$$H(m) = -\sum_{j=1}^{m!} p_j \log p_j = -m! \frac{1}{m!} \log\left(\frac{1}{m!}\right) = -(\log 1 - \log(m!)) = \log(m!)$$

Consequently, we can define a normalised PE by dividing by $\log(m!)$, according to

$$\hat{H}(m) = -\frac{1}{\log(m!)} \sum_{j=1}^{m!} p_j \log p_j. \tag{2}$$

This value allows us to compare the results for different orders as well as to define certain ranges in an application. By comparing with the extremal values, which were already checked above, one can easily verify that if a pattern has probability $p_{\tilde{j}} = 1$ again then $\hat{H}(m) = 0$ and if we have an equal probability distribution of $p_j = \frac{1}{m!}$ then $\hat{H}(m) = 1$. Hence, large values of the normalised PE are a sign of high complexity.

Now we will take a look at the various applications, where the PE is already implemented. As Yan et al. explain in [18], the value can characterise the status of rotary machines. Our field of interest is the application to the analysis of an EEG. It has already been used quite often in research, for example by Olofsen et al. in [8]. They describe the effect of anaesthetics on the human brain activity by applying the PE to the time series obtained by the signal processing of an EEG. In this work, we would like to take up this point and present with sleep stage classification another application of the PE based on the analysis of EEG data. From now on the definition of the PE from Eq. (2) is used.

5 Sleep-Stage Classification Using Permutation Entropy

For sleep analysis it is important to know the specific sleep states, i.e. to be able to "measure" and determine them. There are two main categories of sleep: rapid eye movement sleep (REM) and nonrapid eye movement sleep (NREM). The latter is further divided into sleep stages 1 to 4. Stage 1 is drowsiness, when someone is in between being awake and asleep. During a sleep cycle, stage 2 takes around 40 to 50% of the time, which is when brain waves slow down. In stage 3, δ-waves, which were mentioned in Sect. 2, begin to occur. In stage 4 these waves become dominant. Stages 2, 3 and 4 are called light sleep, deep sleep and very deep sleep, respectively, see [10]. Including the stage of being awake we have a total of 6 stages that can be classified. This goes back to the year 1968, published by Rechtschaffen and Kales [19]. In 2007, the American Academy of Sleep Medicine introduced a new classification by merging stage 3 and stage 4, which is internationally recognised and used, see [20].

Due to its importance, the analysis of sleep is an already widely discussed topic [3–7]. The first three publications apply methods using entropy to an EEG for further investigation, but only [4] uses PE. In the latter publication the method is only applied to three patients, so we wanted to take a look at a larger sample size and look at patients suffering from a sleep disorder as well.

As a motivating example for the correlation between the normalised PE and the sleep stage classification take a look at Fig. 3. There it is illustrated that the PE and sleep stages have a similar course when watched over time.

The normalised PE will now function as the underlying modelling method to analyse the sleep stages from the EEG via simulation. We used a part of the open source CAP sleep database [13] to apply this classification model. Since it was published in 2001, the sleep stage classification by Rechtschaffen and Kales was used. The database is a collection of 108 recordings of people suffering from a sleep disorder, but also including a control group, which are people without any sleep disorder. Every data set included, among other things, at least three EEG signals. We selected 18 patients. Nine of them are n1 to n9, people without any sleep-related condition, and the other nine are ins1 to ins9, patients suffering from insomnia, a sleep disorder. The data set also contained information about the sleep stages of the patients, which are given in a 30 s intervals. Because of this 30 s window we also split the EEG into 30 s windows.

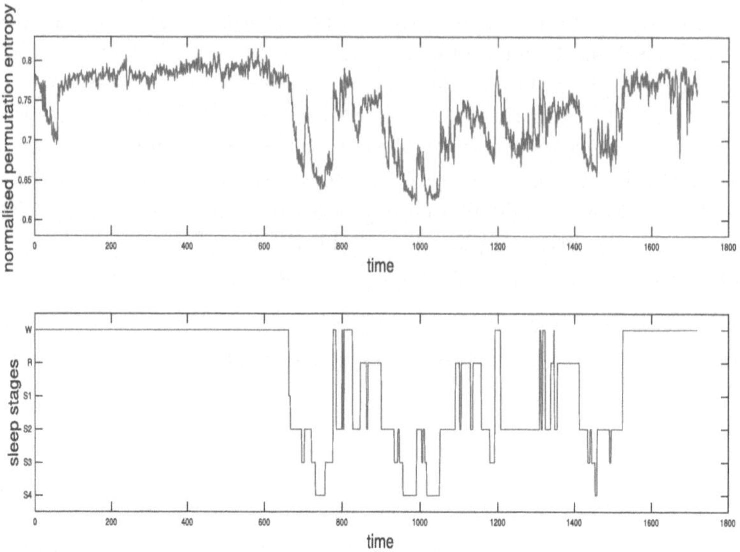

Fig. 3. Course of the normalised PE and the sleep stages over time using the example of the patient ins5.

The implementation was done in MATLAB©, Version 2022b. It is based on the code of S. Berger, published in [11,21], but adapted and extended to the specific needs of this project. The given EEG is resampled to a sampling frequency of 200 Hz and afterwards low-pass filtered at 30 Hz again. The measured data is then partitioned into windows of 6000 data points, corresponding to 30 s of signal duration. Then, the normalised PE is calculated of each of the windows, which associates an entropy value to every given sleep stage classification. We will take a closer look at the correlation between the two. For the calculation, we assigned each sleep stage a value, as displayed in Table 1. The stages are abbreviated as W for wakefulness, R for REM sleep and S1 to S4 for the sleep stages 1 to 4, respectively. Nowadays the non-REM sleep is abbreviated as N instead of S and there are only three stages of this type of sleep.

Table 1. Abbreviations and values of each sleep stage.

Sleep stage	stage 4	stage 3	stage 2	stage 1	REM sleep	awake
Abbreviation	S4	S3	S2	S1	R	W
Value	0	1	2	3	4	5

First we determined which combination of order and time delay gives a good correlation measure. A small investigation for the patients ins1-ins4 and n1-n4 showed that an order of $m = 3$ and a time delay of $\tau = 1$ give the best correlation. There were also good values for order $m = 7$ as well as for time delay of $\tau = 3$ (for orders $m = 3$ and $m = 7$) but a higher time delay led to a lower correlation value. This is already indicated in [22], where it is described that higher time-delays behave the same as lower sampling frequencies.

Each patient had multiple measurement electrodes, so we had to decide which electrode to take the EEG from in order to apply the method. The measurement of the electrodes Fp2-F4 was taken from all patients, except n6, n7 and n9, which showed a strong correlation in all of them. For the patients n6, n7 and n9 we chose the measurement of the electrodes C3-A2 because it also showed a strong correlation.

The data set provided the sleep stage classification only in a certain time range, hence we had to limit the measured EEG to the corresponding time range. There was also a difference in the recording time between the patients, which resulted in a different amount of values. The final range was between a measured time of 6 h 6 min and 14 h 20 min, corresponding to 732 and 1719 values respectively. All simulations and classifications were performed and evaluated in MATLAB©.

6 Results

First, the patients suffering from insomnia are considered and the normalised PE is computed. Then, also its correlation with the assigned values to the sleep stages, which are listed in Table 1, is calculated and shown in Table 2. In order to have a closer look at the sleep stage classification, boxplot diagrams of the normalised PE for each of the sleep stages of each patient were generated with the pre-implemented function of MATLAB©, see Fig. 4.

Table 2. Correlation between normalised PE and sleep stages for data of the patients suffering from insomnia.

Patient	ins1	ins2	ins3	ins4	ins5	ins6	ins7	ins8	ins9
Correlation	78.62%	87.43%	80.79%	82.25%	91.11%	92.36%	91.23%	70.45%	62.39%

Table 2 shows a strong correlation for all nine data sets. The only one with a correlation below 70% correlation is the one of patient ins9. One reason to explain this is due to the large overlap of the values of the sleep stages S4 and S3, as well as S1, R and W. Looking more closely at the boxplot diagrams one can see the respective ranges of the PE of the sleep stages of each patient. Single outliners are marked with a red "+". They are determined by the default settings of the boxplot function of MATLAB© where the whiskers are set to 1.5 times the interquartile range. Thus, any data point that is more than 1.5 times the interquartile range away from its quartile is an outliner.

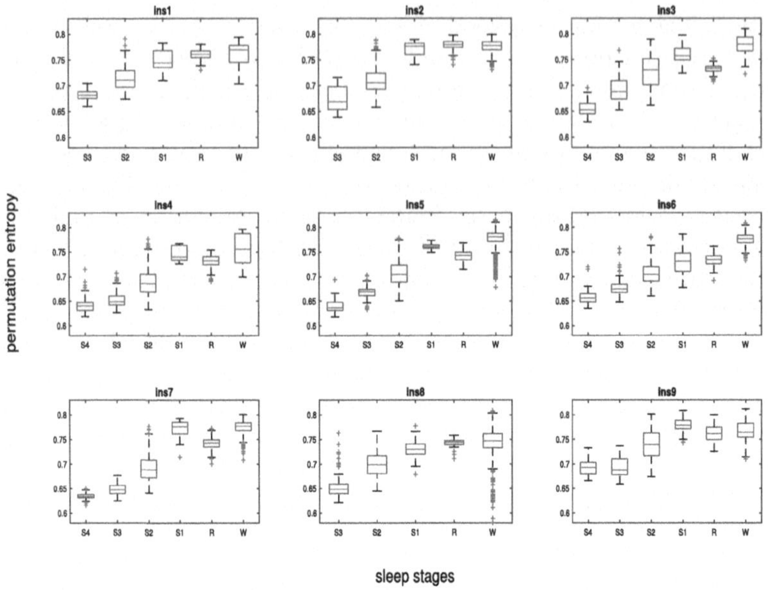

Fig. 4. Boxplots of the normalised PE over the sleep stages of each of the patients suffering from insomnia.

Furthermore, it is visible that the patients ins1, ins2 and ins8 never reach sleep stage 4. Since S3 and S4 are merged in the nowadays used sleep stage classification by the American Academy of Sleep Medicine, this fact is negligible.

For comparison, we now look at Table 3 and Fig. 5, where the correlation value and the boxplots for the patients without a sleep disorder are shown. This time we have a very strong correlation for all studied subjects.

Table 3. Correlation between normalised PE and sleep stages for data of the patients without a sleep disorder.

Patient	n1	n2	n3	n4	n5	n6	n7	n8	n9
Correlation	87.81%	80.96%	79.60%	87.86%	86.32%	81.49%	77.53%	81.21%	86.44%

Overall there is a difference between each range of the PE value, but all are between 0.58 and 0.83. The correlation coefficients are also similar in both groups, with the exception of ins9. What all studies have in common is the steady increase in the PE value as the patients become more awake, i.e. when comparing sleep stages S4-S1 and awake. This applies to both groups, people with and without a sleep disorder. This was also indicated in [4], who showed it for a sample size of three. REM sleep mostly ranges somewhere between S1 and S2.

The range of S2 is also the broadest one. This can be explained by the fact that it describes the phase of light sleep in which we spend 40 to 50% of the time of one sleep cycle, as already mentioned in Sect. 5.

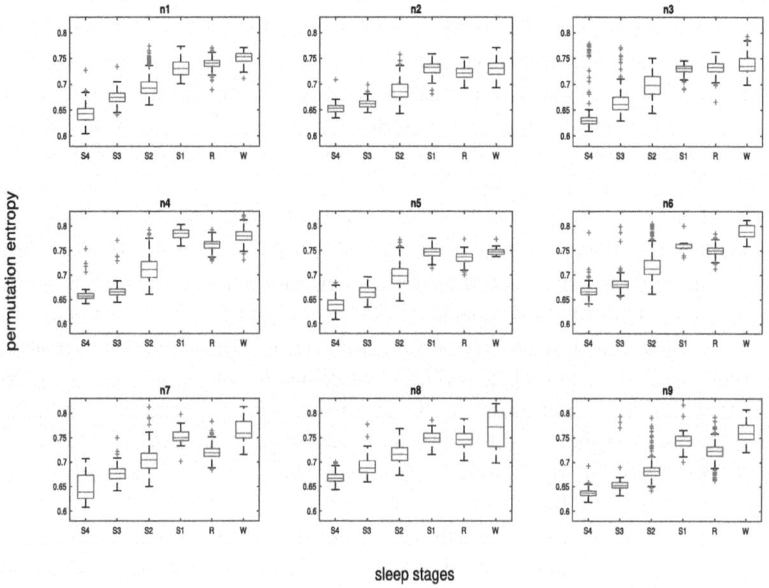

Fig. 5. Boxplots of the normalised PE over the sleep stages of each of the patients without a sleep disorder.

7 Discussion

In our work we used the PE as a modelling method for analysing brain activity. The signal processing in terms of an EEG was described and the method itself was specified in every mathematical detail. Furthermore, the importance of sleep stage classification was motivated and then also performed with the normalised PE for 18 data sets of patients suffering from the sleep disorder insomnia and people without any sleep-related sickness. Within the investigation we showed that there is a strong correlation between the PE and sleep stages for all of the patients. Also a connection between deeper sleep and lower value of the normalised PE was shown, suggesting that it is a good parameter for sleep stage classification.

We looked at one EEG signal per patient, which is a current limitation that we will be working with. Whenever possible, we chose the measured signal Fp2-F4, which had the best correlation value on average among all data sets. Here, different electrodes or the combination of more signals may give a better insight into the classification. Moreover, we applied the method with order $m = 3$ and with time delay $\tau = 1$, as it gave the best results on average. A detailed study of changing these parameters is not done in this work, but will be considered in the future. Regarding sleep disorders, we only selected patients suffering from insomnia as the most common sleep disorder [23]. Perhaps interesting results could be obtained by looking at sleep stage classification of different disorders in consultation with experts in this field.

Other variants of spectral entropy measures work well for sleep stage classification. Examples are the Walsh spectral entropy and Haar spectral entropy [3] and the fuzzy measure entropy [5]. In [6] a different approach is proposed. Deep learning methods for physiological signals are described and it is stated that this method improves more and more over time as more data becomes available. Another perturbation, the multivariate multi-scale weighted permutation entropy, is also proposed in the diagnosis of Alzheimer's disease in [24].

Sleep scoring systems have already been developed for the electrocardiogram, electroencephalogram and electrooculogram. All of these signals contain information regarding the sleep stages [7]. A combination of these signals is likely to give a better insight. This indicates that adding a new parameter, which could be as simple as peripheral pulse or blood pressure, will lead to a better classification. With several measured parameters the stages can then be found by clustering the data points.

Sleep stage classification was the first step in a larger project proposal. It is planned to apply the concept of PE to the analysis of the brain activity during anaesthesia in a surgery. This should allow the anaesthetist to control the level of a patient's loss of consciousness, depending on the PE, among other parameters. In order to find the optimal amount of anaesthesia, i.e., not to give too much or too little anaesthetics, further research in this area is important [8].

In our work, we have learnt some lessons and picked up aspects that we want to investigate further. As the results for the classification of sleep stages met our initial expectations, we intend to continue our work and apply it in the field of anaesthesia for the classification of stages of loss of consciousness during surgery.

References

1. Bandt, C., Pompe, B.: Permutation entropy: a natural complexity measure for time series. Phys. Rev. Lett. **88**(17), 174102 (2002). https://doi.org/10.1103/PhysRevLett.88.174102
2. Shannon, C.: A mathematical theory of communication. Bell Syst. Tech. J. **27**(3), 379–423 (1948). https://doi.org/10.1002/j.1538-7305.1948.tb01338.x
3. Kirsch, M., Monahan, K., Weng, J., Redline, S., Loparo, K.: Entropy-based measures for quantifying sleep-stage transition dynamics: relationship to sleep fragmentation and daytime sleepiness. IEEE Trans. Biomed. Eng. **59**(3), 787–796 (2012). https://doi.org/10.1109/TBME.2011.2179032
4. Mateos, D.M., Guevara Erra, R., Wennberg, R., Perez Velazquez, J.L.: Measures of Entropy and Complexity in altered states of consciousness. arXiv (2017). https://doi.org/10.48550/ARXIV.1701.07061

5. Zhang, Z., et al.: Efficient sleep classification based on entropy features and a support vector machine classifier. Physiol. Meas. **39**(11), 115005 (2018). https://doi.org/10.1088/1361-6579/aae943
6. Faust, O., Hagiwara, Y., Hong, T.J., Lin, O.S., Acharya, U.R.: Deep learning for healthcare applications based on physiological signals: a review. Comput. Methods Programs Biomed. **161**, 1–13 (2018). https://doi.org/10.1016/j.cmpb.2018.04.005
7. Faust, O., Razaghi, H., Barika, R., Ciaccio, E.J., Acharya, U.R.: A review of automated sleep stage scoring based on physiological signals for the new millennia. Comput. Methods Programs Biomed. **176**, 81–91 (2019). https://doi.org/10.1016/j.cmpb.2019.04.032
8. Olofsen, E., Sleigh, J., Dahan, A.: Permutation entropy of the electroencephalogram: a measure of anaesthetic drug effect. Br. J. Anaesth. **101**(6), 810–821 (2008). https://doi.org/10.1093/bja/aen290
9. Hoffmann, K.P., Krechel, U.: Geräte und Methoden der Klinischen Neurophysiologie (EEG, EMG/ENG, EP). In: Kramme, R. (eds.) Medizintechnik, pp. 129–168. Springer, Heidelberg (2007). https://doi.org/10.1007/978-3-540-34103-1_11
10. Sanei, S., Chambers, J.: EEG Signal Processing. Wiley, Hoboken (2013). https://doi.org/10.1002/9780470511923
11. Berger, S., Schneider, G., Kochs, E., Jordan, D.: Permutation entropy: too complex a measure for EEG time series? Entropy **19**(12), 692 (2017). https://doi.org/10.3390/e19120692
12. Zanin, M., Zunino, L., Rosso, O., Papo, D.: Permutation entropy and its main biomedical and econophysics applications: a review. Entropy **14**(8), 1553–1577 (2012). https://doi.org/10.3390/e14081553
13. Terzano, M.G., et al.: Atlas, rules, and recording techniques for the scoring of cyclic alternating pattern (CAP) in human sleep. Sleep Med. **6**, 537–553 (2001). https://doi.org/10.1016/s1389-9457(01)00149-6
14. Cooper, R., Osselton, J., Shaw, J.: Elektroenzephalographie: Technik und Methoden. Fischer (1974)
15. Shannon, C.: Communication in the presence of noise. Proc. IRE **37**(1), 10–21 (1949). https://doi.org/10.1109/JRPROC.1949.232969
16. Titz, S.: Welt der Physik: Entropie (2013). https://www.weltderphysik.de/thema/phaenomene-der-thermodynamik/entropie/. Accessed 7 Feb 2023
17. Popovic, M.: Researchers in an entropy wonderland: a review of the entropy concept. arXiv preprint arXiv:1711.07326 (2017). https://doi.org/10.48550/arXiv.1711.07326
18. Yan, R., Liu, Y,. Gao, R.: Permutation entropy: a nonlinear statistical measure for status characterization of rotary machines. Mech. Syst. Signal Process. **29**, 474–484 (2012). https://doi.org/10.1016/j.ymssp.2011.11.022
19. Rechtschaffen, A., Kales, A.: A manual of standardized terminology, techniques and scoring system for sleep stages of human subjects. United States Government Printing Office, Washington DC (1968)
20. Iber, C., Ancoli-Israel, S., Chesson, A.L., Quan, S.F.: The AASM Manual for the Scoring of Sleep and Associated Events: Rules, Terminology, and Technical Specifications. American Academy of Sleep Medicine, Westchester (2007)
21. Berger, S., Kravtsiv, A., Schneider, G., Jordan, D.: Teaching ordinal patterns to a computer: efficient encoding algorithms based on the lehmer code. Entropy **21**(10), 1023 (2019). https://doi.org/10.3390/e21101023
22. Popov, A., Avilov, O., Kanaykin, O.: Permutation entropy of EEG signals for different sampling rate and time lag combinations. In: 2013 Signal Processing Symposium (SPS), pp. 1–4 (2013). https://doi.org/10.1109/SPS.2013.6623608

23. Van Someren, E.J.W.: Brain mechanisms of insomnia: new perspectives on causes and consequences. Physiol. Rev. **101**(3), 995–1046 (2021). https://doi.org/10.1152/physrev.00046.2019
24. Deng, B., et al.: Multivariate multi-scale weighted permutation entropy analysis of EEG complexity for Alzheimer's disease. Cogn. Neurodyn. **11**, 217–231 (2017). https://doi.org/10.1007/s11571-016-9418-9

Production Systems

Application of Digital Twin Simulation in Deploying AGV/ARV System to Hi-Tech Industry

K. Harikrishnan[1], David Gyulai[2(✉)], and C. Niranjana[3(✉)]

[1] Western Digital Corporation, Bangalore, India
harikrishnan.kanagasabesan@wdc.com
[2] Western Digital Corporation, Amsterdam, Netherlands
david.gyulai@wdc.com
[3] Sri Venkateshwara College of Engineering, Bangalore, India
niranjana3194@gmail.com

Abstract. The application of Automated Guided vehicles (AGVs) in manufacturing environments has become globally popular to enhance process automation. A current trend is that AGVs have become more prominent and the advancement in its technologies places the solution at top of the industry with its sustainability in many complex application environments. For every single portable stage, no matter what their application is, there is an extraordinary interest to model the system in a simulation environment. Various use of simulation tools analyses the routing, scheduling and allocation of models. The application of predictive simulations related to intra-logistics in assembly systems additionally projects a significant role, which pertains to the establishment of the operation of AGVs too. In this paper a non – human intervention approach is discussed wherein AGV/ARV (Automated Robotic vehicles) are used in Surface Mount Technology (SMT) line to replace the traditional operator-based material handling. It deploys optimized utilization with high performance of AGV/ARV with a new dispatching approach. The analysis for increased future production volume of the system is provided.

Keywords: Automated Guided vehicles (AGVs) · ARV (Automated Robotic vehicles) · dispatching · simulation

1 Introduction

The use of Automated Guided Vehicles (AGV) has increasing presence in manufacturing domain. AGV's fundamental goal is the transportation of materials from one point to the next point in a production area. Prior to the execution of the AGV in a real-time environment, the routing, scheduling, dispatching, and positioning can be done in a simulated environment [1]. The manufacturing techniques uses wide range concepts like sensor system, Artificial Intelligence, Digital Twins and so on. It is seen that for improving efficiency of the system integration of all these concept helps attaining optimum level

M. Mujica Mota and P. Scala (Eds.): EUROSIM 2023, CCIS 2032, pp. 221–236, 2024.
https://doi.org/10.1007/978-3-031-68435-7_16

[2–9]. With the demand that AGVs are used by many industries for any kind of manufacturing, intralogistics simulation model is also spreading and becoming a conventional tool in industry. Among all simulation methods, digital twin modelling with discrete event simulation (DES) has high significance and considered efficient approach, as it can model complete real system in a virtual simulation [3].

Digital Twin concept became widely known in industry, due to its capabilities for real time observation, optimization, maintenance, and many other advantages shown on the improvement and evaluation of the growing hi-tech industry. Because of its hierarchy and dynamic nature, it is the backbone of many manufacturing industry [4–13].

It is noted that battery-powered AGVs (B-AGVs) are cheaper than diesel-powered AGVs (D-AGVs). Based on survey, it can be concluded that B-AGVs have simple system model with less maintenance and fuel cost. Likewise, [the noise and emission produced are also on the lower side. However, due to restricted battery limit, the B-AGV is expected to re-energize in time when the battery arrives at threshold value. The charging progression of B-AGV should be reformed to stay away from clog at electrical charging stations. With the use of B-AGV the environmental sustainability goals are higher because usage of D-AGV the pollution and variation in environmental parameters are high [5].

Utilization of smarter AGVs improves the efficiency, furthermore, it prompts layout issues and the demand of effectively assessing the system performance with different designs. As for the routing logic, it is difficult to optimize and define the best path for the AGV, considering number of rows and columns on the shop-floor grid and different number of AGV's further complicate the problem [6]. The most fundamental approach to calculate the optimal path for the AGV's in a static environment is Dijkstra's algorithm. However, there is no solution for collision avoidance in this method. Finally, it has reached to a point that discrete event simulation-based modelling approach with AGVs and ARV is best to design for complex environments [7].

Even though there is a tremendous growth in the development of digital twin technology, discrete-event simulation modelling approach has some shortfalls in decision making, for which data-driven approaches are used to enhance the solution [8–11]. In the paper, the discrete-event simulation modelling approach is combined with various dispatching approaches, like FIFO, Minimum Distance, Minimum Distance + Full load, FIFO + Minimum Distance, FIFO + Minimum Distance + Full load. With analysis of various parameters, FIFO + Minimum Distance has provided the significant result depicting that it takes the best of both FIFO and Minimum Distance. From the expense evaluation, it reached a conclusion that AGV/ARV cost and operator cost is on the higher side however this factor remains positive since there is no occurrence of shortfall of any AGV/ARV like humans [12]. Finally with the use of AGV/ARV 95% no blunder faults will occur, and no troubles will be resulting proving high efficiency. The developed model was validated by using Wilcoxon-test and proved to reach the desired accuracy. The data taken for validation can be Continuous, Skewed distribution and Random sample.

2 Proposed Strategy

A semiconductor factory (SMT line) with the production capacity at 100% is taken for analysis and discussion where logistic are implemented by operators [10]. The estimated plan is to increase the capacity of the production to 175% by 2025. To address this rapid increase in production capacity with future outlook, the existing system [13] of operator-based Material handling is replaced with the sophisticated AGV/ARV based automated system. In this way digital twin model is employed to evaluate the best way of deploying AGV/ARV system and determine the required size to handle different volume of production capacity considering the future needs. In this facility, majorly there are two types of supply system.

1. Magazine supply.
2. Tray Supply.

2.1 Magazine Supply

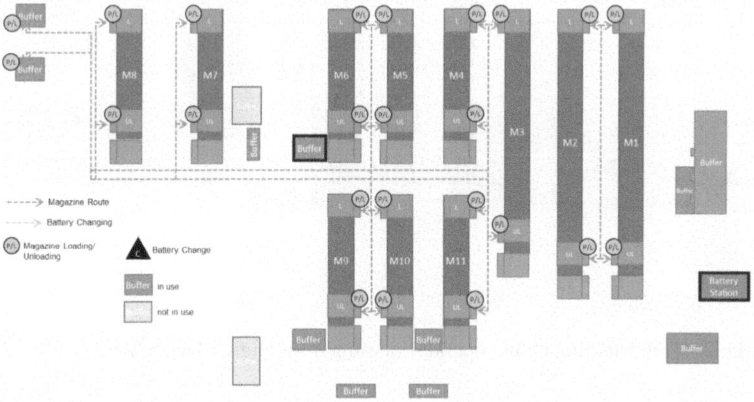

Fig. 1. Current operator handling route of Magazine (assumed for Digital Twin study)

The current operator handling route of magazine within the facility is shown in Fig. 1. Each machine has a load and unload point. At the entrance of the machine, magazine is loaded and at the exit empty magazine is unloaded and moved back to the entrance of the machine, to be loaded for the next batch of product. This cycle corresponds to the continuous machine operating cycle. A battery station indicated in the layout is the proposed location where AGV/ARV goes for change of battery in the layout. Buffer indicates where magazines will be changed if there is a product change over. Overall, we have assumed there are 11 machines within which magazine loading and unloading will be handled.

2.2 Tray Supply

The handling of tray by the operator in current facilities is shown in Fig. 2. The grey line indicates filled tray pick up from exit of the machine to buffer (temporary storage area). Purple line indicates filled tray pick up to the buffer, if buffer is occupied with an empty trolley which can hold these trays. It also demonstrates the case when buffer (indicated in white colour) is occupied, it will be moved to buffer. Red line does the similar process as purple line. Pink line indicates empty tray supply from empty tray buffer (indicated in light blue) to both Buffer (temporary storage for empty tray). Blue line signifies empty tray supply to all the machines in the layout. Green line indicates movement of trolley between buffer to other process. Yellow line proposes the route for change of battery if required for AGV/ARV. Board tray buffer depicts which will supply the tray cover required for covering the filled trays.

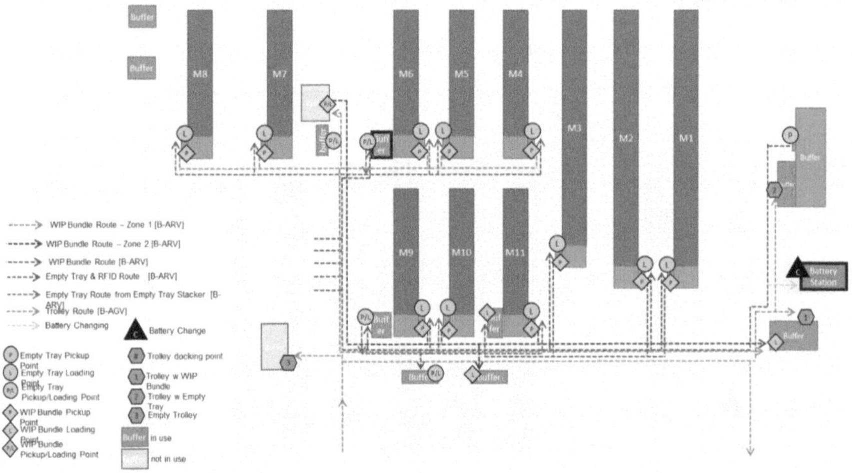

Fig. 2. Tray handling by the operator (assumed for Digital Twin study)

To choose the optimal method of deploying ARV/AGVs in the system, we have proposed five different strategies like FIFO, Minimum distance, Minimum distance + full load, FIFO + minimum distance and FIFO + minimum distance + Full load and figured out the best among the same.

2.3 FIFO Logic

According to this logic as shown Fig. 3.any task that is created first will be assigned to first available AGV/ARV. Task from all the machines will be listed in the central task list, as shown in the Table 1 and Table 2 helps in showing the availability of AGV/ARV and its current operational status.

Table 1. Central task list

Task_name	Task_type	From_station	From_Port	Destination_station	Destination_port	Assigned_ARV	Creation_time
ET_Drop	Main_task	*.Models.XXX_Line.Buffer		*.Models.XXX_Line.ET_Drop	L1		41:51.3
ET_Drop	Main_task	*.Models.XXX_Line.Buffer		*.Models.XXX_Line.ET_Drop	L2		43:22.6
FT_pick	Main_task	*.Models.XXX_Line.FT_Pick	L5	*.Models.XXX_Line.LB_Buffer2			44:26.6
ET_Drop	Main_task	*.Models.XXX_Line.Buffer		*.Models.XXX_Line.ET_Drop	L2		44:52.0
ET_Drop	Main_task	*.Models.XXX_Line.Buffer		*.Models.XXX_Line.ET_Drop	L1		46:17.3
FT_pick	Main_task	*.Models.XXX_Line.FT_Pick	L4	*.Models.XXX_Line.LB_Buffer2			47:00.7
ET_Drop	Main_task	*.Models.XXX_Line.Buffer		*.Models.XXX_Line.ET_Drop	L2		47:00.7
ET_Drop	Main_task	*.Models.XXX_Line.Buffer		*.Models.XXX_Line.ET_Drop	L2		47:12.0
FT_pick	Main_task	*.Models.XXX_Line.FT_Pick	L4	*.Models.XXX_Line.LB_Buffer1			47:53.3
FT_pick	Main_task	*.Models.XXX_Line.FT_Pick	L5	*.Models.XXX_Line.LB_Buffer2			48:05.1
ET_Drop	Main_task	*.Models.XXX_Line.Buffer		*.Models.XXX_Line.ET_Drop	L1		48:12.0

Table 2. Availability of AGV/ARV

Type	Name	CurrentLoad	Maxload	Agv_status	LocalTaskList	Working	Failure
Main_task	XXX1	5	5	active	Task assigned table	TRUE	FALSE
Main_task	XXX2	3	5	active	Task assigned table	TRUE	FALSE
Main_task	XXX3	1	5	active	Task assigned table	TRUE	FALSE
Main_task	XXX4	0	5	active	Task assigned table	TRUE	FALSE
Main_task	XXX5	3	5	active	Task assigned table	TRUE	FALSE
Empty_carry	XXX6	3	5	active	Task assigned table	TRUE	FALSE
Empty_carry	XXX7	5	5	active	Task assigned table	TRUE	FALSE

Table inside table

Task_name	Task_type	From_station	From_Port	Destination_station	Destination_port	Completed	Distance
ET_Drop	Main_task	*.Models.XXX_Line.Buffer		*.Models.XXX_Line.ET_Drop	L2		
FT_pick	Main_task	*.Models.XXX_Line.FT_Pick	L4	*.Models.XXX_Line.LB_Buffer1			

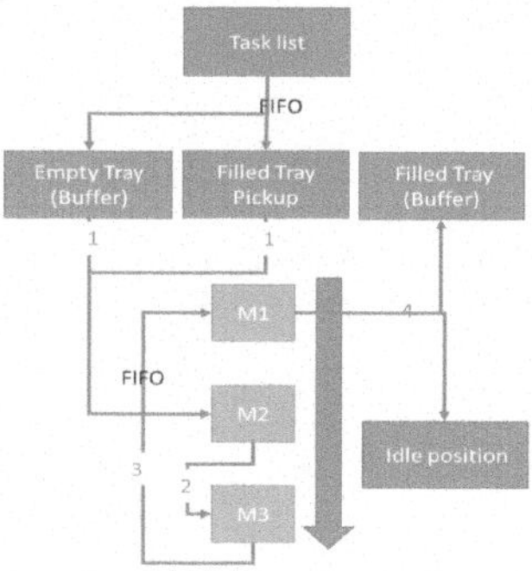

Fig. 3. First in First Out

If the first task listed in the central task list is an empty tray supply, then logic will search for first available AGV/ARV and assign to it followed by which it will check if any other empty pick-up task is available. If available, the AGV/ARV will be loaded to it carrying capacity. If not, then AGV/ARV will revert to buffer and lift the required trays and moves towards destinations as per the list of supply. As the supply is based on FIFO then the point to highlight is the AGV/ARV supply can also happen as seen in Fig. 3. On completion, AGV/ARV will return to idle position. If the first task listed in central task list is filled tray pickup, the logic will work similar as illustrated. Only difference to be noticed is it will go to different machine as per FIFO and carry the filled tray to buffer.

2.4 Minimum Distance Logic

Minimum distance logic works on the principle where AGV/ARV takes the shortest distance to travel such that the allocated task is performed. In this logic the best AGV/ARV is picked to finish the task. The availability of AGV/ARV and central task list is same for all the process shown as Table 1 and Table 2. As per the logic shown in Fig. 4, the available AGV/ARV is selected and based on task list it figures out which task can be performed with minimum distance (logically finding out travel distance of each task in the task list and picking minimum out of it). Upon task selection, the similar task from the task list is figured out (again listed based on minimum distance) and added to the task list of AGV/ARV. If the task isn't available AGV/ARV is loaded to it carrying capacity and the task which is assigned already will be performed. Incase if it's an empty tray carry, it will move to the buffer, supply it to the machine and revert to idle position. If it's a filled tray carry, it will pick up from each position and supply it to the buffer.

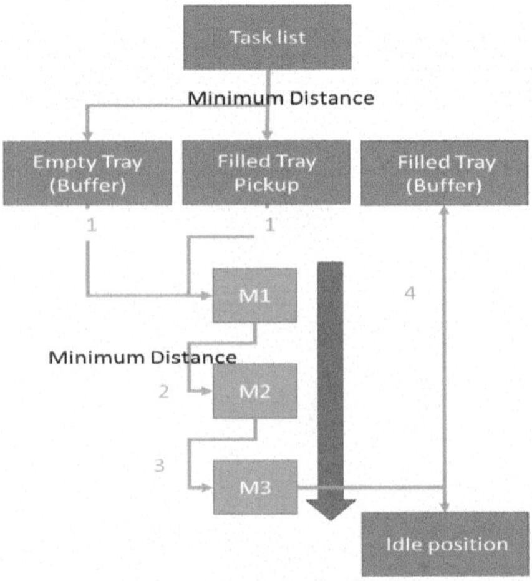

Fig. 4. Minimum Distance Logic

2.5 Minimum Distance Logic + Full Load

This logic shown in Fig. 5 is almost identical to minimum distance logic. Only contrast is that in minimum distance logic + full load, the AGV/ARV is maximum loaded to its fullest carry capacity. Consider if an AGV/ARV can carry out 5 tasks simultaneously, Until the AGV/ARV is assigned with 5 tasks hold on period is continued and it will be dispatched to perform the assigned tasks.

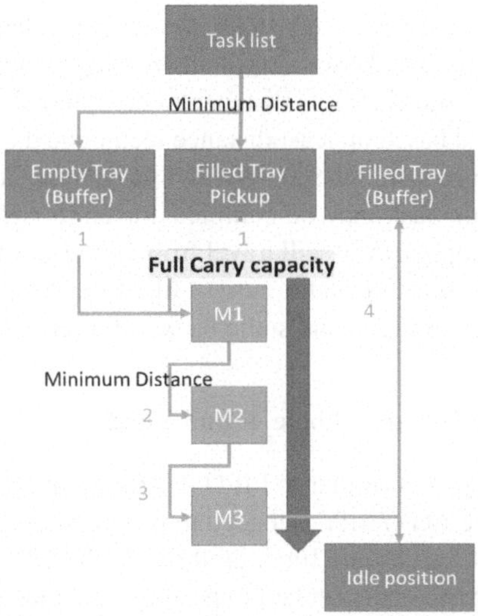

Fig. 5. Minimum Distance Logic + Full Load

2.6 FIFO + Minimum Distance

In this logic as shown in Fig. 6, the benefit of both FIFO and minimum distance logic is utilized. The first task in central task list (Table 1) will be given high priority through FIFO method and for that particular task the availability of AGV/ARV is chosen from the availability table (Table 2). Based on availability the minimum distance that any AGV/ARV takes to perform the task is studied and selected.

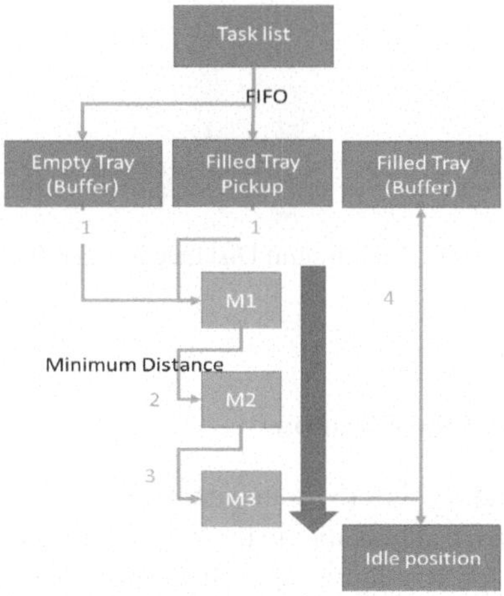

Fig. 6. FIFO + Minimum Distance

After filtering out the available AGV/ARV, we will calculate the distance taken by each against the picked-up task. Based on calculation high priority will be assigned to the AGV/ARV which has travelled less distance to complete the task. The similar task is taken from the table and based on least distance technique the AGV/ARV is assigned with task. The process repeats until the AGV/ARV is filled with its capacity. In few cases it will be observed that the task may be completed even though the AGV/ARV is not fully loaded. In case of empty tray, it will travel from buffer to supply the machine. Post supplying it reverts to idle position and in case of filled tray pick up, it will pick up from each machine as indicated in Fig. 6 and settle back to buffer.

2.7 FIFO + Minimum Distance Logic + Full Load

This logic as shown in Fig. 7 is similar to FIFO + Minimum Distance logic. The found contrast in this logic is the AGV/ARV is loaded to its maximal capacity. Considering an AGV/ARV can carry out 5 tasks at a time, until the AGV/ARV is assigned 5 tasks the wait time is kept on hold and is dispatched to perform assigned tasks.

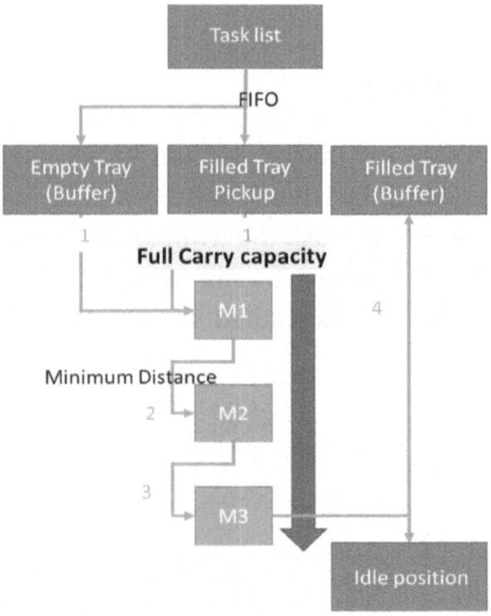

Fig. 7. FIFO + Minimum Distance logic + Full load

2.8 AGV/ARV Route and Specifications

In current system, 12 Operators/shift (5 for magazine, 7 for Tray and total 2shifts/day) are carrying out the task. To arrive at best among the five different strategies, same number of AGV/ARV is considered for this digital twin simulation. Analysing the current layout, to avoid high traffic density and collision between the AGV/ARV following route is planned as shown Fig. 8.

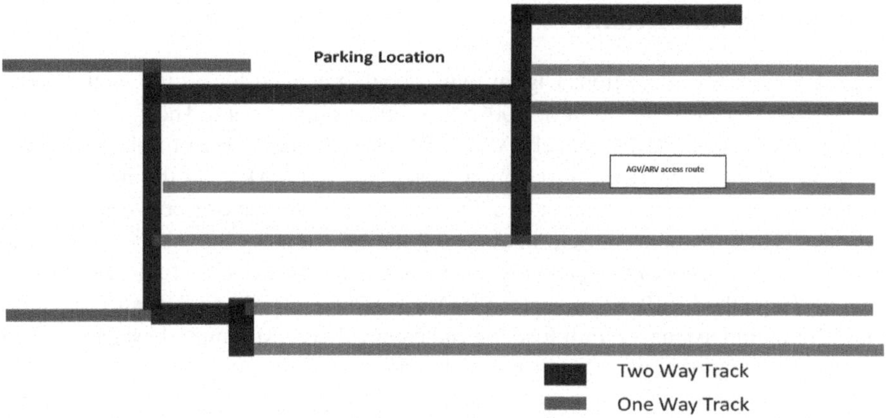

Fig. 8. AGV/ARV Route Layout (assumed for digital twin study)

Route where we anticipate higher movement of AGV/ARV, proposed to have a two-way track with the width of 3 m (Considering AGV/ARV size and parallel passing of the same) and one way track will be proposed to access the machine for supply In and supply out with the width of 1.5 m (Only 1 AGV/ARV can access the route). Table 3 indicates the AGV/ARV specification planned for the Digital Twin study. To be noted that the data mentioned below is based on AGV/ARV general specifications.

Table 3. AGV/ARV Specification

Parameter	Value
AGV/ARV speed	0.3–0.6 m/s
Availability	~95%
Battery Capacity	72 Ah
Reserve	29 Ah
Carry Capacity	1–5

As shown in Table 3, carry capacity of AGV/ARV changes depending on the type of task it performs. If it's a magazine carry, then maximum of 1 magazine can be carried. If it's a tray carry, then maximum of 5 trays can be carried at a time. If the AGV/ARV reaches the reserve status, then it will move to the battery station shown in Fig. 8 and swapping of battery is carried out. The Speed of AGV/ARV varies depending on some conditions like whether it is near the machine or turning around a track or travelling straight in the two-way track or one way track.

3 Experimental Results

As explained in above section, Digital twin model is created for all the method considering the current production at 100% theoretical requirement. The following key performance indicator (KPI) namely AGV/ARV utilization, number of tasks performed by AGV/ARV to support the production, impact of AGV/ARV on the machine utilization and estimated production is achieved or failed to achieve between different strategies. These parameters help in choosing the best strategy suited for the system. Thus AGV/ARV is deployed optimally and helps scaling the same for future production ramp up. From the Fig. 9, S1 indicates FIFO, S2 indicates Minimum Distance, S3 indicates Minimum Distance + Full load, S4 indicates FIFO + Minimum Distance and S5 indicates FIFO + minimum Distance + Full load.

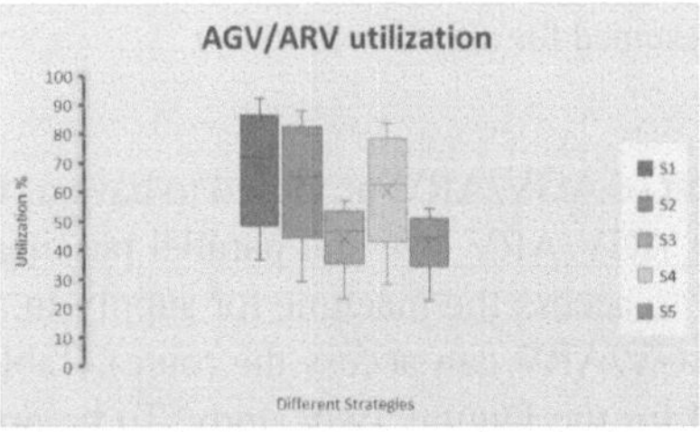

Fig. 9. AGV/ARV Utilization

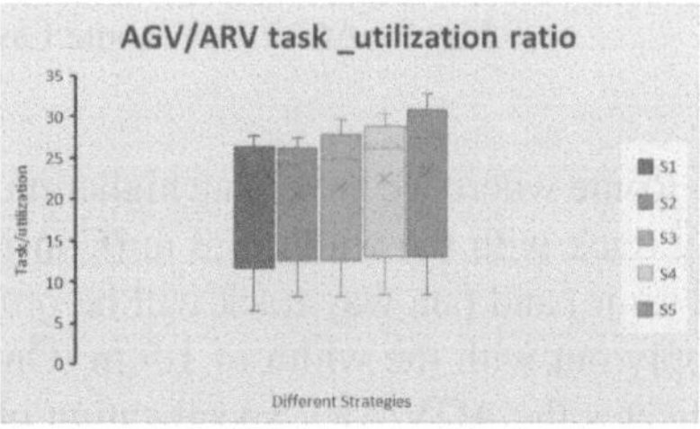

Fig. 10. AGV/ARV task_utilization ratio

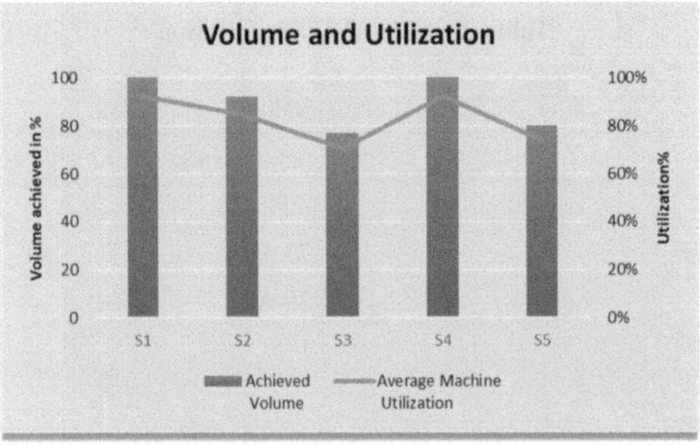

Fig. 11. Volume Analysis

The Goal is to pick a logic that does a greater number of tasks with less utilization (task number/utilization ratio). Concluding based on this, the work of AGV/ARV that produces a high machine utilization and finally production target of 100% is achieved. AGV/ARV utilization graph as seen in Fig. 9 signifies that Range of Utilization of S1 is high compared to that of the other strategies. S5 is the least utilized method. But looking at volume as seen in Fig. 11, S5 has did less compared to S1. So S5 strategy isn't ideal to deploy in the factory. Next to S5 on the least utilization side is S3. S3 has done less volume compared to S1. So, it can be said that S3 also is not a good logic for deployment. Now taking S4 for comparison, it has done same number of tasks as that of S1, and its task/utilization ratio is better than S1 as seen in Fig. 10. S2 is neither less utilized nor task /utilization ratio is high compared to that of any of the strategies. Because of this S2 is not considered further as a valid option. Now looking at the volume obtained and machine utilization, Both S1 and S4 has achieved the planned production and machine utilization on both the cases which is at an average of 92%. So, looking at the overall graphs it is clearly depicted that both S1 and S4 has performed better than other strategies and S4 is the better scenario, considering its task/utilization ratio with S1. With the deep analysis it is concluded that S4 (FIFO + minimum distance) is chosen to be best strategy.

3.1 Model Validation

After the chosen strategy, the next step is to validate the corelation of the model with the actual scenario (baseline). Based on the research, it is advisable to deploy a model which has an accuracy above 95% with the actual data, to best utilize the model for what-if analysis and resource optimization. Table 4 indicates actual data collected from the production facility. It also provides the data collected from each machine when an operator performs a task. (Loading and unloading of magazine or tray).

Table 4. Actual data from production facility

c_TaskId	c_TaskCount	c_TaskNo	c_WorkLoc	c_WorkPort	c_TaskCode	c_TaskCode2	c_CarrierID	c_Priority	d_DueTime	d_SendTime	d_StartTime	d_EndTime	d_ActionTime	d_DoneTime
2.02E+24	2	1	XXXX	E	44171266	131597	aaa	1	2.02E+13	57:17.4	00:49.4	03:59.8	03:57.7	03:59.8
2.02206E+24	2	2	XXXX	A	32571392	65804	bbb	1	2.02E+13	57:16.4	04:00.4	06:59.3	06:57.0	06:59.3
2.02206E+24	2	1	XXXX	10	212860928	74582544	ccc	1	2.02E+13	56:50.3	57:49.0	57:50.5	NULL	57:50.5
2.02206E+24	2	1	XXXX	E	30539778	131596	ddd	1	2.02E+13	49:15.4	49:15.6	51:23.2	51:20.8	51:23.2
2.02206E+24	2	2	XXXX	A	42008577	65805	aaa	1	2.02E+13	49:16.4	51:23.5	54:26.6	54:24.8	54:26.6
2.02206E+24	2	1	XXXX	E	45219842	131597	bbb	1	2.02E+13	45:01.9	52:51.6	55:30.8	55:28.4	55:30.8
2.02206E+24	2	2	XXXX	A	33619969	65804	ccc	1	2.02E+13	45:02.9	55:30.8	58:04.1	58:02.6	58:04.1
2.02206E+24	2	2	XXXX	A	45154305	65805	ddd	1	2.02E+13	45:00.8	58:19.8	00:49.1	00:47.2	00:49.1
2.02206E+24	2	1	XXXX	E	33685504	131596	aaa	1	2.02E+13	45:04.0	58:04.5	58:19.7	58:17.4	58:19.7
2.02206E+24	2	1	XXXX	E	48234496	131776	bbb	1	2.02E+13	43:52.0	48:27.7	48:43.2	48:41.0	48:43.2
2.02206E+24	2	2	XXXX	A	46137344	66000	ccc	1	2.02E+13	43:50.9	48:43.6	51:46.1	51:43.9	51:46.1

Table 5. Data from Digital Twin Model

Task_name	Task_type	From_station	From_Port	Destination_station	Destination_port	Creation_time	AGV_starttime	AGV_Endtime	Load_unload_starttime	Load_unload_Endtime
ET_Drop	Main_task	*.Models.XXX_Line.BUFFER2				1373.696902	1500	1988.144331	1913.144331	1988.144331
ET_Drop	Main_task			*.Models.XXX_Line.ET_Drop	L1	1988.144331	1988.144331	2081.144331	2018.144331	2081.144331
ET_Drop	Main_task	*.Models.XXX_Line.BUFFER2				1632.475354	2100	2188.810997	2113.810997	2188.810997
ET_Drop	Main_task			*.Models.XXX_Line.ET_Drop	L1	2188.810997	2188.810997	2281.810997	2218.810997	2281.810997
ET_Drop	Main_task	*.Models.XXX_Line.BUFFER2				1914.974521	2100	2188.810997	2113.810997	2188.810997
ET_Drop	Main_task			*.Models.XXX_Line.ET_Drop	L1	2188.810997	2188.810997	2449.144331	2386.144331	2449.144331
ET_Drop	Main_task	*.Models.XXX_Line.BUFFER2				2603.872979	3000	3169.627764	3094.627764	3169.627764
ET_Drop	Main_task			*.Models.XXX_Line.ET_Drop	L1	3169.627764	3169.627764	3279.29443	3216.29443	3279.29443
FT_pick	Main_task	*.Models.XXX_Line.FT_Pick	L4			2841.04945	3300	3477.29443	3386.29443	3477.29443
FT_pick	Main_task			*.Models.XXX_Line.LB_Buffer2		3477.29443	3477.29443	3852.627764	3786.627764	3852.627764
FT_pick	Main_task	*.Models.XXX_Line.FT_Pick	L4			2884.116563	3300	3732.961097	3641.961097	3732.961097
FT_pick	Main_task			*.Models.XXX_Line.LB_Buffer2		3732.961097	3732.961097	3852.627764	3786.627764	3852.627764
ET_Drop	Main_task	*.Models.XXX_Line.BUFFER1				3613.120318	3900	4077.423629	4002.423629	4077.423629
ET_Drop	Main_task			*.Models.XXX_Line.ET_Drop	L1	4077.423629	4077.423629	4293.833333	4230.833333	4293.833333

After collecting data from Digital Twin model, the Next step is to learn if the data is distributed normally or not. The normality calculation is done by using QQ plot and Shapiro test where QQ plot confirms visually how the data is spread and Shapiro test statistically validate the same (Table 5).

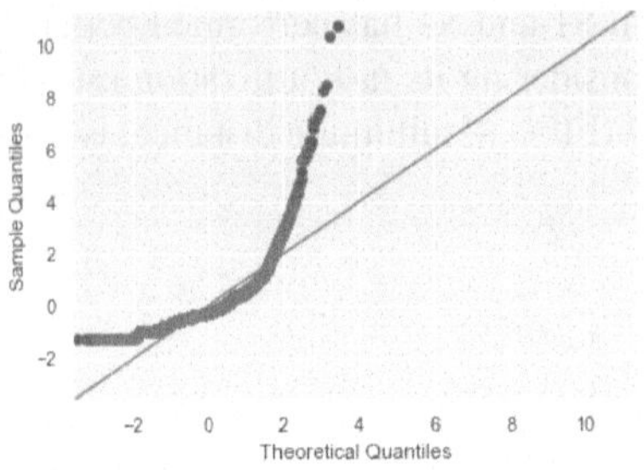

Shapiro Result (statistic=0.72, p value=0.0)

Fig. 12. QQ plot – Actual Data

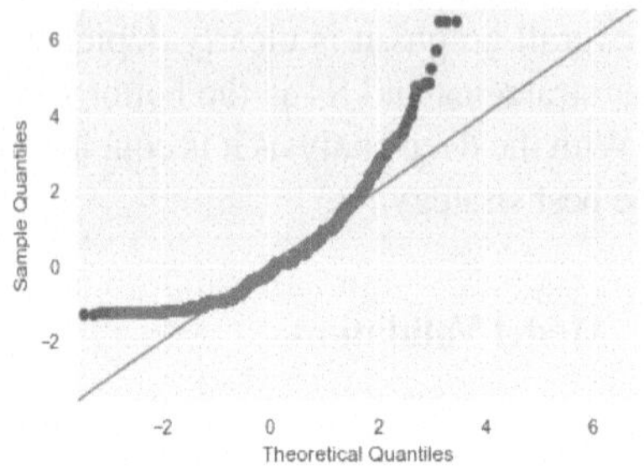

Shapiro Result (statistic=0.88, p value=0.0)

Fig. 13. QQ plot – Model Data

From the QQ plot as seen in Fig. 12 and Shapiro test as shown in Fig. 13 for normality, it is clearly derived that both model data and actual data doesn't follow normality. Wilcoxon t-test is preferred in this case to statistically test both the data. Data validation is carried out graphically as well as using Wilcoxon t-test.

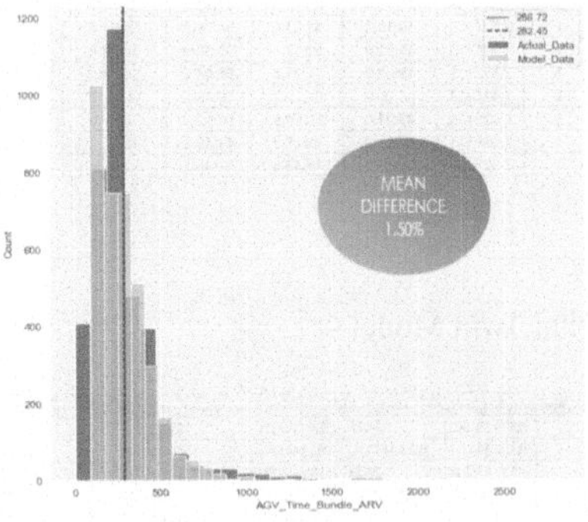

Fig. 14. Histogram

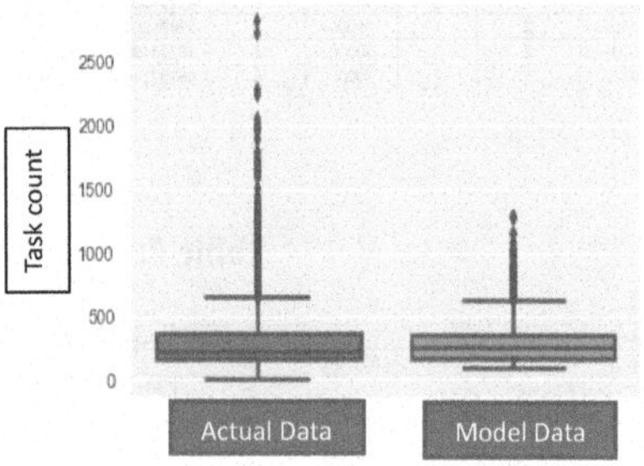

Fig. 15. Box Plot

Overall Sample size of the data collected is for couple of days. In Fig. 14, Cyan indicates data spread of model data and dark green signifies data spread of actual data. From histogram, the overlap of model data over actual looks good with mean Difference of only 1.5%. (Mean of actual data = 286.75 s per task, Mean of model data = 282.45 s per task). In box plot in Fig. 15, dark blue indicates actual data and brown indicates model data. It is clear that interquartile range (IQR) of model data is within the interquartile range (IQR) of actual data. Standard deviation of actual data is 235.59 s and model data is 153.19 s, indicating that though model is accounting for variability of the actual line, some of the outliners aren't accounted as it is difficult to mimic the one-time real issue happening in the production line because of man, machine or material. Overall outcome from histogram and box plot is that the model data is close to actual data.

Wilcoxon Result (statistic = 1850073.0, p value = 0.32)

Wilcoxon t-test p value > 0.05 confirms that there is no significant difference between model and actual data. Overall, model data is closely related to Actual data, confirming the accuracy of Digital twin model for optimization and what-if analysis.

3.2 AGV/ARV Results at 100% Theoretical Requirement

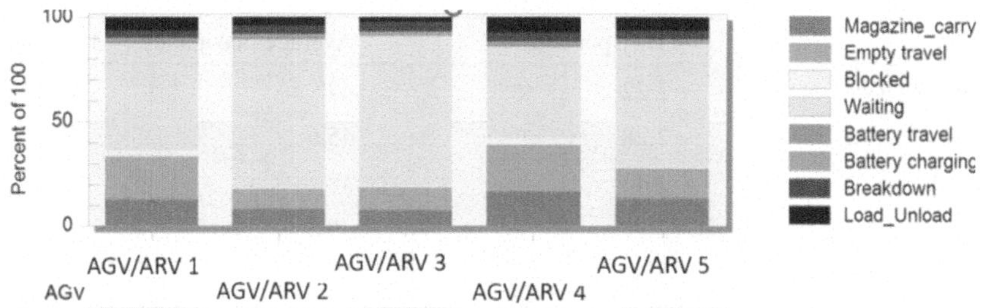

AGV	AGV/ARV 1	AGV/ARV 2	AGV/ARV 3	AGV/ARV 4	AGV/ARV 5
Magazine_carry	12.63%	7.93%	7.12%	16.30%	13.16%
Empty travel	20.44%	9.66%	11.25%	22.48%	14.37%
Blocked	3.08%	0.28%	1.04%	3.18%	0.73%
Waiting	50.87%	71.31%	71.59%	43.57%	58.79%
Battery travel	1.54%	1.37%	0.79%	1.45%	1.01%
Battery charging	1.40%	1.16%	1.16%	1.44%	1.30%
Breakdown	3.50%	4.26%	4.77%	4.34%	4.18%
Load_Unload	6.55%	4.04%	2.28%	7.24%	6.47%

Fig. 16. Magazine AGV/ARV Utilization

Average magazine AGV/ARV utilization is 40.8%as shown in Fig. 16, indicates that at 100% theoretical requirement, it is underutilized and can perform more task if there is an increase in the production capacity. Also, AGV/ARV can be further optimized for the same capacity.

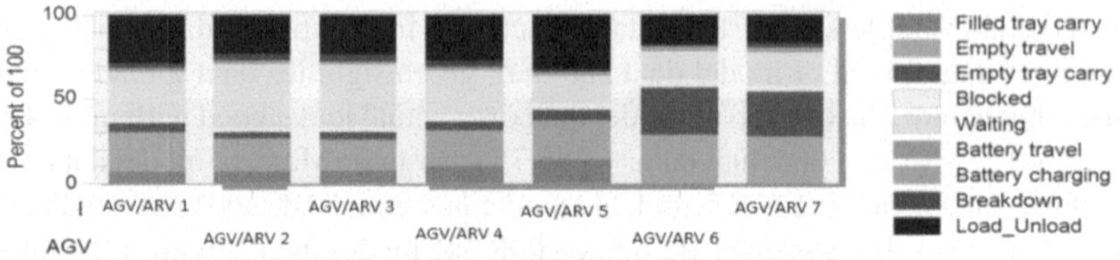

Filled tray carry	6.93%	6.98%	7.30%	10.04%	13.85%	0.00%	0.00%
Empty travel	23.29%	19.63%	18.56%	21.11%	23.38%	28.56%	27.62%
Empty tray carry	5.50%	3.80%	4.44%	5.47%	6.18%	27.91%	26.31%
Blocked	2.97%	2.28%	3.10%	4.08%	3.79%	4.68%	3.54%
Waiting	27.51%	38.03%	36.78%	26.17%	16.34%	17.15%	19.94%
Battery travel	0.56%	0.67%	0.74%	0.76%	0.91%	1.05%	1.27%
Battery charging	1.40%	1.23%	1.29%	1.45%	1.52%	1.58%	1.58%
Breakdown	3.26%	4.20%	4.27%	3.72%	1.66%	1.39%	3.21%
Load_Unload	28.58%	23.19%	23.52%	27.20%	32.37%	17.69%	16.54%

Fig. 17. Tray AGV/ARV Utilization

Average task AGV/ARV utilization is 74% as shown in Fig. 17, indicates that at 100% theoretical requirement, it is moderately utilized. But in case of task carry AGV/ARV, there is a need to check how many trays are carried out during each trip, as in contrast to magazine carry which does only 1 magazine for a trip, 5 trays can be carried out.

Table 6. Evaluation of tray carry by each AGV/ARV

Name	Tray per trip	Average occupancy (%)
AGV/ARV1	2.22	44
AGV/ARV2	2.28	45.6
AGV/ARV3	2.43	48
AGV/ARV4	2.46	49.3
AGV/ARV5	2.6	52
AGV/ARV6	3.69	73.8
AGV/ARV7	3.38	67.8
	2.72	**54.36**

From the analysis shown in Table 6, it is clear that, though utilization is at 72%, average tray carry per trip is at 2.72 (54%), indicating that more trays can be carried out for the same utilization if there is increase in capacity. Also, AGV/ARV can be further optimized for the same capacity.

3.3 AGV/ARV Number vs Volume Analysis

For volume Vs optimized AGV/ARV number, 85% utilization is considered as the thumb rule value for the AGV/ARV utilization. Reason being is that 5% is considered for any maintenance or breakdown in the unit and another 10% is accounted for any external variabilities, which are not accounted for in the digital twin model. so, optimized number is arrived for each volume when the AGV/ARV average utilization reaches above 85%.

Table 7. Volume Production

Volume	AGV/ARV at 85% Utilized	
	Mag AGV/ARV	Tray AGV/ARV
100%	3	5
125%	4	7
150%	4	8
175%	5	8

From Table 7, it is clear that for current production volume 8 AGV/ARV would be sufficient to support the system (4 can be reduced from initial plan of 12 AGV/ARV). If the volume ramp-up to 125%, we would be requiring 11 AGV/ARV. For 150%, it is 12 units and for 175%, it is 13 units.

4 Conclusion

As per the results, discrete event simulation model contributes significantly to improvement in the performance of a complex production system where the existing material handling system is dependent on the operator. It proposes strategy and simulation procedure for AGV/ARV allocation in the existing system. The greatest benefit is that various strategies are tried out without impacting the existing manufacturing facilities and specific problem was solved arriving at the optimum number of AGV/ARV, considering the critical parameters namely speed of the unit, loading capacity, its ideal route and failure data. The effectiveness of the proposed strategy is shown by achieving maximum output with the minimum economic input. Although such complex strategy can be implemented with ease in digital twin model, real environment requires efficient control system to mimic same performance and efficiency in system. Finally, notable outcome of this paper as follows:

1. We arrived at the strategy that solved the problem of unbalanced and unreasonable allocation of task to the AGV/ARV and maximized the task output with less utilization.
2. Existing material handling system of using 24 operators (12/Shift) can be efficiently handled by 8 AGV/ARV at 100% volume (Current production condition of the system)
3. Volume vs AGV/ARV gives a clear indication of how many AGV/ARV to be planned at different production ramp up (125%–175%)

As a result of successful implementation of digital twin method in the present use case, similar approach has be planned to increase system performance in other parts of the facilities.Furthermore, as a part of the future look, the author plans to conduct more detail study by linking the model with the real line production system (on implementation of AGV/ARV), thereby able to understand dynamic system behaviour and finding out efficient feedback for the system. In addition to the proposed distance-based dispatching approach, other solutions involving online scheduling and optimization are sought by using constrained programming and heuristics-based approach.

References

1. Skapinyecz, R., Landschützer, C.: Application of freely moving agvs in a discrete event simulation environment: a first model. Adv. Logist. Syst.-Theory Pract. **16**(2), 5–23 (2022)
2. Begout, P., et al.: Augmented reality authoring of digital twins: design, implementation and evaluation in an industry 4.0 context. Front. Virtual Reality **3**, 918685 (2022)
3. Jungmann, M., et al.: Real-time activity duration extraction of crane works for data-driven discrete event simulation. In: 2022 Winter Simulation Conference (WSC). IEEE (2022)
4. Borucka, A., Guzanek, P.: Application of the digital twin concept in assessing the readiness of production systems (2022)
5. Hu, H., Yang, X., Jin, J.: Battery-Powered Automated Guided Vehicles Scheduling Problem in Automated Container Terminals for Minimizing Energy Consumption. Available at SSRN 4229476
6. Singh, M., et al.: Digital twin: origin to future. Appl. Syst. Innov. **4**(2), 36 (2021)
7. Rožanec, J.M., et al.: Automotive OEM demand forecasting: a comparative study of forecasting algorithms and strategies. Appl. Sci. **11**(15), 6787 (2021)
8. Melesse, T.Y., Di Pasquale, V., Riemma, S.: Digital Twin models in industrial operations: state-of-the-art and future research directions. IET Collab. Intell. Manuf. **3**(1), 37–47 (2021)
9. Maksymenko, K., et al.: Towards the myoelectric digital twin: ultra fast and realistic modelling for deep learning. bioRxiv (2021)
10. Rasheed, A., San, O., Kvamsdal, T.: Digital twin: values, challenges and enablers from a modeling perspective. IEEE Access **8**, 21980–22012 (2020)
11. Melesse, T.Y., Di Pasquale, V., Riemma, S.: Digital twin models in industrial operations: a systematic literature review. Procedia Manuf. **42**, 267–272 (2020)
12. Barricelli, B.R., et al.: Human digital twin for fitness management. IEEE Access **8**, 26637–26664 (2020)
13. Tao, F., et al.: Digital twin-driven product design framework. Int. J. Prod. Res. **57**(12), 3935–3953 (2019)
14. Testi, D.: A statistical research on the impact of lean and industry 4.0 on companies in northern Italy (2019)
15. Viharos, A.B., Németh, I.: Simulation and scheduling of AGV based robotic assembly systems. IFAC-PapersOnLine **51**(11), 1415–1420 (2018)

Reward Calculation in Real-Time Scheduling Based on Simulation and Q-learning

Shufang Xie(✉), Tao Zhang, and Oliver Rose

Universität der Bundeswehr München, 85577 Neubiberg, Germany
shufang.xie@unibw.de

Abstract. The precondition for reinforcement learning to solve real-time scheduling problems is that the defined reward function must guarantee that the objective of the scheduling is achieved while the cumulated reward is maximized in the algorithm. This requires that the reward function must directly derive from the objective. Since jobs are unknown in real-time scheduling, an overall objective is not able to know before the concerned time horizon ends. As a substitute, the whole horizon is divided into several small periods. A period-wide objective is calculated after each period. It is considered to be the collective reward of the decisions in the period. Then the collective reward is distributed to every decision in this period. The value function will be updated according to these individual rewards. Based on this idea, five methods are proposed to distribute the reward and evaluated by experiments in a single-machine model.

Keywords: Real-time Scheduling · Simulation · Q-learning · Collective Reward · Reward distribution

1 Introduction

On shop floors, once a machine becomes idle and meanwhile, more than one job is waiting in the queue, operators need to select one job to be processed first so that a long-term objective (like minimizing average cycle times) can be achieved. We call this selection real-time scheduling (shown in Fig. 1). The operators usually follow some predefined decision rules, like FIFO (First In First Out), SPT (Shortest Process Time), CR (critical ratio), and so on, to select jobs.

Usually, several candidate rules are given in the predefining process, and simulation is often applied for evaluating and selecting the rules from the candidates. Since the number of candidates is in most cases limited in the process but the information which may influence the decision-making is quite large, the selected rules may not be as good as expected. In this study, we are going to use simulation in a different way to overcome this problem.

If we consider each decision-making instance of real-time scheduling individually, all decisions make up a sequence. Real-time scheduling turns out to be finding the sequence of decisions that results in a good performance of the system. This type of decision-making is called sequential decision-making. Q-learning is known to be an

M. Mujica Mota and P. Scala (Eds.): EUROSIM 2023, CCIS 2032, pp. 237–248, 2024.
https://doi.org/10.1007/978-3-031-68435-7_17

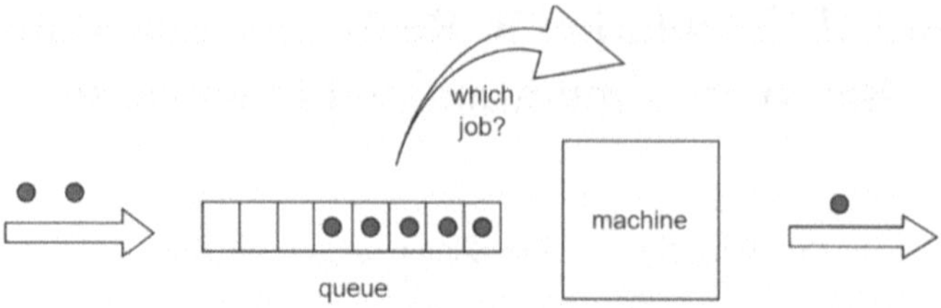

Fig. 1. Real-time scheduling problem

appropriate way to solve these kinds of decision-making problems. Q-learning [1] is one of the most-used algorithms in reinforcement learning [2], which is essentially based on the Markov decision process (MDP) [3], which aims to find an optimal action-value function to take the best action once the state changes. The decision-maker (agent) during the Q-learning tries to acquire knowledge from scratch by interacting with the system (environment): Once a decision needs to be made (action), the decision-maker will take one action according to her current knowledge, and after a while, the system will return a reward for that decision. The current knowledge will be updated according to that reward and the updated knowledge will be utilized in the next decision-making. The goal of the decision-maker is to maximize the received reward in the long run. In the algorithm, the system or environment is often replaced with a simulator. During the simulation, the simulator asks the decision-maker for decisions and the decision-maker makes the decision and responds to the simulator. The simulator will take the decisions and return the reward to the decision-maker accordingly. The decision-maker gains knowledge according to the rewards and uses the new knowledge to make later decisions. The decision-maker will be able to make decisions in a real system after this offline learning phase is done.

Applying that idea to scheduling, each action represents selecting a job from the queue. The value of an action reflects the job's priority, determined by both the current state of the environment and the action itself. The crucial questions revolve around quantifying states and actions, as well as defining reward functions. In a recent paper of ours [4], we quantified states and actions and assessed their impact on algorithm performance through numerous experiments. In the study described here, we will propose various methods for calculating rewards and evaluate them experimentally.

2 State of the Art

Using reinforcement learning to solve real-time scheduling can be traced back to about 20 years ago. It has become more and more popular in recent years. In this section, we will review some papers and mainly focus on reward calculation.

Aydin and Öztemel [5] introduced an intelligent agent-based dynamic scheduling system comprising two distinct components: the agent and the simulated environment. The agent dynamically selects the most suitable priority rule based on real-time shop conditions, while the simulated environment executes scheduling tasks using the chosen rule. The agent undergoes training via an enhanced reinforcement learning algorithm during the learning phase, enabling it to subsequently make scheduling decisions. Rewards are computed using a heuristic that considers the selected rule, incorporating both the size of the relevant queue and its mean slack time. Ou et al. [6] also approached the real-time scheduling challenge as a reinforcement learning task, employing the Q-learning algorithm. They emphasized the critical role of defining a reward function in constructing a learning model. To assess the learning efficacy of the Q-learning algorithm, they conducted simulation experiments utilizing five distinct reward functions derived from various interpretations of the production system. Simulation results indicated that learning performance differed across reward functions, with only the reward function aligned with a deeper comprehension of the system demonstrating superior performance compared to others. Like Aydin and Öztemel, determining the reward from the understanding is a heuristic too. However, calculating the reward based on the heuristic indicates that the agent is learning knowledge from the heuristic. The best case of the learning will be that the agent performs as same as the heuristic does. Wang and Usher [7] sought to develop a dynamic control policy using reinforcement learning for single-server scheduling with multiple products. This dynamic policy optimizes a cost function incorporating WIP inventory, backorder penalty costs, and setup costs while ensuring adherence to productivity constraints for each product. Their dynamic policies demonstrated superior performance compared to several cyclic policies. The experiments show that for a single machine problem with different reward functions different system objectives can be achieved. But it is difficult to know whether a reward function is the best fit for a certain objective. Gabel and Riedmiller [8] applied reinforcement learning to tackle a job-shop scheduling problem. They viewed job-shop sequencing problems as sequential decision processes and assigned an adaptive agent to each resource. These agents make dispatching decisions independently and refine their decision-making through trial and error using a reinforcement learning algorithm. They defined the reward to be the sum of the reward that is associated with the resources and jobs. Since all jobs must be known in advance, this definition is only suitable for static scheduling. Zhang et al. [9] minimized mean weighted tardiness in an unrelated parallel machine scheduling problem with reinforcement learning by constructing elaborate state features, actions, and the reward function. The state features and actions are defined fully utilizing prior domain knowledge. Minimizing the reward per decision time step is equivalent to minimizing the schedule objective, i.e., mean weighted tardiness. This reward calculation is only suitable for static scheduling too.

From the study above, we can see that the reward calculation has a big influence on the learning procedure. Different reward functions are defined for different objectives of scheduling. In static scheduling, the reward calculation is quite easy. Because all jobs are given in advance, an overall objective can be calculated whenever a decision is made, and the overall objective can be directly considered to be the reward. However, this won't work in real-time scheduling in which jobs arrive randomly. We will never know an overall objective before the time horizon ends. In this case, some studies use heuristics to calculate the reward, but this learning conducted by the heuristics can only result in an agent which performs as well as the heuristics. It is unnecessary to train the agents and we can just use heuristics instead if they perform well. However, obtaining good heuristics is very hard. This is also the reason why reinforcement learning is introduced.

In this study, we will try to find a generic and easy way to calculate the reward for all kinds of objectives.

3 Real-Time Scheduling Based on Simulation and Q-learning

3.1 Simulation and Reinforcement Learning

In reinforcement learning, agents always interact with the environment. When a decision needs to be made, the environment provides the current state and action set to the agents and requests them to choose one action from the set. After one action is selected, the environment will take the action and return a reward to the agent at the next decision-making. For real-time scheduling, we cannot let the agent interact directly with the real system in the learning phase, because it is too risky to take the decision made by the agent. Instead, a simulation model is built from the real system. The agents interact only with the simulation. Whenever the agent finishes the learning and has been tested in the simulation, it can be connected to the real system and make decisions. The agent can also continue to learn with the real system.

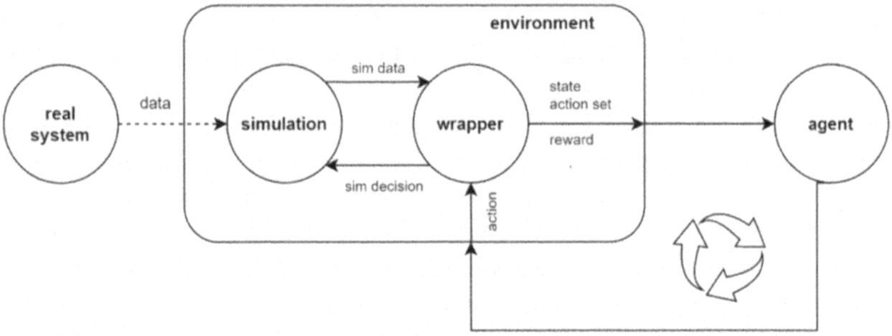

Fig. 2. Simulation and reinforcement learning

In addition, the simulation generates huge amounts of data. Not all of them are needed or directly needed by the agent. Thus, there is a gap between the simulation and the environment. A wrapper is added to bridge this gap, shown in Fig. 2. According to the definitions of the state, action, and reward, it extracts or calculates the state, action set, and reward from the simulation data at decision-making points. How to define the state, action, and reward is quite tricky in real-time scheduling. For the definition of states and actions, please refer to our papers [4] and [12]. This paper will focus on the reward function.

3.2 Real-Time Scheduling with Q-learning

As we said before, the selection of a job from the queue relies on the state of the production system at that time. In our algorithm, a value called Q value is calculated for each state-action pair (job) in the learning phase. The Q values are then used in practice to select the job – more explicitly, the state-action pair with the biggest Q value will be selected.

A general learning procedure of the real-time scheduling based on simulation and Q-learning is shown in Fig. 3. We run as many simulation replications as possible. In each replication, once we need to select a job from the queue, an ε-greedy algorithm is applied. The action that has the biggest Q value (initialized randomly) will be selected with the probability ε. With probability 1- ε, the action is selected randomly. When we come to the next decision point, the reward for the last selection will be calculated, and the Q value of the last state-action pair is updated according to the formula in the bottom block in Fig. 3. The procedure terminates when the Q values converge after enough replications.

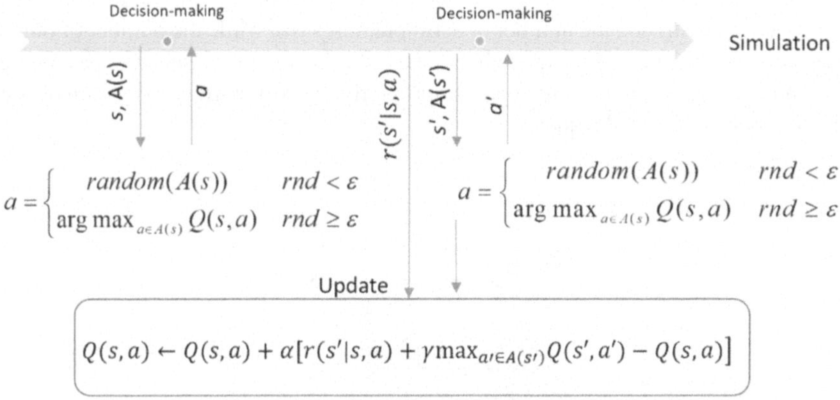

Fig. 3. A general procedure of Q-learning

If the state and action spaces are quite large, it is practically impossible to visit each state-action pair during the simulation runs. We must generalize the Q value. Usually, a function of state-action pair with some parameters is adopted. Each feature of the state and action receives a parameter (weight). In the update block, these parameters are updated rather than the Q value. This is mostly done by supervised learning. For lack of space, for a detailed description of the Q-learning and the parameterized value function, please refer to Watkins and Dayan [1] and Hado et al. [10].

3.3 Deep Q-learning with Featured Action Space

In a real-time scheduling scenario, the available actions correspond to the jobs currently in the queue. The quantity of actions directly reflects the number of jobs awaiting processing. As jobs enter the system at random intervals and the queue undergoes continuous changes, attempting to encompass all available jobs within the action set can quickly lead to an overwhelming volume, particularly over extended periods due to the ongoing influx of new jobs. Furthermore, since job arrivals are unpredictable, establishing fixed actions becomes unfeasible. To deal with these challenges, we restructure the action space into a more organized format called a featured action space. This helps us categorize actions based on their unique characteristics.

To accommodate different types of action spaces, we need to customize our neural network structures accordingly. This also applies to the learning algorithms we use. In scenarios with discrete action spaces, we employ a single neural network. It takes the state as input and generates Q values for all potential actions as outputs. Each output corresponds to a specific action. For continuous action spaces, we typically utilize two neural networks. The first network takes the state as input and outputs a distribution of possible actions. We then sample actions from this distribution. The second network takes both the state and the sampled action as inputs and outputs the corresponding Q value. In the case of the featured action space, both the state (s) and action (a) are input together into the neural networks, producing the value Q (s, a) as output, as illustrated in Fig. 4. In this setup, the number of inputs is determined by the sum of the dimensions of the state space (m) and the number of action features (n), while the number of outputs is fixed at 1. Here, x represents the value of each dimension of the state (s), and y refers to the value of the action (a)'s features. Importantly, the structure remains independent of the number of jobs in the queue.

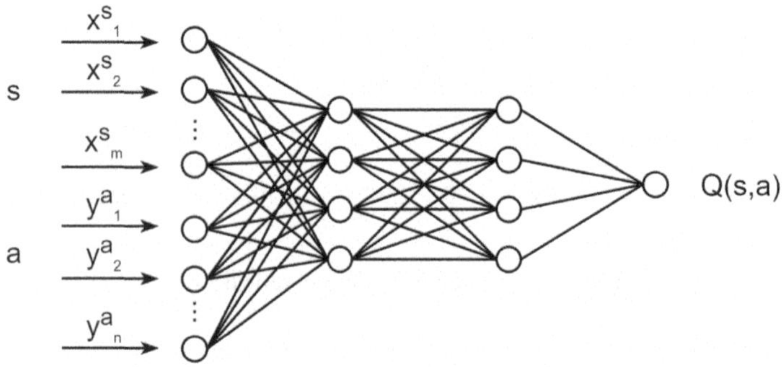

Fig. 4. Structure of neural network for featured action space

When choosing a job from the queue, we simply compute the Q value for each action and select the one with the highest value. For estimating future rewards, we input the next state along with each action from the next action set into the neural network, then choose the highest Q value as the future reward. However, this approach necessitates additional computation, which is its drawback.

4 Reward Calculation

4.1 Difficulties

The objective of Q-learning is to maximize the cumulated rewards. The objective of real-time scheduling is to optimize a KPI over a time horizon, for example, to minimize the average cycle time in six months. To use Q-learning to solve the real-time scheduling, we must define a reward function to make sure that the objective of the scheduling is achieved while the cumulated reward is maximized in the time horizon.

Before the study, we tried to calculate the immediate reward according to the current queue length, slack of the selected jobs, and so on. But the results are not so good. The reason is that the reward calculated in these ways is not directly related to the objective of the scheduling. We cannot guarantee that the objective of the scheduling is achieved while the cumulated reward is maximized. Thus, in the study, we try to calculate the reward directly from the objective of the scheduling. We have tried to use a so-far-objective (the objective achieved so far) to calculate the reward, but the results are also not good. The reason is that the so-far-objective reflects only the performance of the decisions before, not the performance of the current decisions.

Besides, the objective of the scheduling is mostly the long-term KPIs of the manufacturing system. We do not know one decision's contribution to the long-term KPIs. How to give each decision an objective-related reward is the main issue.

4.2 Collective Reward in a Period

After a job is selected and starts to be processed on a machine, a reward is requested by the algorithm for updating the Q value. As we said before, it is hard to know how good the single selection is in the long run at that time. Luckily, we can know how good a series of selections is after a certain period. We can observe the KPIs in a shorter period for several decisions (like the average cycle time in one week in which 100 decisions are made). This short-term KPI can be used as the reward for decisions made in this period. Since the reward is obtained by all decisions in the period, we call it a collective reward.

For this reason, the whole simulation run is divided into several small periods, like KP1, KP2, and KP3 shown in Fig. 5. The concerned KPIs of the schedule are calculated for each period. An overall reward for all selections in one period can be computed from the KPIs in that period: If the objective is to maximize a KPI, the reward is the value of the KPI; otherwise, the reward is the reciprocal of the KPI.

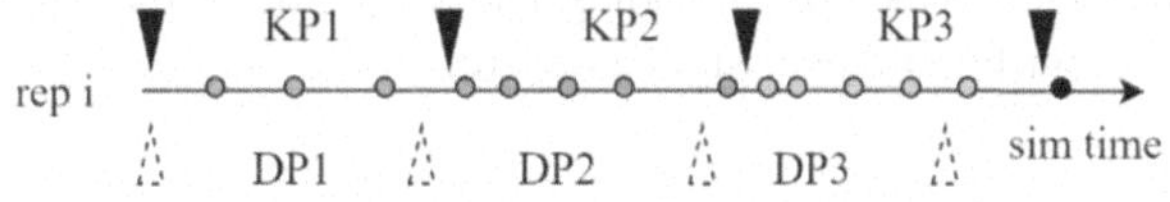

Fig. 5. Overall reward calculation in each period

The minimal length of the period must meet the condition that the concerned KPIs can be calculated in this period. For example, if the objective is to minimize the cycle time and the period is too short, no jobs are finished in that period. We will not be able to calculate the cycle time for that period and consequently, there is no way to obtain the reward. On the other hand, if the length of the period is the same as the simulation run, only one period is formed. This special case will be separately treated in the following experiments. Besides, after a job is selected, there is a delay until the selection eventually affects the reward. Thus, the time window is shifted back a little bit when we are collecting the state-action pairs in a period, for example, DP1, DP2, and DP3 shown in Fig. 4. Thus, the decisions are made in DPs (decision periods) while their rewards are calculated from KPs (KPI periods). Regarding the shifting, we always move the DPs back by the length of the maximal processing time. If the last decision is still included, move to the time that the last decision was made and exclude the last decision, because we need to look ahead one step to update the value of state-action pairs.

4.3 Distribution of Collective Reward

Since the collective reward is for all the selections (state-action pairs), to update the Q values, we need to distribute the collective reward to each selection. Usually, lots of decisions are made in one period. Based on the sequence or timing that the selections occur in the period, the following methods are proposed by us to distribute the reward:

- only the last state-action pair is rewarded,
- even distribution of the reward,
- weighted distribution of the reward,
- the most frequent state-action pair receives the reward, and
- the n most frequent state-action pairs are rewarded.

If only the last state-action pair is rewarded, the other pairs will receive a "zero" reward. This is like the rewarding process in the Grid World Game where agents only get a reward when they reach the destination. If they are halfway, they do not get any reward. The reward can also be distributed by a given weight if their contribution to the reward is different. The weight can be either the occurring times of a pair in the period or the number of selections to the end of the period. If all pairs have the same weight, the reward is evenly distributed. The reward can also be assigned to the n most frequent pairs because these pairs contributed more to the reward. The same as before that the reward can also be distributed by a given weight on the most frequent pairs. If n is 1, only one pair is rewarded.

The assigned reward is considered to be the immediate reward for each state-action pair. At the end of each KP period, once the collective reward is calculated and distributed to each state-action pair in the corresponding DP period, the values of the state-action pairs are updated according to the assigned reward.

5 Experiments and Results

The reward-calculation methods proposed in the previous section will be investigated by experiments in this section.

5.1 A Single Machine Model

Single-machine scheduling has already received substantial attention in the last decades because of its importance as a building block for more generalized and complex problems. Moreover, in a multi-machine environment, the schedules may be made around a bottleneck, or sometimes the entire production process may be integrated and treated as a single machine for scheduling purposes [11]. Thus, a single-machine model with the objective to minimize the average cycle time is used as the experimental environment. In the single machine model, two products P1 and P2 are produced with processing times of 2 and 6 min, respectively. Jobs of both products arrive randomly and the interarrival times follow an exponential distribution with a mean value of 10 min. To keep the model as simple as possible, other features which are not mentioned here, like breakdowns, setups, and so on, are all ignored in the model.

5.2 Scenario Design

Table 1 shows the eleven scenarios that we have designed for the experiments. According to the length of periods, there are two groups: 1) the period length is equal to the simulation length; 2) the period length is smaller than the simulation length. For the even distribution of the reward, no matter how many times a state-action appears in a period, we only take account once. Contrarily, in the occurring-time-weighted distribution, the more occurred state-action pairs will get more rewards. If the most frequent state-action pair gets the reward, the other state-action pairs won't get any reward. In the step-number-weighted distribution, the state-action pairs which appear later will obtain more reward.

Table 1. Eleven scenarios for the experiments

Scenario	Period Length	Method
1	< sim length	Only the last state-action pair is rewarded
2		Even distribution of the reward
3		Occurring-time-weighted distribution of the reward
4		The most frequent state-action pair gets the reward
5		The 3 most frequent state-action pairs are rewarded (evenly)
6	= sim length	Only the last state-action pair is rewarded
7		Even distribution of the reward
8		Occurring-time-weighted distribution of the reward
9		Step-number-weighted distribution of the reward
10		The most frequent state-action pair gets the reward
11		The 10 most frequent state-action pairs are rewarded (evenly)

5.3 Experiment Settings

In the learning phase, for each scenario, we run the simulation 10,000 times, and each run for 7 days. In Scenarios 1-5, the period length is 1 day, i.e., the whole simulation is split into 7 periods. In each scenario, the value function is approximated by the neural network as mentioned in Sect. 3.3. Weights are randomly initialized. The optimizer uses the Adam algorithm where the learning rate exponentially decays with an initial value of 0.01. The discount ratio γ is 0.9 and ε is set to 0.3 in the ε-greedy algorithm. The trained neural network is stored after the learning phase and loaded to use in the application phase. In the application phase, we have also run the simulation 10,000 times for each scenario.

5.4 Experimental Results

After each run in the application phase, two KPIs are collected for each product: the number of finished jobs and the average cycle time of the finished jobs. The summary of the two products is the number-of-finished-jobs-weighted-average value. When all replications for a scenario are done, the average of the replications is calculated. The results of all scenarios are listed in Table 2. To make the difference on the average cycle time more obvious, a bar chart is also shown in Fig. 6.

From Table 2 and Fig. 6, we can find out that Scenarios 7 and 2 result in minimal average cycle times with values of 13.38 and 13.40 min, while Scenarios 1 and 6 take up the maximal average cycle time with 14.13 and 14.33 min. Both Scenarios 7 and 2 are evenly distributing the overall rewards. Scenarios 1 and 6 only give the rewards to the last decisions. Since the results are very close, it is hard to judge the other scenarios.

Table 2. Average cycle times and number of finished jobs

Scenario	P1		P2		Summary	
	Finished Jobs	Avg. CT	Finished Jobs	Avg. CT	Finished Jobs	Avg. CT
1	1,003	12.30	1,013	15.94	2,017	14.13
2	1,009	11.52	1,000	15.30	2,009	13.40
3	1,003	11.61	1,003	15.74	2,005	13.67
4	1,000	11.57	1,009	16.01	2,009	13.80
5	1,004	12.05	1,008	15.80	2,012	13.93
6	1,010	12.38	1,012	16.28	2,023	14.33
7	1,003	11.46	1,004	15.29	2,007	13.38
8	1,007	11.65	1,002	15.64	2,008	13.64
9	1,010	11.68	1,003	15.70	2,012	13.68
10	1,012	11.98	1,005	15.74	2,017	13.86
11	1,008	11.83	1,006	15.91	2,014	13.87

The reason why rewarding only the last decisions works fine in the Grid World Game but does not work in our real-time scheduling problems is that the last decisions or final states are the same in the Grid World Game. Through updating, the rewards can easily pass to the previous decisions backward. In real-time scheduling problems, the final states even do not exist and the last decisions in a period are always changing. The rewards of the previous decisions cannot be updated that often. The reason why evenly distributing the overall rewards performs better than others is that we run a huge number of replications. Even though it is possible that the same decision gets a very big reward in one replication, but a very small reward in another replication, after a huge number of replications, the differences between decisions will become obvious. The decisions appearing in "good" replications more often will receive more rewards. In the application phase, we will also run a huge number of replications to evaluate the methods and we always take the decisions with the biggest rewards. The probability that these decisions build "good" replications is high.

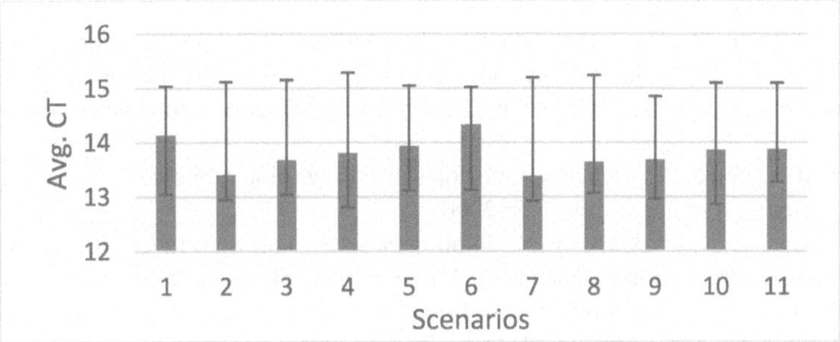

Fig. 6. Average cycle times from all scenarios

6 Conclusions

When Q-learning is used to solve real-time scheduling problems, obtaining rewards from the environment or simulation becomes tricky. The rewards indicate how good the decisions made were for certain system states, and the goal of Q-learning is to maximize the cumulated rewards in the long run. To solve real-time scheduling problems, Q-learning must ensure that the scheduling objectives are optimized while the cumulated rewards are maximized. This implies that the reward functions should be directly derived from the objectives. However, the objectives of real-time scheduling always focus on some long- or mid-term KPIs, i.e., the performance of a sequence of decisions, rather than on individual decisions. It is difficult to return an objective-related reward just after a decision is taken. In this study, a bridge between the rewards and the objectives is built. The whole simulation is divided into several small periods. A period-wise objective is considered to be the collective reward of the decisions in one period. If the objective is to minimize some KPIs, the collective reward can simply be the reciprocal of the KPIs. Since this applies to all kinds of objectives, the reward function is no longer objective-specific. The collective reward is then allocated to every decision. The value function is updated according to these individual rewards. According to the experimental results, we suggest distributing the collective reward evenly. In addition, the essential of reinforcement learning is trial-and-error which needs each state-action pair to appear again and again.

This can only be guaranteed by running a very hung number of replications. Used in a big model, it will require more computing power. This approach can also be used to solve other sequential decision-making problems where the immediate reward is unknown or hard to be obtained, but KPIs can be calculated easily.

References

1. Watkins, C.J., Dayan, P.: Q-learning. Mach. Learn. **8**, 279–292 (1992)
2. Sutton, R.S., Barto, A.G.: Reinforcement Learning: An Introduction. MIT Press, Massachusetts (2018)
3. Puterman, M.L.: Markov Decision Processes: Discrete Stochastic Dynamic Programming. Wiley, Hoboken (2014)
4. Xie, S.F., Zhang, T., Rose, O.: Online single machine scheduling based on simulation and reinforcement learning. In: Putz, M., Schlegel, A. (eds.) 18th ASIM Fachtagung Simulation in Produktion und Logistik, Chemnitz, pp. 59–68. Verlag Wissenschaftliche Scripten, Auerbach (2019)
5. Aydin, M., Öztemel, E.: Dynamic job-shop scheduling using reinforcement learning agents. Robot. Auton. Syst. **33**(2), 169–178 (2000)
6. Ou, X., Chang, Q., Chakraborty, N.: Simulation study on reward function of reinforcement learning in gantry work cell scheduling. J. Manuf. Syst. **50**, 1–8 (2019)
7. Wang, Y.C., Usher, J.M.: Application of reinforcement learning for agent-based production scheduling. Eng. Appl. Artif. Intell. **18**(1), 73–82 (2005)
8. Gabel, T., Riedmiller, M.: Adaptive reactive job-shop scheduling with reinforcement learning agents. Int. J. Inf. Technol. Intell. Comput. **24**, 14–18 (2008)
9. Zhang, Z., Zheng, L., Li, N., Wang, W., Zhong, S., Hu, K.: Minimizing mean weighted tardiness in unrelated parallel machine scheduling with reinforcement learning. Comput. Oper. Res. **39**(7), 1315–1324 (2012)
10. Hado, H., Arthur, G., David, S.: Deep reinforcement learning with double Q-learning. In: Proceedings of the Thirtieth AAAI Conference on Artificial Intelligence (AAAI 2016), Phoenix, Arizona, pp. 2094–2100 (2016)
11. Gupta, S.K., Kyparisis, J.: Single machine scheduling research. Omega **15**, 207–227 (1987)
12. Xie, S.F., Zhang, T., Rose, O.: Real-time scheduling based on simulation and deep reinforcement learning with featured action space. In: Feng, B., Pedrielli, G., Peng, Y. (eds.) Proceedings of the 2022 Winter Simulation Conference, Institute of Electrical and Electronics Engineers Inc., Piscataway (2022)

A Method for Efficient Simulation of Production Lines Through Software-Supported Cooperation Between Production and Simulation Experts

Walter Wincheringer[1](✉) and Marec Kexel[2]

[1] University of Applied Sciences, Konrad Zuse Str. 1, 56075 Koblenz, Germany
wincheringer@hs-koblenz.de

[2] AcuroSim GmbH, Simrockstr. 14, 65187 Wiesbaden, Germany

Abstract. Production simulations are associated with considerable effort, among other things due to the knowledge transfer between the domain expert and the simulation specialist. For small and medium-sized companies, this often represents an economic hurdle when using the simulation. This article presents a method for software-supported collaboration between the production expert and the simulation specialist, which leads to a significant reduction in effort. In this way, the advantages of the simulation can be used economically even with low optimization potential.

Keywords: production · DES · procedure model for a simulation study

1 Introduction

Because of prevailing megatrends, e.g. sustainability, digitization, AI, Smart Factory, production is exposed to a rapidly changing environment. In order to be able to withstand the competitive pressure, holistic optimization of production and logistics processes are required [1].

A possibility for holistic optimization in production that has been established in practice is the value stream mapping method [2]. This is a static snapshot that aligns the production process to customer requirements with the help of various parameters, such as customer takt time, process cycle time, degree of utilization, inventory range and flow rate, and thereby identifies waste [3]. Based on the consideration of average parameters, the value stream mapping method can only insufficiently depict the dynamics in the production process. Therefore, the application of the value stream mapping method fails, especially in the case of branched, parallel material flows and short cycle times [4].

In order to optimize the dynamics in production and the changing influences of employees, material, machine, method, etc. on the production process, the use of discrete-event simulation (DES) has proven itself [5].

However, simulation studies in production are usually complex, despite a large number of established, generally applicable simulation tools. This is particularly due to the

M. Mujica Mota and P. Scala (Eds.): EUROSIM 2023, CCIS 2032, pp. 249–260, 2024.
https://doi.org/10.1007/978-3-031-68435-7_18

fact that the practical knowledge and experience of the application area, the so-called domain knowledge of the production expert, has to be transferred to the simulation specialist for the specific task [6]. Only because of knowledge transfer is the simulation specialist able to create a task-specific simulation model in the simulation environment, which depicts the real, dynamic production process with sufficient accuracy.

The effort associated with knowledge transfer usually means that simulation is only used in production if there is considerable economic potential associated with it. This is often the case in larger companies with multi-shift production and high repetition rates.

For small and medium-sized companies, which often only produce in one shift, the high cost of simulation studies represents an entry barrier [7]. This contribution presents a method how a software-supported synthesis of the production and the simulation expert can accelerate the execution of simulation studies in production.

2 Requirements and Procedure

Production systems are heterogeneous depending on the frequency of repetition, the necessary manufacturing technologies and the respective production control logic. Therefore, in the development of the method, the production line (flow principle) is focused.

However, based on the experience of a development and industrial partner, the *AcuroSim GmbH*, the production lines should not only be strictly interlinked, but should include both a constant and an unsteady material flow. Thus, continuous conveyors with material buffers, driverless and guided transport systems as well as material stores (supermarkets) must be taken into account [8].

It should also be possible to map pull (demand-driven) and push (consumption-driven) controls. In addition, material flows must be branched and able to run in parallel. These requirements are based on different application knowledge of a value stream simulation tool with low granularity [9], which failed in practice when different material flows and control logics were used within an interlinked production.

Twelve simulation studies that were not satisfactorily carried out with the value stream simulation tool were analyzed in their entirety. Further evaluations of a value stream-based simulation and its material flow-specific limits are documented in [10], among others.

In a further step, eight finished simulation studies in production, which were carried out according to the process model of VDI 3633 [11], were examined, see Fig. 1. Here, there is a limitation to the phases of system analysis up to implementation, which results in a model that can be experimented. The data acquisition and processing as well as the verification and validation (V&V) were also examined.

In particular, the system analysis and the creation of a concept model require close cooperation between the production expert and the simulation specialist [6]. Another focus of the close cooperation between the domain expert and the simulation specialist is data acquisition, its processing and interpretation. Therefore, the methodology to be developed should support these phases of a simulation study in particular (system analysis, data acquisition and processing) and lead to a reduction in effort.

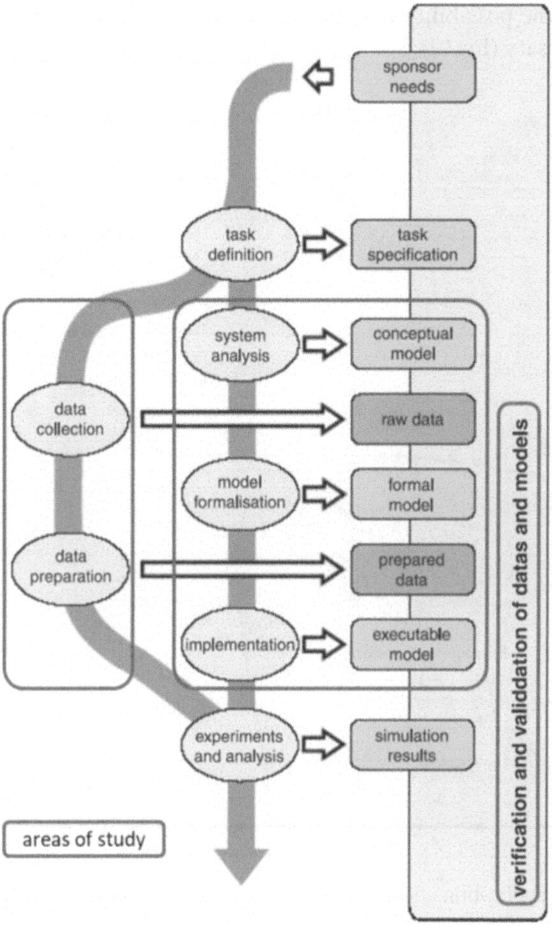

Fig. 1. Procedure model of the simulation according to VDI 3633 [11].

Furthermore, the implementation should be accelerated with the help of defined simulation modules for production and material flow processes, which can be positioned and linked in the WITNESS Horizon simulation environment using drag-and-drop.

The use of specific simulation kits, which are based on application-specific reference models and are implemented in a user-friendly simulation environment, is a proven approach to reducing the effort involved in creating application-specific simulation models [12].

The advantage of saving time by using pre-programmed simulation modules goes hand in hand with the disadvantage of the application-specific limitation to the area of application. There are different value stream simulation tools on the market that partially map the dynamic components in production; e.g. technical malfunctions, set-up processes, fluctuating cycle times. However, the mapping of parallel material flows or special control logic is not possible. Therefore, the simulation kit to be developed should also offer the possibility that the simulation specialist can add specific process logic, etc. if necessary (Fig. 2).

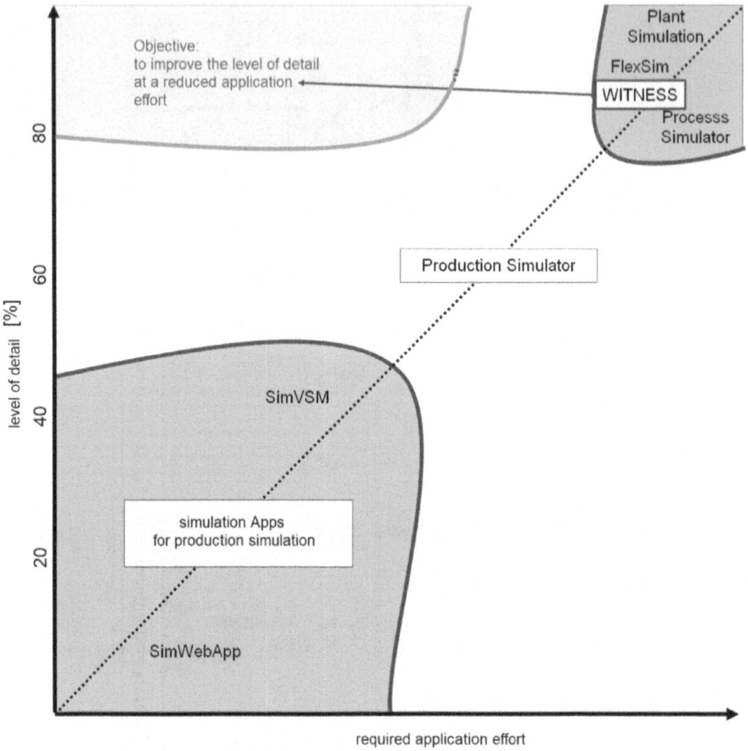

Fig. 2. Required application effort versus level of detail depending on the Simulation tools.

The development is not about developing another value-stream-oriented simulation module set that can be used independently by a production expert. Such building blocks for simple production processes are available on the market [10].

The development is about the collaborative, efficient cooperation (synthesis) of the two experts, the production expert and the simulation specialist, in order to efficiently build simulation models for complex production and material flows.

3 Methodology of Software-Supported Synthesis

The methodology pursued the approach of including a domain expert from the production area and the simulation specialist equally in the implementation phases of a simulation study in order to reduce the individual effort of those involved. In classic simulation studies, according to the process model to VDI 3633 [11], the task-specific system analysis is carried out by the simulation specialist in cooperation with the production expert. The simulation expert then summarizes his findings in a concept model, which he then verifies together with the production expert. Since only the simulation specialist, and not the production expert, is familiar with the simulation environment, he has to carry out the model formalization and implementation in the simulation environment. Only the executable simulation model can be verified and validated together with the production expert.

A significant way of reducing the effort is that with the help of a simulation module kit, the phases of system analysis, model formalization and implementation do not run separately sequentially, but are integrated and carried out jointly between production experts and simulation specialists.

This is made possible by an advantageous division of labor in the modeling and implementation process, as a result of which the two experts only have to draw on their existing knowledge and available resources without having to familiarize themselves intensively with the subject of their counterpart beforehand (Fig. 3).

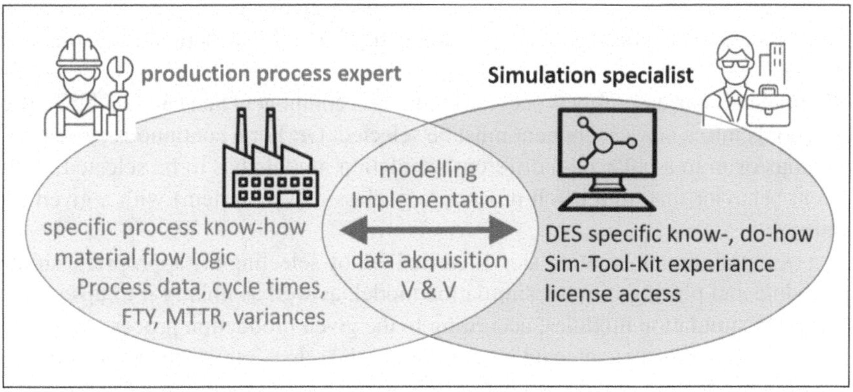

Fig. 3. Devision of labour between production expert and simulation specialist.

In the system analysis phase, the production expert must describe the process of the production line step by step to the simulation specialist. At the same time, the simulation specialist creates the simulation model step by step by selecting the appropriate simulation module from the previously defined simulation module box (see Fig. 4) and placing it in the simulation model. The previously coordinated task specification of the simulation study is taken into account and, if necessary, real process steps in the model are summarized in one module or mapped in more detail by several simulation modules (adjustment of the granularity of the simulation model to the goals of the simulation study).

The respective simulation module has a specific behavior that can be configured via parameters and thus adapted to the specific application. In this step, it is not necessary to specify processes and set-up times or quality parameters and distributions, as these are subsequently determined by the production expert (data acquisition) and recorded in a data table (data preparation).

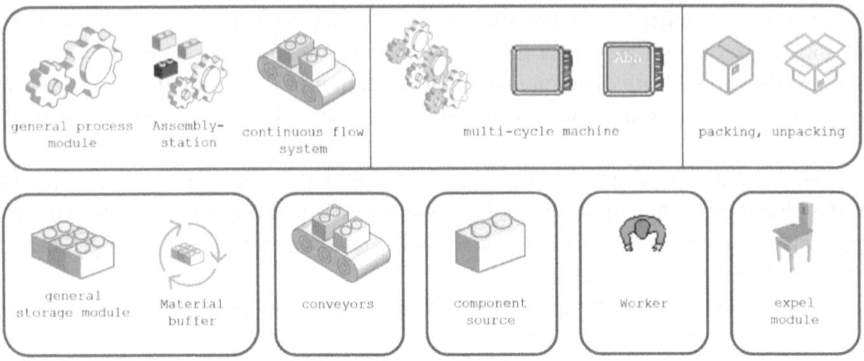

Fig. 4. Simulation model box with specific tools.

This step is all about selecting the appropriate simulation module based on the production process description together – between the production expert and the simulation specialist.

Example: For an assembly process, a simulation component that can combine different materials into a new component must be selected. Or: For a continuous system (e.g. continuous oven in a bakery), a different simulation module has to be selected, with a different behavior than in a batch process (e.g. classic oven system) with a given load quantity.

An essential implementation step is carried out by selecting the appropriate simulation module and placing it in the simulation model, as well as linking it to upstream or downstream simulation modules, according to the given production process.

If the behavior of the selected simulation module does not correspond to the real behavior in production, which can occur in particular with specific process logic of an individual production process, the simulation specialist has the option of adapting the simulation module to the specific needs.

Example: A test system within a linked production line repeats the test, depending on the test result - which leads to an increased cycle time - or immediately classifies the tested product as scrap. It can also happen that a product arrives at the testing station that was previously detected as defective (example: a component is missing). In this case, no test is carried out on the test system and the product is passed through without being tested, which leads to a reduction in the cycle time.

Such peculiarities occur again and again in practice and make it clear how important it is that the simulation modules are not closed, but can be adapted to the specific needs of the simulation task. Despite this customization option, the advantage of fast, simple implementation through the use of simulation modules remains in the course of joint modeling between production experts and simulation specialists, since in this phase the special features of the production step are only documented with a corresponding comment. The simulation specialist can adapt the simulation model in terms of programming later.

The respective simulation modules are linked with conveyor technology modules, material storage or other process steps to form a simulation model of the production line to be mapped. The simulation modules for material transport, for buffers or storage, also have parameterizable behavior and can thus be adapted to specific production requirements (Fig. 5).

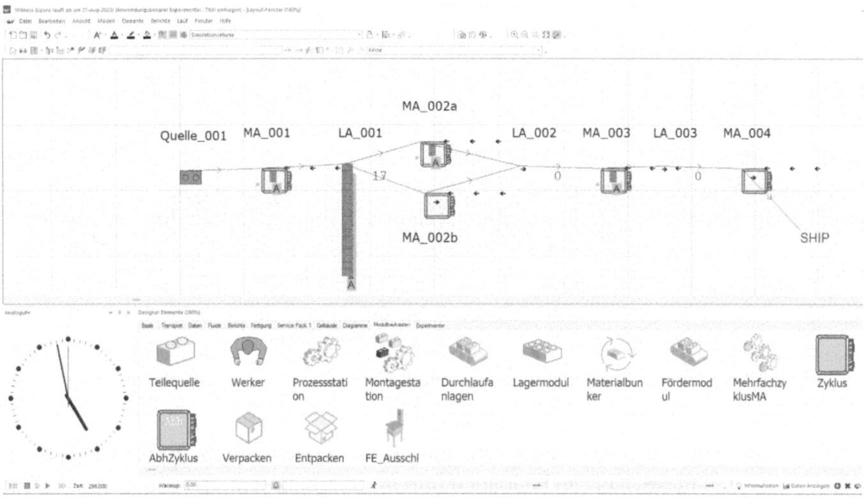

Fig. 5. Simulation kit, simple simulation model, WITNESS Horizon.

4 Parameterization of the Simulation Modules and the Simulation Model

However, for reasons of complexity, the parameterization of the simulation modules does not take place in the common modeling time and not in the simulation environment. On the one hand, this is because that the parameterization data is almost exclusively known to the production expert and the simulation specialist can only make a small contribution here. On the other hand, the production expert usually has no knowledge of how to operate the simulation software. In addition, mostly for cost reasons, there is no license access to the simulation software by the production expert.

Therefore, the simulation tool kit was structured in such a way that after arranging and linking the simulation modules in the simulation environment, an Excel spreadsheet containing all the module-specific parameter fields can be exported. The export table can be expanded or reduced at any time by adding further simulation modules.

After the data table has been exported, the simulation specialist will discuss it with the production expert and the various selection menus, including those for distributions, will be explained. For example, various set-up processes can be assigned to a production process depending on an order lot size or an operating time. The individual set-up process can in turn be linked to workers and be subject to a temporal distribution, depending on the qualifications of the worker. Something similar can be mapped for maintenance or quality aspects. In order to make the process of data collection as time-efficient as possible, various input assistants were programmed, which automatically fill in further fields based on the entries made or suggest possible entries (selection menus, etc.). In addition, other functions in the background ensure the plausibility and completeness of the entered data and parameters and thus a smooth simulation run (Fig. 6).

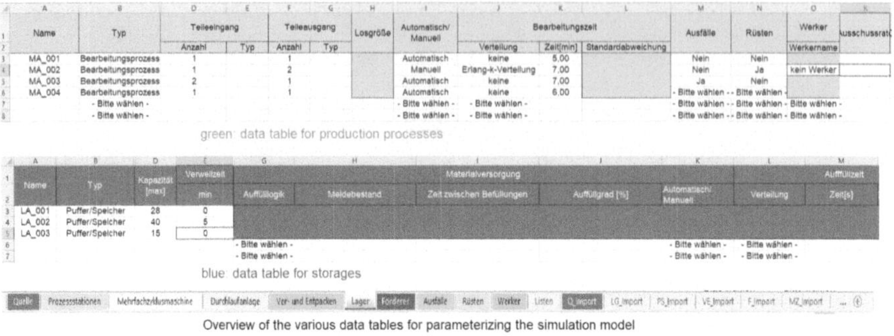

Fig. 6. Overview about data tables, sample for production processes and storages.

After all parameters, distributions, etc. have been entered into the Excel tables by the production expert, these data can be loaded into the simulation model using an interface. Immediately afterwards, the first simulation run can be carried out in order to do an initial verification and validation of the simulation model jointly by the production expert and simulation specialist. If errors are detected in the simulation model, further simulation modules can be added, existing ones deleted or adjusted. The simulation specialist can also adjust parameters directly in the simulation environment.

If you create the export table again, all entered data will be exported. Missing data can be added via Excel or existing data can be adjusted by the production experts. This is followed by another import and further verification and validation.

If the simulation model corresponds to the production system to be mapped with sufficient accuracy, the optimizations can then be carried out with the help of simulation experiments. The Excel table with different parameter variants can also be used for this. For this purpose, the adapted table is saved with an experiment-specific file name. This makes it possible to carry out a wide range of simulation experiments without great effort.

If an adaptation of the simulation model is required for the experiments, for example changing the process or increasing capacity through an additional machine, the model can be adapted and saved under a new model file name. There is a 1:n relationship between the respective simulation model and the parameter files, which means that a specific simulation model can have several parameter files, with each parameter file standing for a specific simulation experiment.

5 Evaluations, Diagrams and KPIs

In order to be able to evaluate the results or findings of a simulation experiment, a presentation of the results in the form of time courses, diagrams and key figures is essential. In order to reduce the effort involved in a simulation study in production here as well, the simulation modules already contain extensive evaluation logic in addition to the selectable control rules, so that the simulation specialist does not have to spend any time programming to display the simulation results (Fig. 7).

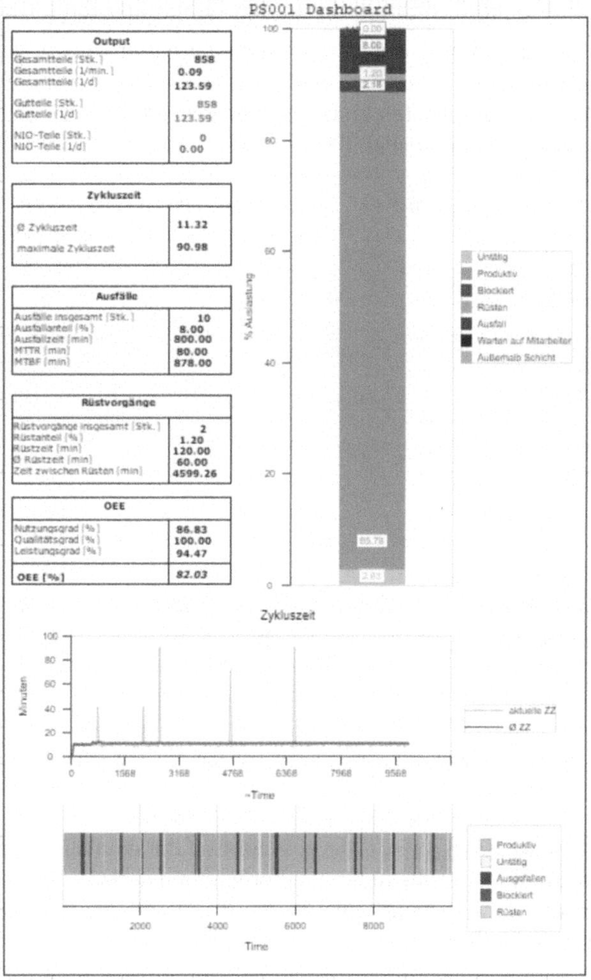

Fig. 7. Example dashboard of the "process station" in the simulation model.

A key performance indicator in production is the Overall Equipment Effectiveness (OEE value). This value is the product of the degree of utilization, the degree of performance and the degree of quality of a production facility [13]. The OEE value is of interest both for individual process stages/simulation modules and for the entire system/the entire simulation model. Therefore, the simulation modules were programmed in such a way that these key figures can be recorded and displayed both for each simulation module and for the entire simulation model.

In addition, parameters such as average cycle time, technical failures (MTTR, MTBF), quality indicators and throughput are important at a production station. This also includes the percentage shares of the production process with regard to productive, idle, blocked, waiting for operators, setting up, etc. as well as their representation over time.

Furthermore, module-specific key figures for the evaluation of simulation results are of great importance. When evaluating a material buffer or a storage module, the utilization over time as well as average and maximum values are important in order to identify possible optimizations. For conveyor modules, the average residence time and the throughput are also required (Fig. 8).

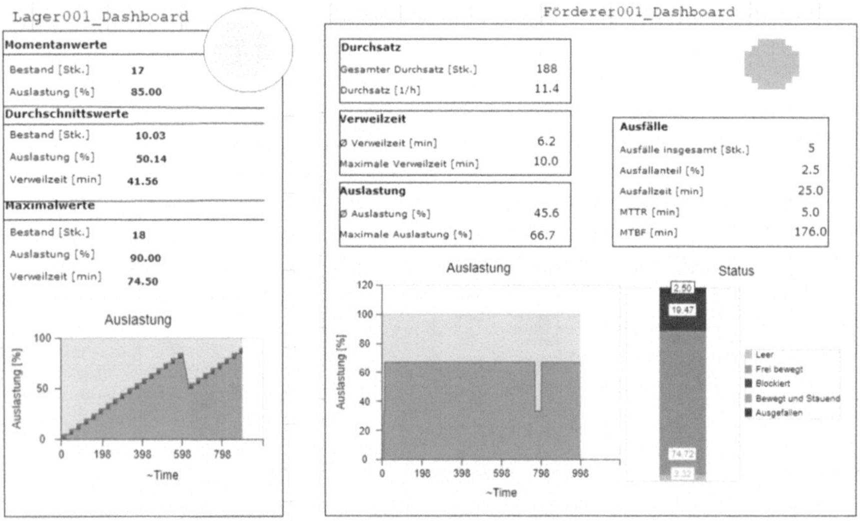

Fig. 8. KPI dashboards of storage and conveyer module.

Nevertheless, if necessary, there is the possibility of the simulation specialist expanding the presentation of the results in terms of programming. Here, too, the openness of the simulation tool kit is advantageous compared to the proprietary simulation tools offered on the market.

6 Benefits and Results

The programmed simulation modules, including the KPI-based presentation of results and the link to the Excel table for data and parameter acquisition (export, import), can be described as a simulation tool kit that enables software-supported, cooperative and efficient collaboration (synthesis) of the production expert and the simulation specialist in a simulation study.

The system analysis and model formalization phases (according to VDI 3633) can be advantageously combined proportionately with the implementation phase. The work content is divided between the production expert and the simulation specialist in such a way (joint modeling & implementation and V&V, separate parameterization) that interactions are reduced and the implementation effort is minimized.

This significantly accelerates the execution of simulation studies in production. The first couple of practical applications of the simulation kit have resulted in a cost reduction of 25–40%, depending on the complexity of the task. Table 1 shows a comparison of the total effort in working days, the production expert and the simulation specialist, from various simulation studies in production, with and without the use of the simulation kit.

Table 1. Total effort (production expert & simulation specialist) of simulation studies in production, with and without Sim-Kit (in man-days).

Simulation project phase, VDI 3336	without Sim-Kit	with Sim-Kit
system analysis	8 - 13	4 - 9
model formalisation	2 - 5	0 - 2
data collection & preparation	4 - 12	3 - 7
implementation	7 - 15	3 - 6
verification & validdation	2 - 5	2 - 4
experiments	2 - 6	2 - 6
total	**25 - 56**	**14 - 34**

The time savings increase with further simulation studies and the same actors, since the production expert does not have to familiarize himself with the processes and the data collection table.

Thus, the implementation of simulation studies is also economically advantageous for production optimization in small and medium-sized companies with a lower optimization potential.

References

1. Kern, W.: Modulare Produktion. Springer Fachmedien, Wiesbaden (2021)
2. Meudt, T.; Rößler, M., et al.: Wertstromanalyse 4.0, ZWF Zeitschrift für wirtschaftlichen Fabrikbetrieb. Jahrgang **111**, 6 (2016)
3. Erlach, K.: Der Weg zur schlanken Fabrik. 3. Auflage. Springer, Heidelberg (2020). https://doi.org/10.1007/978-3-662-58907-6
4. Kaiser, J., Urnauer, C., Reif, M., Metternich, J.: Wertstrombasierte Erstellung von Simulationsmodellen, ZWF Zeitschrift für wirtschaftlichen Fabrikbetrieb Jahrgang **11**, 1 (2019)
5. März, L., Krug, W., Rose, O., Weigert, G.: Simulation und Optimierung in Produktion und Logistik, Praxisorientierter Leitfaden mit Fallbeispielen. Springer, Heidelberg (2011). https://doi.org/10.1007/978-3-642-14536-0
6. Gutenschwager, K., Rabe, M., Spiekermann, S., Wenzel, S.: Simulation in Produktion und Logistik. Springer, Heidelberg (2017)
7. Rudel, S.: Prozess-Simulation in KMU, Dissertation, Fakultät für Wirtschaftswissenschaften der Universität Duisburg-Essen, Essen (2016)

8. ten Hompel, M., Schmidt, T., Dregger, J.: Materialflusssysteme. 4te Auflage, Springer, Heidelberg (2018)
9. Wincheringer, W., Preiß, T.: Simulation von Wertstromprozessen mit Hilfe eines Modulbaukastens. Productivity: Kompetenz in Produktion und Logistik, Band 22, Heft 3, pp. 19–22, GITO - Verlag für Industrielle Informationstechnik und Organisation, Berlin (2017)
10. Sohny, T.: Referenzmodell basierend auf der Wertstrommethode zur Bewertung von automatisierten Materialflusssystemen der Produktion in der Angebotsphase, Dissertation, Fakultät für Maschinenbau, ITPL, Technische Universität Dortmund, Cuvillier Verlag, Göttingen (2023)
11. VDI Richtlinie 3633: Simulation of systems in materials handling, logistics and production - fundamentals. Part 1, Beuth Verlag, Berlin (2014)
12. Wenzel, S.: Referenzmodelle für die Simulation in Produktion und Logistik. Society for Computer Simulation, Ghent (2000)
13. VDI Richtlinie 2870: Lean production systems. Basic principles, introduction, and review. Part 1, Beuth Verlag, Berlin (2012)

Use of Simulation to Determine the Production Capacity of Plastics Line in MSMEs

Yazmin Galván-Ortiz(✉), Susana C. Téllez-Ballesteros, and Ricardo Torres-Mendoza

Facultad de Ingeniería, Universidad Nacional Autónoma de México (UNAM), Av. Universidad 3000, Ciudad Universitaria, Coyoacán, Cd. Mexico, Mexico
yazort.210@gmail.com

Abstract. Currently, Industry 5.0 gave an unexpected turn to Production, since the new concept of "Technology as an aid and not as a threat" makes labor necessary for production to be efficient through current technologies, which makes an important resilience in production lines.

This document gives us a visibility of how to optimize production lines by means of simulation for future decision making in a MSMEs. In this case the main objective is to reduce line production capacity but maintaining process output. The simulation is about reduce from three to two machines process and supply the current demand.

The analysis based on structural, operational, and numerical data is the fundamental basis for the simulation analysis. We followed the Harrell-Ghosh-Bowden simulation procedure, which allows analyze, compile, and propose the solution of an event in a practical and straightforward way. In this paper we evaluate the decision scenarios proposed by the stakeholders in a production line of MSMEs, and we suggest the scenario with the best results using simulation tools to improve the operation.

The simulation process let improve the production line without affect the real operation process and it could apply on this small business case.

Keywords: discrete event simulation · line production · capacity · MSMEs

1 Introduction

Simulation is a tool for the analysis and evaluation of new and existing systems. The process simulation requirements are many; some of them are the knowledge of systems theory, documentation and process improvement, stakeholder communications, programing knowledge, and project management (Harrell, et al., 2011).

Goyal et al. (2018) reported a literature review on business development services in micro, small and medium-sized enterprises (MSMEs). They classified the types of needs related to MSMEs in competition, regulation, innovation, and operation. In addition, the authors concluded about the challenges faced by business development service providers such as market imperfections, customer-related choice offerings, scarcity of human and financial resources, and ethical dilemma.

M. Mujica Mota and P. Scala (Eds.): EUROSIM 2023, CCIS 2032, pp. 261–276, 2024.
https://doi.org/10.1007/978-3-031-68435-7_19

We believe the operating problems of MSMEs can find a solution by putting simulation tools into practice. In this paper we evaluated the decision scenarios have proposed by the stakeholders in a production line of MSMEs, and we suggest the scenario with the best results using simulation tools to improve the operation. As a business development service in the production operation, we integrate simulation tools into the decision process of due to bottlenecks and waste within their production lines.

1.1 State of the Art

The simulation can mix with other methodologies to make improvements in the systems, El-Haik and Al-Aomar (2006) described how Simulation mix with Lean Six Sigma and called Simulation-Based Lean Six-Sigma and Design for Six-Sigma.

Voigt, et al. (2015) developed a model-based fault localization in bottling plants and refers to bottle production lines to make them efficient by means of simulation, crediting that currently there are only a few plant-specific solutions (based on statistical calculations or artificial neural networks) for the automatic diagnosis of bottling plants. Therefore, the use of simulation tools makes it possible to meet specific study objectives.

Industry 4.0 allows flexible interactions with technologies for discrete event simulation development, A flexible and open environment for discrete event simulations and smart manufacturing, García, et al. (2021) supports the feasibility of discrete events executed in a simulation landscape for optimization and functional interface.

Adopting new Industry 4.0 technologies helped to increase productivity and avoid the economic collapse of companies. Currently, Industry 5.0 gave an unexpected turn to Production, since the new concept of "technology as an aid and not as a threat" makes labor necessary for production to be efficient through current technologies, which makes an important resilience in production lines. Industry 5.0: A Transformative Vision for Europe, Based on Dixson-Declève et al. (2022) studies emphasizes the various sources of resilience that markets must present to continue to innovate within the framework of innovative technologies (Fig. 1).

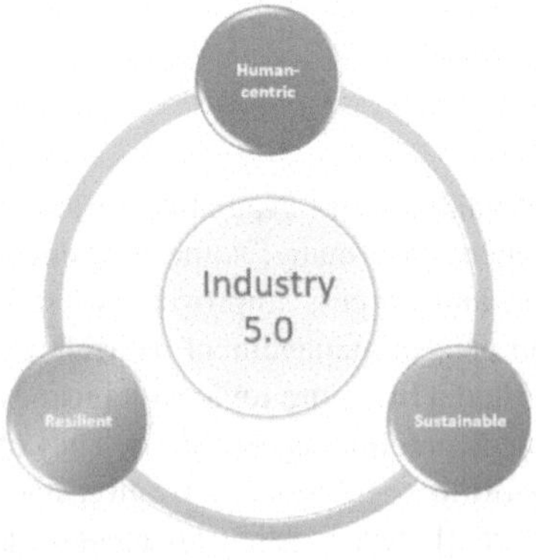

Fig. 1. Industry 5.0 relationship, Source: Breque, et al. (2021)

Derived from the COVID, many companies closed due to lack of income and the loss of human lives, for industry 4.0 was a new adaptation of technologies that did not favor 100 percent of these uncertainties, so Industry 4.0 and production recovery in the covid era, Cugno (2022) made a study of the 3 years of pandemic in manufacturing companies to measure the significant impact it had.

The methodology describe in this paper is discrete event simulation applied to line production in a MSMEs. The simulation allowed us to evaluate the performance of the line production and have findings about the process with the develop of a model base and testing different scenarios before applying the change at the system.

1.2 Problem Statement

The company manufactures and produces juices in plastic containers, taking care of the production of its own containers and thus ensuring that the product meets high quality standards and that its contents remain fresh even without refrigeration.

The company wants to know the production capacity of the lines and based on the demand to see if the orders of containers are satisfied; this leads it to review its installed capacity and if it is feasible (and profitable) to buy or remove a bottle production line to know what maximum capacity it can have and how to minimize production times (with only 2 machines).

The following proposals are proposed for the company's requirement:

- 2 production lines, 1 working shift (12 h).
- 2 production lines, 2 working shifts (12 h each)
- 2 production lines, 1 working shift (12 h) and 3 packing areas.

The project is to demonstrate what would happen and how it would affect the production of the product.

The goal of the project is to avoid bottlenecks and improve processes to reduce back to a three to two production line scenario and eliminate waste within the plant.

2 Material and Methods

To achieve the objective, the Harrell-Ghosh-Bowden (2000) simulation procedure was used, which consists of 6 steps to analyze, compile, and propose the solution of an event in a practical and simple way.

- Step 1: Define objective, scope, and requirements. Define the purpose of the simulation project and what the scope of the project will be. Requirements need to be defined in terms of resources, time, and budget for carrying out the project.
- Step 2: Collect and analyze system data. Identify, gather, and analyze the data defining the system to be modeled. This step results in a conceptual model and a data document on which all can agree.
- Step 3: Build the model. Develop a simulation model of the system.
- Step 4: Validate the model. Debug the model and make sure it is a credible representation of the real system.

- Step 5: Conduct experiments. Run the simulation for each of the scenarios to be evaluated and analyze the results.
- Step 6: Present the results. Present the findings and make recommendations so that an informed decision can be made.

Steps 1 to 5 will be analyzed and referenced in Sect. 3. Model construction, while step 6 will be seen as part of Sect. 4 where we will have the results of the original simulation and the improvement proposals for this case.

A simulation project involves describing the elements of system, this could consist of erroneous assumptions which must later be abandoned, reformulation of the problem objectives, and repeated evaluation and redesign of the model. However, if this iterative process does properly, it results in a simulation model which properly assesses alternatives and enhances the decision-making process (Harrell−Ghosh−Bowden, 2000).

3 Building the Model

The company manufactures and produces juices in plastic containers, taking care of the production of its containers themselves and thus ensuring that the product meets high quality standards and that its contents remain fresh even when not refrigerated.

To achieve the objective, the simulation procedure was used (Garcia Martinez, R. F., et. al. 2021), which consists of 6 steps to analyze, compile, and propose the solution of.

3.1 Step 1: Define Objective, Scope, and Requirements.

To support the company's decision making by always looking at the efficiency and optimization of its plastic production lines and based on information and proposals that we provide with the help of the simulation and in a term no longer than one month to be able to invest or not in the new project.

To achieve the company's objectives, it is necessary to take time, make site visits and obtain information provided by the owner and operators to understand their main pain within the production lines and what requirements must be covered to improve their efficiency.

The aspect to answer within this framework are main interest in improving the target process, the motivation for changing his production lines, kind and characteristic of ideal scenarios, and actual production demand.

3.2 Step 2: Define Objective, Scope, and Requirements.

The collection step begins by defining the simulation data to be used. We are going to classify the simulation data in structural data, operational data, and numerical data.

Structural Data. The objects in the production system to be simulated are the structural data. These objects are inputs entities, processes, and flow process. We began recording each activity, and its description. However, for reasons of order, classification and for simulation purposes in FlexSim®, We grouped activities into four main groups (Combined, crushing process, injection process and packaging of plastic bottles). At table 1 is describe the activities and necessary machine for made, also the process group used at the simulation model.

Table 1. FlexSim® groups.

Activity	Process group name	Machine (Machines number)
Pick up and transport a sack to the shredding area Place 70% of raw material and 30% of recycled material	Combined	Mixer (1 machine)
Place the mixture in the shredder Shred the mixture Empty and deposit the mixture in a barrel	Grinding process	Shredder (1 machine)
Move the barrel from the crushing area to machine 1 Transfer mixture from the canister area to the cone of machine 2 Empty the mixture into the cone Start up and work the injection machine	Injection process	Injection machine (3 machines)
Place the bag on the scantling machine by a second operator Insert 110 bottles per bag Seal bag with tape Place label to the bag Take the bag to warehouse Transport to another machine to plastic bottle packaging	Plastic bottle packaging	Warehouse (1 machine)

Figure 2 represents the base model system to be simulated. Input entities such as raw material and recycled material enter the system as a source resource. The mixed material is to be combined and shredder by a process facility. The mixture is then processed in the injection and packaging facility. And finally, the bag of bottles goes to the final warehouse.

Operation Data. In the case for operational data the lay out of the company is used to observe the location of the machinery and raw material to optimize transfer times and avoid unnecessary crossings in material transport, see Fig. 3.

Numerical Data. We use quantitative data to help us describe insight into the system. Numerical data is quantitative information about the system elements. As arrival time is entities data; the production capacity is numerical data about the facilities.

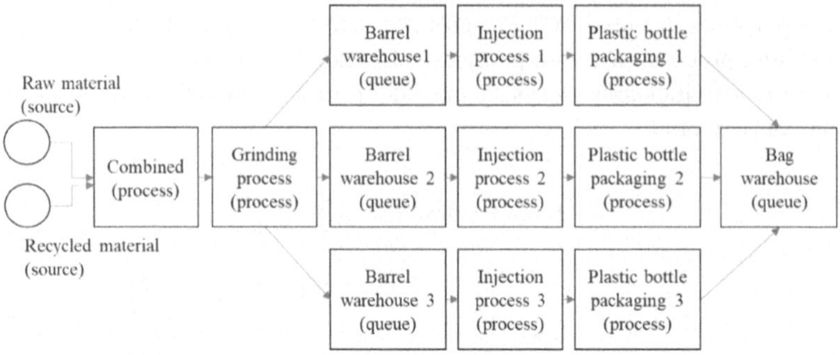

Fig. 2. System flow diagram, Source: own elaboration.

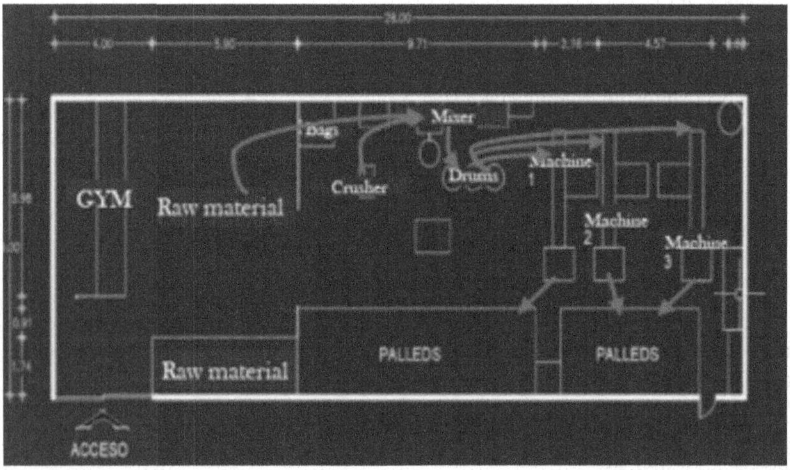

Fig. 3. Production system lay out, Source: own elaboration.

Numerical data is obtained through records and time studies. This data is used to turn raw data into valuable and actionable information. They serve to understand the operation of the system and its context, allowing more effective decision-making. These data are presented at Table 2, grouped by type of material, quantity, and process times.

Table 2. Numerical data

Material	Quantity	Time
Raw material	3 tons	Each month (22 working days)
Bottles production plan	Production average rate: 22,500 bottles per day, between the 3 machines Rejection rate (±2% by defective ones) Bottle weighs: 22g	25 kg of material per day per machine
6 cans of material (each can of 25 kg)	105 kg of virgin material (70%) 45 kg of recycled material (30%)	Daily for 3 machines
Injection molding machines	3 machines	Permanent
Shredder	1 machine	Permanent
Material canisters casting	3 machines	Permanent

When recording the times of each activity (prior to grouping) we corroborated that the data were independent by performing a visual independence test with the scatter plot. Since, if the graph follows a pattern or trend line, the data are dependent; and if there are random points, the data are independent. Figures 4, 5, 6, 7, 8, 9, 10 and 11 display the data time scatterplots for each process group. We show the process group is having an independent process because the scatterplots are random points.

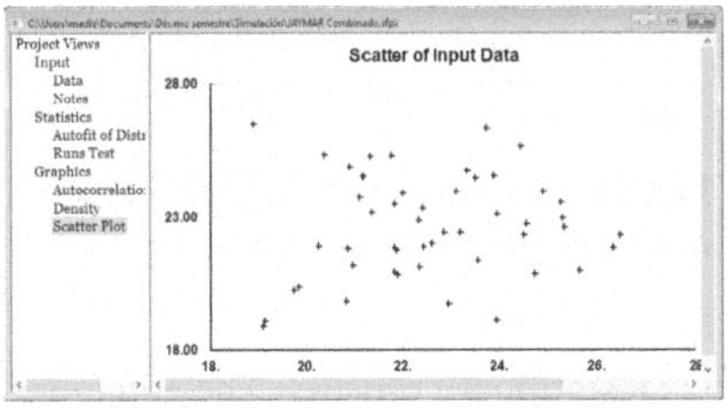

Fig. 4. Scatter plot in combined process, Source: own elaboration

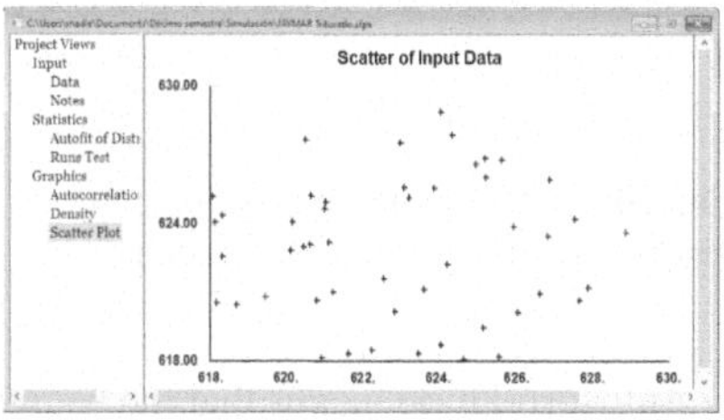

Fig. 5. Scatter plot in grinding process, Source: own elaboration

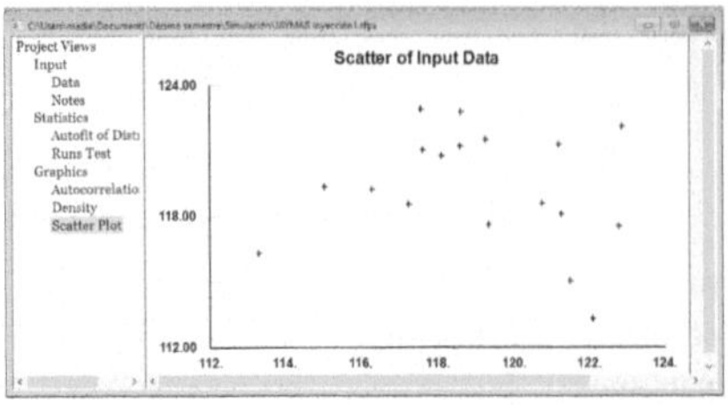

Fig. 6. Scatter plot in injection molding process machine 1, Source: own elaboration

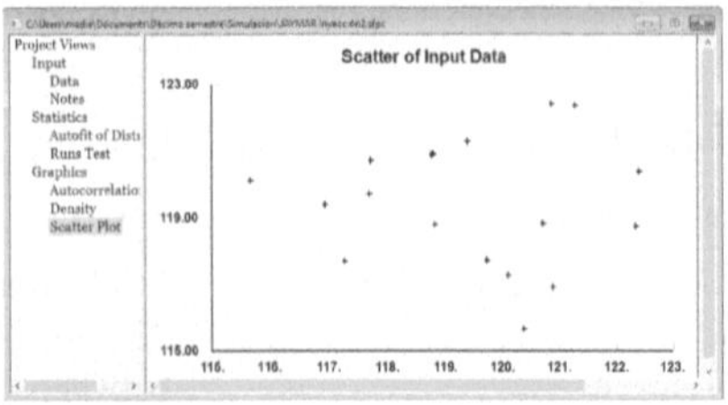

Fig. 7. Scatter plot in injection molding process machine 2, Source: own elaboration

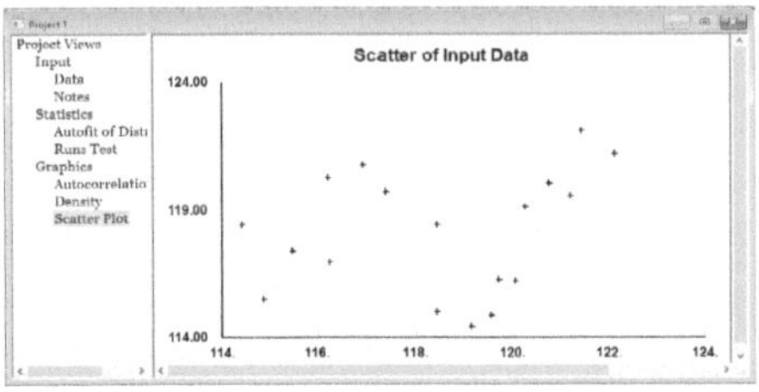

Fig. 8. Scatter plot in injection molding process machine 3, Source: own elaboration

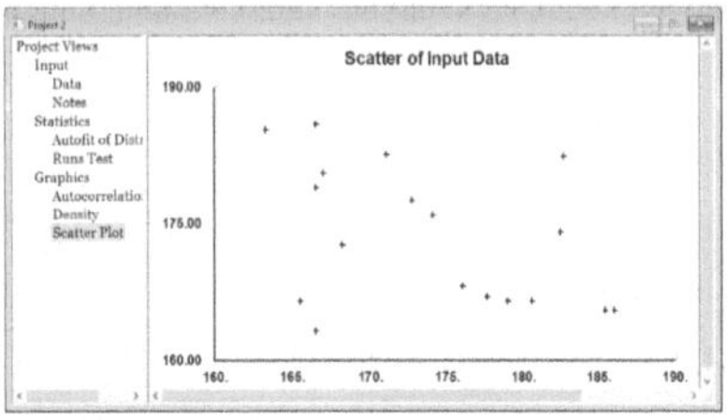

Fig. 9. Scatter plot in packing process machine 1, Source: own elaboration

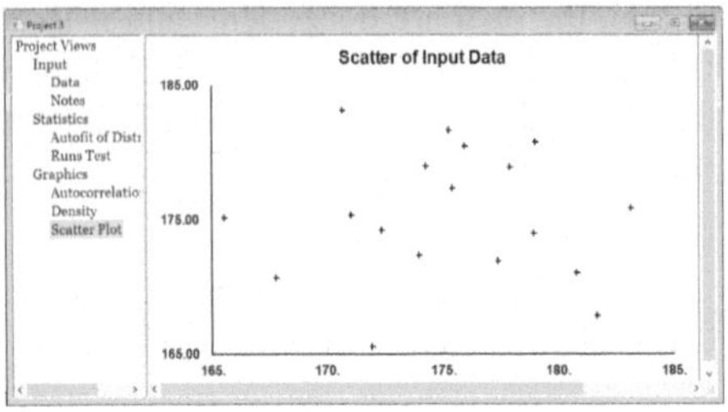

Fig. 10. Scatter plot in packaging process machine 2, Source: own elaboration

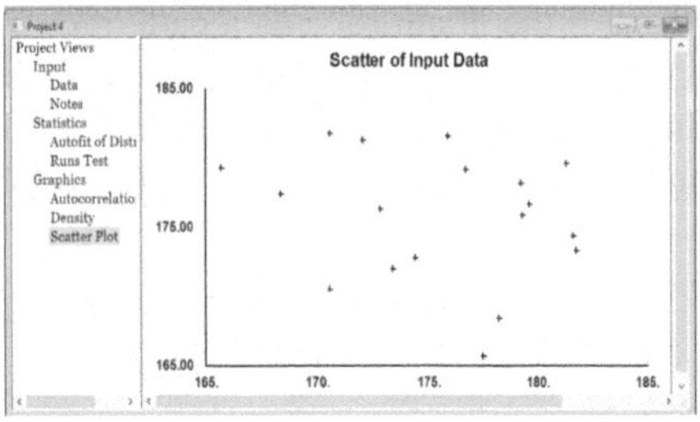

Fig. 11. Scatter plot in packaging process machine 3, Source: own elaboration

Autocorrelation means that the source data is a stationary process. The autocorrelation value could be between 1 and −1, if the autocorrelation is closer to zero, indicating little or no correlation. The Table 3 show the autocorrelation in the positive and negative directions of the group processing time.

Table 3. Autocorrelation Process group facilities.

Process group name	Correlation positive direction	Correlation negative direction)
Combined process	0.134	−0.268
Grinding process	0.178	−0.306
Injection molding process machine 1	−0.129	−0.166
Injection molding process machine 2	0.203	−0.041
Injection molding process machine 3	0.238	−0.209
Plastic bottle packaging Process Machine 1	0.247	−0.438
Plastic bottle packaging Process Machine 2	0.241	−0.505
Plastic bottle packaging Process Machine 3	0.039	−0.223

After having tested the numerical independence and the correlation analysis, we define the conversion of numerical data into information suitable for the simulation. From using numerical data as we have, we choose applying a goodness-of-fit tests method for having suitable data simulation.

The goodness-of-fit test helps us, based on the production study time (time taking in the plastic bottle production process), to know what distribution the data follow and thus to obtain information to simulate the production, see Table 4.

Table 4. Autocorrelation Process group facilities.

Process group name	Distributions tested	Distribution valuation	Selected distribution
Combined process	Normal (22.6, 1.89) Uniform (18.9, 26.5) Exponential (18.9, 3.69) Lognormal	Do not rejected Do not rejected Reject Reject	Normal (22.6, 1.89)
Grinding process	Lognormal (558, 4.16, 0.0457) Normal (623, 2.95) Uniform (618, 629) Exponential (618, 4.78)	Do not rejected Do not rejected Do not rejected Reject	Normal (623, 2.95)

(*continued*)

Table 4. (*continued*)

Process group name	Distributions tested	Distribution valuation	Selected distribution
Injection molding process machine 1	Normal (119, 2.57) Lognormal (−1.09e+0.003, 7.09, 0.002) Uniform (113, 123)	Do not rejected Do not rejected Do not rejected	Normal (119, 2.57)
Injection molding process machine 2	Lognormal (−634, 6.62, 0.00237) Normal (119, 1.79) Uniform (116, 122)	Do not rejected Do not rejected Do not rejected	Lognormal (−634, 6.62, 0.00237)
Injection molding process machine 3	Uniform (114, 122) Normal (118, 2.37) Lognormal (−635, 6.62, 0.00315)	Do not rejected Do not rejected Do not rejected	Uniform (114, 122)
Plastic bottle packaging Process Machine 1	Lognormal (160, 2.45, 0.602) Exponential (163, 10.2) Normal (173, 7.34) Uniform (163, 186)	Do not rejected Do not rejected Do not rejected Do not rejected	Lognormal (160, 2.45, 0.602)
Plastic bottle packaging Process Machine 2	Normal (175,4.64) Lognormal (−1.03e+0.003, 7.1,0.003) Uniform (166, 183)	Do not rejected Do not rejected Do not rejected	Normal (175, 4.64)
Plastic bottle packaging Process Machine 3	Normal (176,4.52) Lognormal (−1.03e+0.003, 7.1, 0.003) Uniform (166, 182)	Do not rejected Do not rejected Do not rejected	Normal (176, 4.52)

3.3 Step 3: Build the Model

We use FlexSim® simulation software to represent the production system. The operating conditions are a 12-h working day (starting at 8:00 a.m.). Hygiene rules are the use of masks and hats. Safety regulations are the use of safety shoes, vests, and belts.

Simulation objectives.

- Closing of the operation on the third line production machine.
- Increase production with only two Injection molding process machine.

Decision variables.

- Injection molding process machine in operation
- Packing process machines in operation.

State variables.

- Operation time
- Work shifts number.

Response variables.

- Finished production.

Three simulation scenarios are proposed where the current state of operation is analyzed, with an operation of one extra machine and one less, to know the feasibility of the decision that can be taken within the company, showing the following results.

3.4 Step 4: Validate the Model

Based on the collected data of times and production plan, simulate the production process of plastic bottles considering the following: two bottles are produced per injection molding machine, the units are in kilograms, and the 2% of the production is defective (this was already considered in the simulation). The Fig. 12 show the model base simulate at FlexSim®.

A total of 150 [kg] of raw material enters at the system per day, which is divided into 3 lines distributing an equal amount, of 50 [kg] to each line, remembering that we have only one mixer and one crusher, which ensure that the raw material is homogeneous. In each line the mixed is introduce into the bottle injection process, once this process is finished, bottles will be packaged in bags (with 110 bottles).

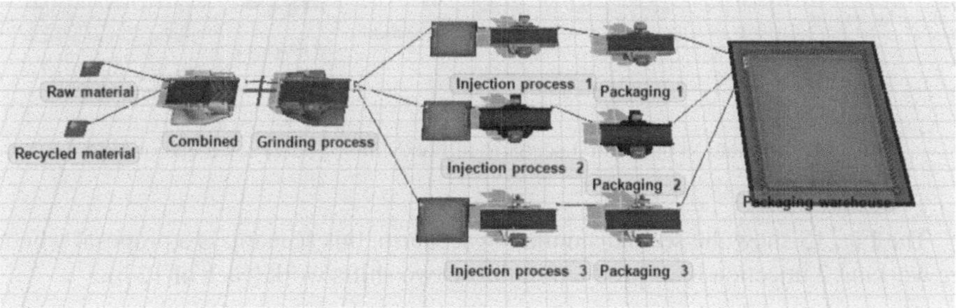

Fig. 12. FlexSim® simulation design, Source: own elaboration

The base model was review by the stakeholder of the process. The process owner was agreed with the results of the base model results. Then we follow withe the experiment evaluation at process simulation step 5, see Fig. 13.

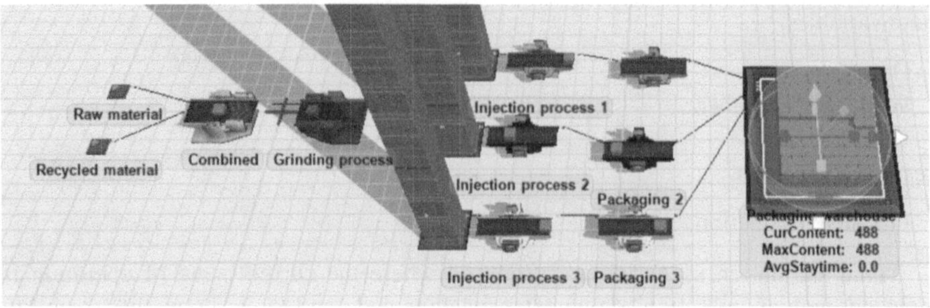

Fig. 13. Simulation of the base model at FlexSim®, , Source: own elaboration

3.5 Step 5: Conduct Experiments

For the problem raised, we make some proposals that will help your efficiency and guide you to a good decision-making process:

The first proposal is to simulate the 2 lines that you want to maintain during only 1 shift (defined as 12 h). These scenarios let prove the tendential results of decision to close of the operation on the third machine, see Fig. 14.

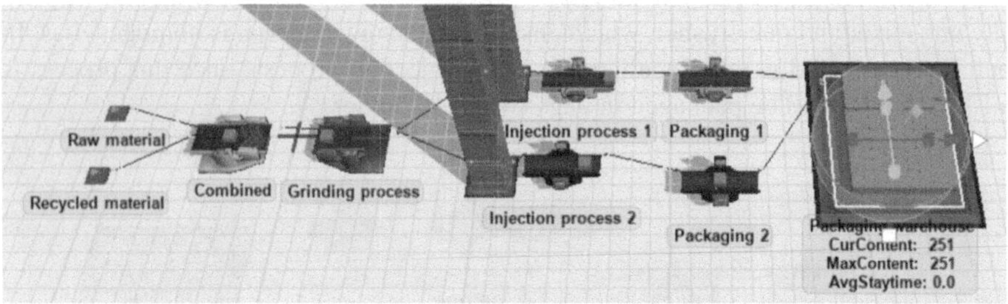

Fig. 14. Simulation of the first scenario at FlexSim®, Source: own elaboration

The Fig. 15 show the second simulation scenario, this scenario is a proposal where are working 2 injection lines production throw two shifts work (each of 12 h).

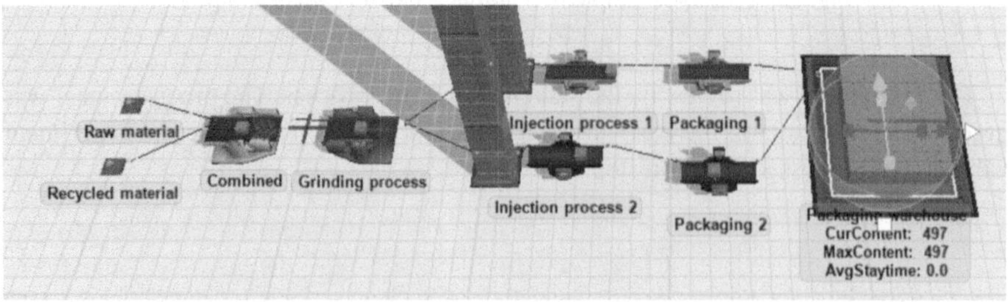

Fig. 15. Simulation of the second scenario at FlexSim®, Source: own elaboration

The third scenario is to maintain our proposal with 2 injection lines and 1 shift (12 h defined), in addition we leave 3 packaging areas. This would be the best proposal if we were looking to produce more and sell line 3 (except, of course, the packaging area). When proposal three is proposed, the idea is to make the process in the packaging area more efficient, and this is reflected in a minimum backlog, since the time in which the bottles are removed allows reducing the bottleneck generated by leaving the bottle in the injector (Fig. 16).

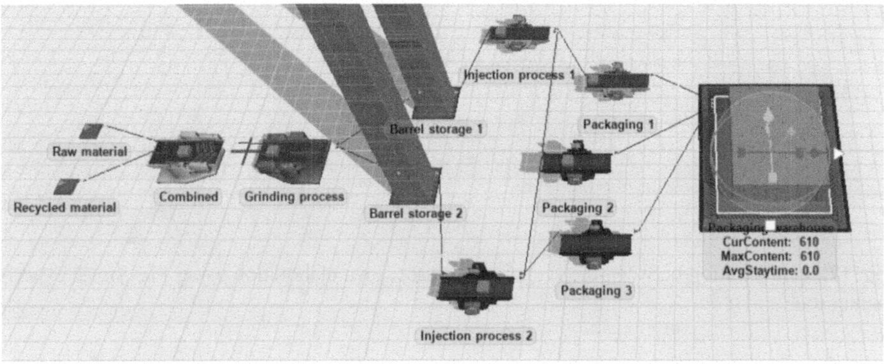

Fig. 16. Simulation of the third scenario at FlexSim®, Source: own elaboration

Table 5 shows the way the decision and state variable are changing between each simulation scenario using FlexSim ®. At the end of the Table 5 we can see the variable response with only one replication of the simulation results.

Table 5. Conduct experiment results.

Variables	Description variable	Base model	First scenario	Second scenario	Third scenario
Decision variables	Injection molding process machine in operation	Three	Three	Two	Two
	Packing process machines in operation	Three	Two	Two	Three
State variables	Operation time	12 h	12 h	24 h	12 h
	Work shifts number	One turn	One turn	Two turns	One turn
Response variables	Produced kilograms	488 kg	251 kg	497 kg	610 kg

We tested with the FlexSim® simulation experimenter tool. The experiment tool aims to perform system replication in the same execution of the simulation. Figure 17 shows the comparation between the base model and the three tested scenarios.

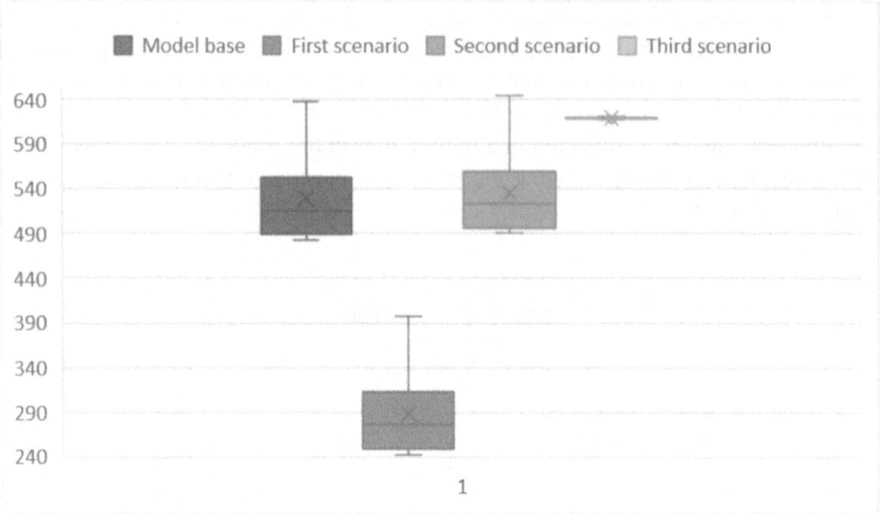

Fig. 17. Comparation between the scenarios, Source: own elaboration using FlexSim® experimenter tool.

4 Results and Discussion

The experimenter's tool allowed knowing a result of thirty simulation models. We were able to identify the maximum and minimum final production among the thirty replicates. And we are 95% sure that the final production will be within a production confidence interval. Tables 6 shows statistical result of the scenario using experimenter tool.

Table 6. Statistical scenarios results

Scenario	Mean (95% confidence interval)	Sample standard deviation	Minimum observation	Maximum observation
Model based	529.1 ± 16.3	43.5	483	638
First scenario	288.4 ± 16.2	43.5	242	398
Second scenario	535.3 ± 16.1	43.2	490	645
Third scenario	619.267 ± 0.324	0.868	617	621

Having said the above, if we use the simulation results as we have at Table 7, we can see for model base that we get 529 [kg] of finished product, that is, if we divide this result by 22 [g] that a bottle weighs, we get a result of approximately 24045 bottles (as finished product without shortages). This means that 218 bags with 110 bottles each are packed and 65 are instore.

Table 7. Production scenarios

Model results	Base model	First scenario	Second scenario	Third scenario
Mean produced kilograms	529	288	535	619
Produced bottles	24045	13091	24318	28136
Bag of 110 bottles	218	119	221	255
Bottles without packing	65	0	8	86
Percentage of target production		−45.56%	1.13%	17.01%

At first scenario whit removing the production line and not establishing alternative, there is a backlog of 45.56%, which makes it unprofitable due to the high demand required. The second scenario reaches a production of 1.13% more than the initial scenario. This scenario consists of eliminating a production line with an extra shift of work. The third proposal is to produce more and sell line 3 (except the packaging area). We observe that the production increases 17.01% more in a daily work of 12 h. This would be the best proposal for reduce a work shift.

5 Conclusion

The line production simulation was tested with three scenarios. If stakeholder wants to close the operation of the third production line (which includes injection molding process 3 and packaging 3), the process must operate twice as long to have almost the same final output. If the stakeholder wishes to close the operation of the third production line (including the injection molding process 3 and packaging 3), and you also must have the same work shift, the system process should remove only the injection molding process 3 and keep the packaging process 3 for having the same final output.

When the simulation was carried out with simulation, we could identify the best option, and let the stakeholder makes final decision. This project let the MSMEs knowing the simulation benefits and describe the process simulation tool that could be using for future implementation. The simulation process let improve the production line without affect the real operation process and it could apply on this small business case. The MSMEs line simulation process tested the change of operation between the needs of the owner's business. However, we consider that it might be necessary to apply other tools such as bottlenecks production and simulation-based Lean Six-Sigma, which we can apply for later publications.

References

Breque, M., De Nul, L., Petridis, A.T.: Industry 5.0: towards a sustainable, human-centric and resilient European industry. Publications Office of the European Union, Luxembourg (2021)

Cugno, M., et al.: Industry 4.0 and production recovery in the Covid era. Technovation **114** (2022)

Dixson-Declève, S., et. al.: Industry 5.0: A transformative vision for Europe. ESIR Policy Brief No. 3. Publications Office of the European Union, Luxembourg (2022)

El-Haik, B., Al-Aomar, R.T.: Simulation-Based Lean Six-Sigma and Design for Six Sigma. Wiley, New Jersey (2006)
FlexSim® Homepage. https://www.flexsim.com/es/. Accessed 02 Feb 2023
Garcia Martinez, R.F., et. al.: A flexible and open environment for discrete event simulations. Interactive Design and Manufacturing. Int. J. Interact. Design Manuf. (IJIDeM) 509–524 (2021)
Goyal, S., Sergi, B., Esposito, M.F.: Business development services for micro, small and medium enterprises – literature review of past trends and future directions. World Rev. Entrep. Manage. Sustain. Dev. **14**(3), 312 (2018)
Harrell, C., Ghosh, B., Bowden, R.T.: Simulation Using ProModel. The McGraw–Hill (2000)
Voigt, T., Flad, S., Struss, P.F.: Model based fault localization in bottling plants. Adv. Eng. Inform. **29**(1), 101–114 (2015)

Enhancing Smart Manufacturing Systems: A Digital Twin Approach Employing Simulation, Flexible Robots and Additive Manufacturing Technologies

Romão Santos[1](✉), Cláudia Rocha[1], Rui Dias[1], and João Quintas[2]

[1] INESC TEC - Institute for Systems and Computer Engineering, Technology and Science, Porto, Portugal
romao.f.santos@inesctec.pt
[2] IPN, Instituto Pedro Nunes, Coimbra, Portugal

Abstract. A new generation of manufacturing systems is emerging through the adoption of new policies to overcome future crises highlighted by constant social, environmental, and economic concerns. The rise of so-called smart manufacturing is noticeable. However, new risks to humankind are being introduced, and, more than ever, science and technology are required to guarantee the future sustainability and resilience of our manufacturing systems. This research presents a Digital Twin approach resorting to simulation models with embedded intelligence to transform efficient manufacturing systems and react to complex and unpredictable circumstances. The methodology covers production scheduling incorporating flexible robots, internal logistics supervision contemplating planning and control of mobile robots, and capacity management. The method demonstrates the potential of integrating Additive Manufacturing technologies to quickly react to production needs. The developed strategy was enforced and assessed in an industrial experiment, exhibiting its robustness and promising application. The attained results were very encouraging, highlighting its potential extension to more complex industrial systems.

Keywords: Digital Twin · Smart Manufacturing · Discrete Event Simulation · Real-Time Decision-Making · Addictive Manufacturing

1 Introduction

Industry 4.0 covers digital technologies and automation of industrial processes in order to improve the production efficiency, quality, and flexibility of the manufacturing industry. Nowadays, the fourth-revolution concepts have been deployed, integrating different technologies such as robotics, autonomous systems, the Internet of Things (IoT), distributed and cloud computing, and multi-agent systems in a smart factory fashion. This change of paradigm allows an inter-connected environment that enables real-time feedback and a high level of adaptability in smart manufacturing systems [1].

M. Mujica Mota and P. Scala (Eds.): EUROSIM 2023, CCIS 2032, pp. 277–291, 2024.
https://doi.org/10.1007/978-3-031-68435-7_20

The manufacturing industry is increasingly evolving to cope with the technological market needs, seeking more productive, optimized, flexible, and streamlined processes. Internal logistics tasks such as carrying objects, which were commonly performed by humans, are being automatized by mobile vehicles without requiring infrastructure changes. Automated Guided Vehicles (AGVs) that follow a pre-determined route and Autonomous Mobile Robots (AMRs) that autonomously move resorting to real-time path planning are capable of driverless materials transportation. Moreover, robotic manipulators can also be incorporated to allow the handling of objects [2, 3].

Altogether, service or product innovations require large investments in both research and development. Hence, one of the challenges addressed by Industry 4.0 is on how to reduce these costs through a more lean and rapid prototyping approach. The development of the new product could virtually be conducted without the need for its own production line [4]. A set of technologies such as 3D printing of objects contributed to coin the term Additive Manufacturing (AM) [5]. The AM paradigm has been developed in conventional industry with the goal to facilitate the creation of exclusive products and their addition into production without changing traditional lines, thus creating a significant reduction in production costs of prototype products and in assessing their viability.

Taking direct examples from the industry, companies from the moulding and plastic industry face the need to increase their production capacity, mainly characterized by products with great variability in terms of complexity and in terms of quantities produced [6]. The production process of plastic parts can be divided into two sub-processes: plastic injection and post-injection processing. Currently, the injection process is highly automated due to the use of robotic gantry systems that unload the plastic parts from the injection machines. Post-injection part processing, however, has a suboptimal level of automation due to the lack of flexibility of traditional robotic solutions to adapt to the variability in the complexity of producing a multi-product line [7]. Nowadays, the key to dealing with the different complexities of a large product catalogue is to keep the level of automation deliberately low in post-press part processing. Hence, combining all above mentioned technologies aim to contribute to a more flexible and robust AM, increasing the traceability of products and optimize productive tasks and internal logistics.

The combination of recent sophisticated mobile vehicles (AMRs and/or AGVs) with AM technology in the scope of smart manufacturing systems seems to be a useful opportunity. However, the control, monitoring and optimisation of this technology is still complex and demand for new or improved digital technologies. In this regard, the aim of this work is to provide procedures on how to boost simulation models as a component of a Digital Twin (DT) architecture and benefiting from its main advantages such as virtual testing and evaluation of what-if scenarios, analysis of complex systems under uncertainty and assessment to Key Performance Indicators (KPIs). The main contributions of the presented work rely on: i) a conceptual architecture enhancement based on Open Scalable Production Systems - OSPS framework [8, 9]; ii) a detailed technical description of a simulation base model implementation using Flexsim[1] simulation engine to automatically load real-time data and support decision making on using AM technology

[1] https://www.flexsim.com/pt/.

deal with possible bottlenecks or undesired KPIs; iii) guidelines on how to apply the proposed methodology to real industrial environments.

The structure of this research can be summarized as follows: Sect. 2 describes the background and related literature in terms of solutions dealing with complex manufacturing systems and allowing the integration of mobile robots, AM technologies, DT, Industrial IoT and Discrete Event Simulation (DES). Also, in this section, gaps and similar research works are presented to further analysis and exploitation. Section 3 presents the Simulation-Based Digital Twin methodology focusing on two levels – conceptual and technical – and explores details of a simulation base model and its connections to external services. Section 4 identifies an industrial experiment, representative of an automotive industry real use case and serves as the environment for methodology validation. Finally, Sect. 5 describes the main conclusions and future work.

2 Background and Related Literature

This section provides a comprehensive overview of the existing research in the field of the topic at hand. It serves as a foundation for the research study by presenting the current state of knowledge, identifying gaps in understanding, and highlighting areas of potential exploration. In this sense, studies of digital twins are presented with the associated concerns and opportunities. Also, the DT concept is then related to AM and automated vehicles to identify benefits from this integration.

Michael Grieves introduced the DT concept in a presentation about Product Lifecycle Management (PLM) with the title "Conceptual Ideal for PLM". After this first informal introduction, the author formally described, in his white paper [10], DT as a virtual representation of a real object with data flowing in both directions and allowing convergence and real-time synchronization of virtual and physical spaces. The DT in its basis defines mirroring a product. However, the literature shows that processes (e.g. manufacturing) can also be subjects of virtual space reproduction and achieve similar benefits [11]. In the manufacturing subject, DTs have been used to deal with challenging situations by optimising operations, revealing bottlenecks, and taking advantage of simulated scenarios to forecast manufacturing systems' performance [12]. Also, several examples appear on literature of DTs being applied to production planning and scheduling [13, 14]. Although DTs have been demonstrated relevant improvements in manufacturing challenges, there are still gaps to overcome. Sharma et al. [15] highlight the wide deviation between theoretical and practical DTs. Additionally, the important role of IoT and its technology maturity dependency is mentioned and searching for practical IoT platforms becomes even more relevant. Recent years have witnessed the emergence of a wide variety and heterogeneity of IoT support solutions [16]. Despite some standardisation efforts in the IoT domain, there are no universally agreed standards for IoT platforms architecture [17]. However, a common feature of IoT platforms is to support interoperability between disparate systems, supporting the secure exchange of information even if they natively use different communication protocols [18]. To this end, IoT platforms include a diverse set of functional components that provide services aimed at eliminating the direct interaction between systems and applications.

Moreover, there are still opportunities for improvement when implementing DTs incorporating AM technology. Bartsch et al. [19] discuss the need for standardization and consistent use of terminology to connect different systems, along with the development of practical frameworks and interfaces. The digital twin offers a novel solution for addressing several additive manufacturing challenges, including process simulation, monitoring, and control. The process simulation is used by Filz et al. [20] for planning material supply using AGVs and describing the advantages of testing different scenarios with associated velocity and the number of available vehicles. Furthermore, Popper et al. [21] present a concept on how to integrate take advantage of integrated systems, which include data, models and processes and identify the need of developing new methods, models, and simulations that can calculate "what if" scenarios. This capability is necessary to enable timely reactions to changing framework conditions. Also, Leng et al. [22] made available a review of research studies applying the digital twin to enable smart manufacturing systems (SMS). Here, it is highlighted the relevance of using DES for process flow designing and planning of a SMS.

Observed results in the above-referred works clearly show the relevance of digital twins in addressing manufacturing systems problems. Hence, a real-time decision-making framework, taking value from innovative solutions such as AM and automated vehicles, seems quite relevant to explore. Moreover, to the best of the authors' knowledge, the current literature does not have any work on a DT methodology with combination of mobile robots, AM technology and DES to optimise production scheduling and balancing in real-time.

3 Simulation-Based Digital Twin Methodology

In the flexible production system illustrated in the previous sections of this research, collaborative robotic systems should be seen as resources that can be allocated as production needs adjust, integrating seamlessly with the production system and its operators. In this chapter, the adopted methodology emerges, initially a conceptual explanation of the main components of the approach is presented. Subsequently, technical information regarding the simulation base model, connectors to the industrial IoT (IIoT) Platform and corresponding implementation, is revealed.

3.1 High-Level Digital Twin Architecture

To reach a smart manufacturing system that permits an AM cell to be used as backup resource, a Digital Twin architecture (Fig. 1) is presented. The base of the architecture is the OSPS framework [8, 9], where the central actor is the Advanced Plant Model (APM). The main adjustments of the OSPS framework are related to the introduction of AM cells and new functionalities for the decision support system based on discrete event simulation.

The introduction of IoT technologies in the industrial context plays a crucial role in trying on one hand to get information coming from the shop floor devices and on the other to guarantee the mediation and aggregation of information among the various systems used in the context of an industrial environment [23]. Normally, data is routed to a message broker that promotes asynchronous communication between the various systems through the publication and subscription of data streams. The message broker acts as an intermediary that receives information from various services ("Publishers") in a given channel which is then made available to the services that subscribe to that channel ("Subscribers") [24].

An IIoT platform has also been provided, which ensures asynchronous communication between components whose native communication mechanisms are different. The IIoT platform was built using open-source technologies such as Apache Kafka[2] and included custom connectors that make it possible to provide MQTT[3], Kafka, OPC-UA[4] and ROS[5] interfaces. As an example, and illustrated in Fig. 1, the APM uses the IIoT platform's Kafka interface to interact with the robotic resources, which in turn connect to the IIoT platform through its ROS interface. The AM cell aggregates a fused deposition modelling (FDM) 3D printer (Creality Ender-6), a computational unit acting as controller for the 3D printer running Octoprint with an MQTT plugin, a robotic arm (Franka Panda 6 DoF robot with gripper) and a computational unit acting as controller for the robotic arm (Aaeon Up Squared Pro), and additional sensors for visual inspection for the 3D printer (Logitech Webcam) and robot arm (Intel Realsense D435i). In addition, the AM cell includes a LAN/WAN router to establish the local network that connects the 3D printer and robotic arm, and grants connection to the internet for communicating with the APM.

Although the general architecture is presented here, the focal point of this research is to explore the simulation component and its contribution to the manufacturing system KPIs' improvement. Besides production scheduling scenarios' evaluation, the simulation model also includes heuristics to automatically perform production rescheduling and jobs allocation to the AM cell. The conditions to trigger and sequence jobs in the AM cell can be summarized as follows:

- due dates delays;
- products with a low demand;
- stock disruptions;
- machines or resources unavailability or maintenance.

[2] https://kafka.apache.org/.
[3] https://mqtt.org/.
[4] https://opcfoundation.org/about/opc-technologies/opc-ua/.
[5] https://ros.org/.

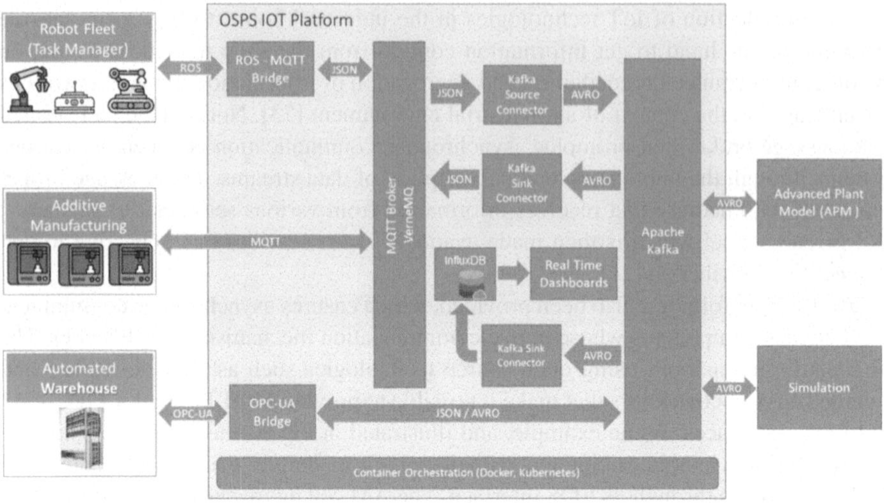

Fig. 1. Digital Twin main components and protocols, adapted from [8, 9]

Heuristics, working as rescheduling rules, perform based on production KPIs (e.g. throughput, capacity utilisation, lateness) resulting from multiple simulations (using experiment replications to obtain more accurate results). They are designed in a routine enabling them to adapt based on the true-to-life data generated during the test runs. Greedy algorithms such as "Earliest deadline first" or "Shortest processing time first" are tested in the simulation model runs.

3.2 Simulation Base Model - Technical Approach

Manufacturing systems involve many interacting components such as machines, workers, material handling systems, and other resources, which results in a high degree of system complexity. This complexity can lead to unexpected bottlenecks, low efficiency, and other operational problems. DES is commonly used to simulate the behaviour of manufacturing systems, allowing decision-makers to evaluate different scenarios and optimize the system's performance. Moreover, DES can facilitate the identification of potential bottlenecks, inefficiencies, and other operational issues in the system, enabling decision-makers to implement changes or improvements to the system in a cost-effective way. In this sense, this section describes the main technical details of the digital twin focusing on a simulation base model and its components that will allow decision-makers to automatically assess better solutions. Also, the interaction of the simulation engine with real-time data from the IoT platform is exhibited.

In terms of the simulation engine, Flexsim, a commercial DES software package, is used. It provides a graphical user interface (GUI) that allows users to create models by assembling 3D objects, machines, conveyors, operators, and other components. It also supports the integration of external data sources and real-time connections to enable dynamic decision-making in response to changes in the environment. In this regard, this work examines how these external connections can be deployed, improved, and replicated. Figure 2 points out the main components of the proposed architecture of a simulation base model and its link to a wrapper service, represented by a file programmed in python language.

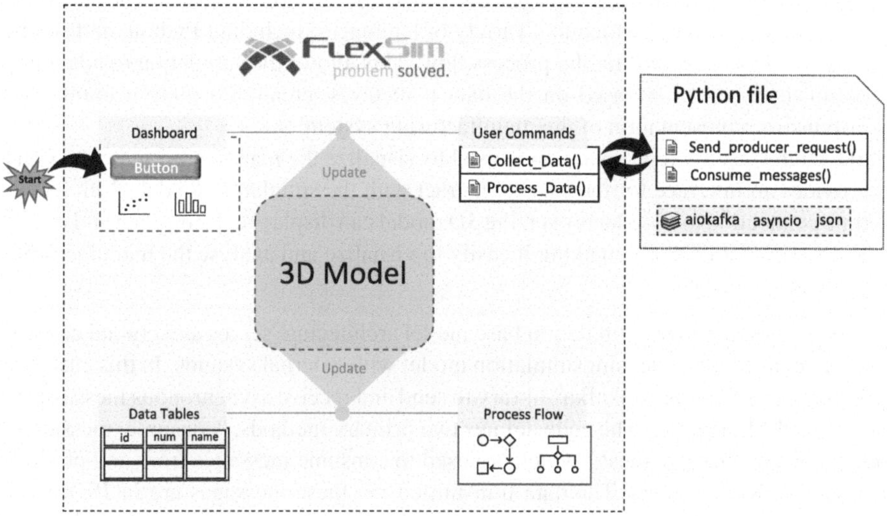

Fig. 2. Simulation Base Model Architecture based on Flexsim and external connections.

The role and features of the elements presented in the simulation base model architecture can be described as follows:

- **Dashboard** - A interface to monitor and control the simulation, providing real-time or simulated information, reports, and graphical analysis tools. Furthermore, using a custom button (with Flexscript programmed code), a set of user commands and Flexsim functions are invoked to reset, load data and run the model.
- **User Commands** - Custom user coding functions used to send instructions or adjust the simulation model. Many possibilities are available such as to modify model parameters, start or stop a process, or change the model state. Furthermore, user commands can be sent to the simulation model from external services, triggering specific actions or changing the model's behavior. Particularly, two user commands are essential in this process:

  – "Collect_Data()" - establish a connection to a Python file with two methods and allows real-time data collection and return as output a JSON string with complex data structure embedded.
  – "Process_Data()" - receives the output of the previous user command and converts JSON data into Flexsim objects and data tables information. As output, it fulfils source data tables of process flow, triggers processes and creates 3D objects corresponding to a snapshot of the real system.

- **Data Tables** - Provide an interface to access the simulation data generated during runtime, which can be analysed, manipulated, and exported for further processing.
- **Process Flow** - A visual representation of the logic and flow of the simulation model. It allows the user to organize, control and make decisions based on different outcomes. Also, custom Flexsim steps and logic can be used to develop heuristics. The heuristics programming can be written in a variety of languages, including Python, and can be called and executed within the process flow. This allows the simulation to adapt and optimize in real-time based on the data it receives, creating a more accurate and responsive representation of the manufacturing system.
- **3D Model** - A 3D representation is used to visualize the manufacturing process and provides an interface for the user to interact with the simulation model. With a real-time connection to a Kafka broker, the 3D model can display real-time data and events before simulation starts, making it easily to visualize and analyse the manufacturing process as it happens.

The Python file in the simulation base model architecture serves as a crucial component that connects the Flexsim simulation model with external systems. In this case, the Python code utilizes the aiokafka[6] library to send and receive asynchronous messages to and from a Kafka broker. The code defines two primary methods, "consume_messages" and "Send_producer_request," which are used to consume messages from and produce messages to Kafka topics. The data transmitted via these messages are in JSON format, which allows for easy processing and analysis by the simulation model. The use of asynchronous messaging with Kafka provides several benefits, including high scalability, fault tolerance, and low latency. By utilizing asyncio[7] and aiokafka libraries, the Python script can send and receive messages without blocking the main thread, allowing for high-performance and responsive communication with external systems. This is particularly important for real-time manufacturing systems, where timely data processing and analysis are crucial for efficient operations. Overall, the Python file serves as a critical component in the simulation base model architecture, enabling seamless communication and integration with the IIoT platform.

Considering the description and specification of the Flexsim simulation base model, integrated in a Digital Twin architecture, the novel component of this research, Fig. 3 exhibits more detail regarding the Flexsim simulation base model main process flow. Here, a set of tokens is created to manage, optimise and trigger the execution of a set of sub-flows and embedded custom scripts. Also, this figure represents the main control of the manufacturing process in different levels of detail (production orders, standard operation procedures and manufacturing tasks) and it is relevant to highlight that this can be exported to a user library and reused in other Flexsim simulation models of other industrial case studies without major changes.

[6] https://aiokafka.readthedocs.io/.

[7] https://docs.python.org/3/library/asyncio.html.

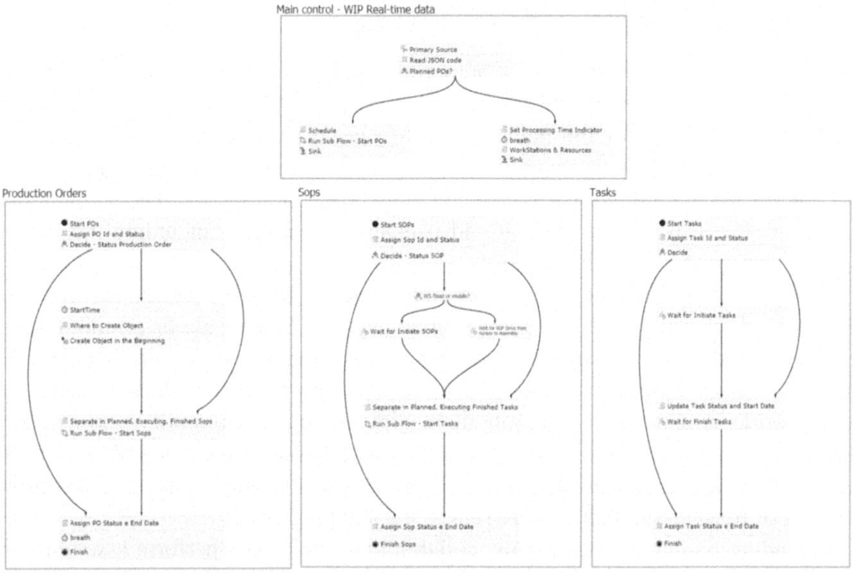

Fig. 3. Flexsim base model main process flow.

This section has presented the methodology used to develop a digital twin and a real-time optimisation of manufacturing processes using Flexsim, external connections and taking advantage of heuristics and AM resources. To validate the concept and explore further real use-case scenarios, an industrial experiment was assessed.

4 Industrial Experiment

This section describes the application context of this study which is a thermoplastic injection moulding company whose activity is focused on the Automotive Industry. Currently, the company needs to increase its production capacity, characterised by products with associated high variability in terms of complexity and quantities produced. To overcome this challenge and considering the impossibility of using the real production system, an industrial simplification of the process was taken into account. An industrial laboratory was chosen for the demonstration of the developed solution. It disposes of equipment and automation technology that allows recreating the production process of the pilot company. Also, the choice of the laboratory facilities allows the development of the solutions in a controlled environment without interference with production, ensuring not only the compatibility and applicability of the solution to the real context, but also contributing to its generalization and scalability to other application sectors.

4.1 Manufacturing Process and Configuration

In this experiment, the production of automotive plastic components was divided into two sub-processes: moulding and assembly. Injection moulding is the most widely used manufacturing process for the manufacture of plastic parts, however this study also attempted to analyse the potential of integrating additive manufacturing technologies to react quickly to the production needs of a flexible production system. The second phase of production consists of processing and packaging operations including mechanical assembly, screwing and quality control.

The physical environment also incorporates a vertical warehouse, an additive manufacturing cell as well as a collaborative robotic system capable of performing internal logistics, assembly and packaging tasks. The additive manufacturing cell includes a 3D printer and a fixed robotic manipulator, and the assembly tasks include a vertical warehouse, a workstation and a mobile robotic manipulator capable of building kits of parts collected from the vertical warehouse and used as a basis for the assembly tasks. The assembly line includes a workstation and has allocated an AMR capable of transporting the boxes of finished product that are sent to the shipping area.

Depending on demand variability, collaborative robots can perform assembly and packaging operations side-by-side with human operators. When demand is low, human operators are sufficient to handle the assembly and packaging operations, easily achieving the desired production throughput. However, when demand increases, human operators should be supported by collaborative robots capable of performing repetitive or physically demanding tasks, thus improving both ergonomics and production efficiency. Similarly, at peaks in demand, a fully robotic solution could be used to achieve the desired production levels.

4.2 Performed Tests and Preliminary Results

The developed solution includes the creation of a digital twin representation model, which supports the traceability of products and by-products. Figure 4 - Flexsim simulation base model of the conducted experiment. Figure 4 presents the Flexsim 3D model and the main dashboard of the conducted industrial experiment. In this figure, the main components of the above-described experiment are modelled, where a set of mobile robots and an AM cell, composed by a 3D printer and a collaborative robotic arm are used in internal logistics and manufacturing processes.

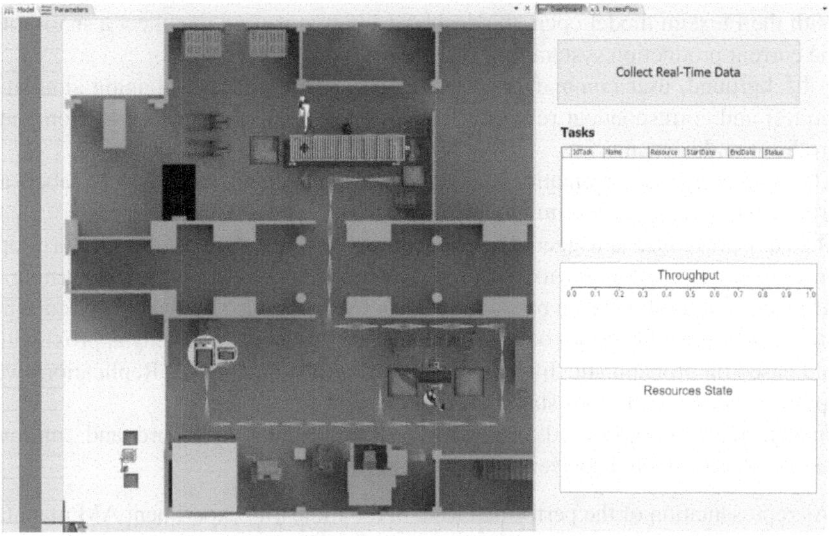

Fig. 4. Flexsim simulation base model of the conducted experiment.

The digital twin aims at virtual representation of production data in order to achieve absolute interoperability being integrated with both the manufacturing execution system and shop floor equipment. The digital twin representation provides the detail of the current production Work-In-Process (WIP) to be processed by the decision support system, allowing simulations of production scenarios and response to the occurrence of faults and/or delays in the production line. Finally integrated optimisation procedures can suggest rescheduling of new production orders.

After a setup phase and taking into account the process described above, a set of steps can be examined and supported the execution of the industrial environment process:

1. A set of production orders with tasks associated to resources are inserted in a production schedule based on a pre-defined demand;
2. The APM ensures the production schedule execution and starts assigning tasks to resources and workstation or cells throughout time;
3. After receiving a task, the AM cell starts printing the product associated to the production order;
4. As well as every workstation and resource, while performing jobs, the AM cell sends its status and relevant information to APM;
5. After finalizing the part 3D printing, it is placed into the vertical warehouse and movements to each stage of the process are performed by an AGV that send geographic position and status to APM;
6. In real-time the product placement, movements and status of resources and workstations or cells are presented in the 3D environment of APM.

At any stage of these steps, the decision-maker or automatic disruptive events (machines failure, deviation of typical processing time of a task, resources unavailability, obstacles' identification) can trigger the simulation base model as follows:

1. With the Flexsim model open, the dashboard button is used to collect a snapshot of the current production system;
2. In background, user commands are activated and messages containing simulation request and correspondent response from APM are sent through the Python Kafka producer and consumer;
3. After receiving this information, data is converted to data tables, objects, labels and process flow logic in Flexsim model;
4. The start procedure is triggered and associated heuristics are used to perform optimised allocation of tasks to resources, re-scheduling operations and turning on AM jobs to permit increase production capacity and reduce bottlenecks occurrence. Stochastic processing times of tasks were used to ensure closer reality approximation and based on probabilistic distributions (fit using historical data). Replications were applied to ensure solutions statistical confidence;
5. Finally, KPIs of optimized solutions are provided in dashboard and improved decisions are gathered and sent to APM.

As representation of the performed tests in the industrial experiment AM manufacturing cell, Fig. 5 a) shows the robotic arm and the 3D printer and b) shows the APM interface with the digital representation of the AM cell environment.

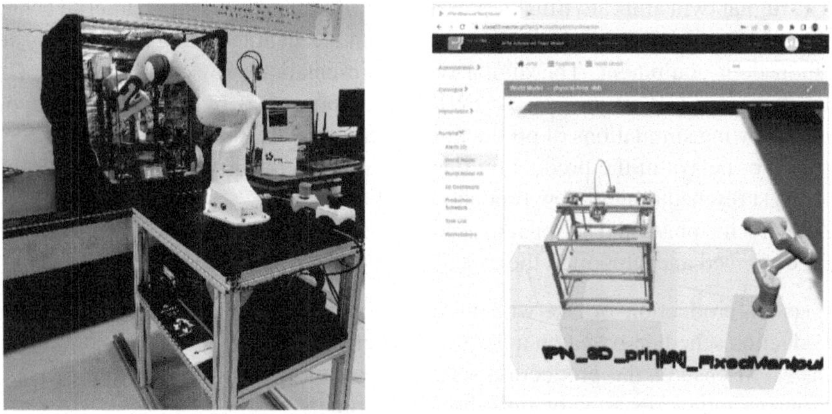

(a) (b)

Fig. 5. a) AM cell of industrial experiment b) APM user interface of AM cell

After performing tests in this environment, results in terms of communication, integration and performance gains were assessed and validated. Preliminary results demonstrated that the Flexsim simulation base model was able to communicate, load current manufacturing process and perform manufacturing orders changes in the real system. These changes (mainly affected by modification of the production orders sequence and job allocation) were updated in APM through the IIOT platform connection and permitted the accomplishment of the pre-defined orders due dates by making use of available capacity and enhancement of KPIs such as resources utilisation.

5 Conclusions and Future Work

Nowadays, moulding and plastic automotive industries are looking for innovative production processes in order to assure high product variability and quantities produced. Additive manufacturing has become a popular method for fabricating some complex components. However, the low interoperability of the production systems combined with the lack of real-time monitoring and control during the AM process can result in low efficiency. For this purpose, a digital twin-based solution for real-time monitoring and optimisation of processes, such as AM technologies and automated vehicles, was developed. Initially, the general architecture of the methodology is presented with relevant details in terms of software and protocols. Furthermore, given the impact of the process simulation in the proposed approach, details of a simulation base model are provided. This includes an integration diagram from the Flexsim simulation model perspective with an explanation of each relevant component and library that permit asynchronous data exchange between the DES component and real-time service such as the APM.

Following the proposed approach, a real automotive industrial experiment with a special highlight on an AM cell is described. After the methodology was applied to the experiment, preliminary results were obtained and demonstrated to be reasonable to validate the overall integration, communication, and data exchange strategy. In conclusion, the proposed work contemplates an improved strategy based on a DT architecture and with particular relevance to discrete event simulation, to take advantage of recent systems to deal with production scheduling and balancing.

Nevertheless, further development is currently being carried out to improve the functionalities and, as the preliminary results were obtained using simple optimisation strategies. It is crucial to explore new scenarios and comparisons to other strategies. As future work, an application in more complex industrial environments seems to be fundamental to achieve the real impact of the proposed methodology. New AI-based methodologies can be added to the digital twin aiming to produce even more optimised solutions. Moreover, the integration of additional mobile robotic units can enhance the solution, increasing the system's actors and assessing the complete case study from the product creation to the preparation to deliver it to the client. Finally, despite the methodology being applied to the automotive sector, using the standards and procedures referred in this research, other sectors would benefit from a replication of this approach.

Acknowledgments. This work is financed by the ERDF - European Regional Development Fund, through the Operational Programme for Competitiveness and Internationalisation - COMPETE 2020 Programme under the Portugal 2020 Partnership Agreement within project PAC, with reference POCI-01-0247-FEDER-046095.

References

1. Javaid, M., Haleem, A., Singh, R.P., Suman, R.: An integrated outlook of cyber-physical systems for Industry 4.0: topical practices, architecture, and applications. Green Technol. Sustain. **1**(1), 100001 (2023)

2. Steclik, T., Cupek, R., Drewniak, M.: Automatic grouping of production data in Industry 4.0: the use case of internal logistics systems based on automated guided vehicles. J. Comput. Sci. **62**, 101693 (2022)
3. Bach, S.-H., Yi, S.-Y.: An efficient approach for line-following automated guided vehicles based on fuzzy inference mechanism. J. Robot. Control **3**(4), 395–401 (2022)
4. Kim, H., Lin, Y., Tseng, T.-L.B.: A review on quality control in additive manufacturing. Rapid Prototyp. J. (2018)
5. Chong, L., Ramakrishna, S., Singh, S.: A review of digital manufacturing-based hybrid additive manufacturing processes. Int. J. Adv. Manuf. Technol. **95**, 2281–2300 (2018)
6. Ganesh Sarvankar, S., Yewale, S.N.: Additive manufacturing in automobile industry. Int. J. Res. Aeronaut. Mech. Eng. **7**(4), 1–10 (2019)
7. Roels, E., et al.: Processing of self-healing polymers for soft robotics. Adv. Mater. **34**(1), 2104798 (2022)
8. Arrais, R., et al.: Application of the open scalable production system to machine tending of additive manufacturing operations by a mobile manipulator. In: EPIA Conference on Artificial Intelligence, pp. 345–356 (2019)
9. Santos, R., Toscano, C., de Sousa, J.P.: A new simulation-based approach in the design of manufacturing systems and real-time decision making. In: 17th IFAC Symposium on Information Control Problems in Manufacturing INCOM 2021, vol. 54, no. 1, pp. 282–287 (2021)
10. Grieves, M.: Digital twin : manufacturing excellence through virtual factory replication - a whitepaper by Dr. Michael Grieves. White Paper, no. March, pp. 1–7 (2014)
11. Kritzinger, W., Karner, M., Traar, G., Henjes, J., Sihn, W.: Digital twin in manufacturing: a categorical literature review and classification. IFAC-PapersOnLine **51**(11), 1016–1022 (2018)
12. Roy, R.B., et al.: Digital twin: current scenario and a case study on a manufacturing process. Int. J. Adv. Manuf. Technol. **107**(9–10), 3691–3714 (2020)
13. Negri, E., Berardi, S., Fumagalli, L., Macchi, M.: MES-integrated digital twin frameworks. J. Manuf. Syst. **56**, 58–71 (2020)
14. Wang, Y., Jiang, Z., Wu, Y.: Model construction of material distribution system based on digital twin. Int. J. Adv. Manuf. Technol. **121**(7–8), 4485–4501 (2022)
15. Sharma, A., Kosasih, E., Zhang, J., Brintrup, A., Calinescu, A.: Digital twins: state of the art theory and practice, challenges, and open research questions. J. Ind. Inf. Integr. **30**(November 2020), 100383 (2022)
16. Fortino, G., Savaglio, C., Spezzano, G., Zhou, M.: Internet of things as system of systems: a review of methodologies, frameworks, platforms, and tools. IEEE Trans. Syst. Man Cybern. Syst. **51**(1), 223–236 (2021)
17. Guth, J., et al.: A detailed analysis of IoT platform architectures: concepts, similarities, and differences. In: Di Martino, B., Li, K.-C., Yang, L.T., Esposito, A. (eds.) Internet of Everything. Internet of Things, pp. 81–101. Springer, Singapore (2018). https://doi.org/10.1007/978-981-10-5861-5_4
18. Noura, M., Atiquzzaman, M., Gaedke, M.: Interoperability in internet of things: taxonomies and open challenges. Mob. Netw. Appl. **24**(3), 796–809 (2019)
19. Bartsch, K., Pettke, A., Höbert, A., Lakämper, J., Lange, F.: On the digital twin application and the role of artificial intelligence in additive manufacturing: a systematic review. J. Phys Mater. **4**(3) (2021)
20. Filz, M.-A., Herrmann, C., Thiede, S.: Simulation-based data analysis to support the planning of flexible manufacturing systems. Simul. Notes Eur. **30**(4), 131–137 (2020)
21. Popper, N., et al.: Methods for integrated simulation-10 concepts to integrate. Simul. Notes Eur. **32**(4), 225–236 (2022)

22. Leng, J., Wang, D., Shen, W., Li, X., Liu, Q., Chen, X.: Digital twins-based smart manufacturing system design in Industry 4.0: a review. J. Manuf. Syst. **60**(May), 119–137 (2021)
23. Boyes, H., Hallaq, B., Cunningham, J., Watson, T.: The industrial internet of things (IIoT): an analysis framework. Comput. Ind. **101**, 1–12 (2018)
24. Teixeira, S., Arrais, R., Dias, R., Veiga, G.: On the development and deployment of an IIoT infrastructure for the fish canning industry. Procedia Comput. Sci. **217**, 1095–1105 (2023)

An AI-Enabled Simulation: Applying Neural Network in a Flexible Flow Shop Problem

Soeb Hussain, Yash Nema, and Mohammad Dehghanimohammadabadi(✉)

Department of Mechanical and Industrial Engineering, Northeastern University, 360 Huntington Avenue, Boston, MA 02115, USA
m.dehghani@northeastern.edu

Abstract. Simulation modeling is integral to the successful implementation of Digital Twin. As Digital Twin models become increasingly sophisticated, it is crucial to enhance the capabilities of simulation modeling by incorporating AI-enabled techniques. By leveraging the power of AI, such as deep learning and artificial neural networks, simulation modeling can be enhanced to provide more accurate and efficient predictions. This paper applies a neural network model within a simulation environment to facilitate decision-making in an assembly line setting. After training, the intelligent data-driven neural network model selects the server in real-time for each job. The Simio Neural Network tool is utilized to collect data, train the neural network model, and conduct experiments. According to the experimental results, the trained neural network model outperforms the baseline model solution.

Keywords: TwinAI · AI-enabled Digital Twin · Simio Neural Networks

1 Introduction

Industry 4.0, has brought about significant changes in the way that manufacturing businesses operate and compete. At the heart of this revolution is the rapid growth of data generated from a multitude of sources, including sensors, machines, and connected devices. This increase in data has created a pressing need for Artificial Intelligence (AI) based Machine Learning (ML) techniques that can analyze and extract insights from these large and complex datasets [1,2]. The ability to effectively leverage these algorithms can have a profound impact on a company's ability to make informed decisions, improve operations, save costs, and stay ahead of the competition [3–5]. Manual analysis or traditional optimization techniques are impractical due to the scale of manufacturing data. It is recommended to use simulation software and suitable data mining techniques to verify various test cases and optimize before deployment. The goal is to find the optimal solution, and this solution needs to be tested to ensure its efficacy [6].

M. Mujica Mota and P. Scala (Eds.): EUROSIM 2023, CCIS 2032, pp. 292–306, 2024.
https://doi.org/10.1007/978-3-031-68435-7_21

A digital twin is a virtual replica or simulation of a physical object or system, such as a part, product, or process integrated with real-data [7]. When it comes to processes, a digital twin consists of three components: a physical process, a virtual process, and connected data that links the physical and virtual processes [8]. The physical and virtual reflections of a digital twin can be kept in sync to make sure both processes have the same property. They can be fully integrated to keep the data in sync and updated. Digital twins not only depend on historical data but also collect data from deployed models in real-time. The biggest difference between simulation and digital twin is that digital twin can not only be used to test and optimize the product but it can also be used to understand defects, issues, and edge cases that can be tested and introduced in the physical system [9]. Furthermore, a digital twin is continually refreshed with data from various sources, including ERPs. As a result, unlike simulations where models remain static, digital twin models are dynamic and continually updated to reflect changes in the corresponding real-life system.

Discrete Event Simulation (DES) is a simulation technique that models the behavior of a system as a sequence of discrete events in time. In the context of Digital Twin implementation, DES can be used as a decision-making tool [10] to simulate the behavior of the physical system in a virtual environment, allowing for analysis of the system's behavior and performance in various scenarios [11]. Integrating the simulation environment of DES with AI is crucial to enhance its capability. This integration allows for the creation of an AI-enabled digital twin that can leverage the strengths of both approaches: the ability of the simulation model to provide a detailed representation of the real-world system, and the power of AI algorithms to provide decision support and insights [12]. Therefore, this integration can provide the following benefits:

- **Improved Accuracy:** Simulation generates an abundance of high-quality data from a controlled environment with known inputs and outputs. This allows neural network training with high accuracy.
- **Pattern Identification:** Utilizing the wealth of data generated by the simulation model under various conditions, neural networks can identify patterns and common behaviors within the system.
- **Informed Decision-Making:** simulation modeling captures the system's behavior in detail, allowing neural networks to learn the behavior of complex systems and make accurate predictions to support decision-making.

This paper aims to demonstrate a showcase where a neural network model is applied to add intelligence to the simulation environment. Therefore, this paper has the following contributions:

- Encourage the practical use of AI in simulation modeling and provide insights on how AI can enhance simulation modeling efficiency.
- Illustrate an example of utilizing a neural network model in a discrete simulation setting to support decision-making.
- Perform experiments to compare the performance of an AI-driven simulation model utilizing neural networks with the baseline solutions.

To accomplish this, an flexible flow shop model is created in Simio. The model utilizes a trained neural network as a *selection rule* to assign jobs to machines in a manner that maximizes efficiency and reduces makespan, which is the total time required to complete all jobs.

The remainder of this paper is structured as follows: Sect. 2 discusses the new trend of integrating simulation software with intelligent and AI-enabled models, as well as notable works in the literature where neural network is combined with the simulation environment for prediction modeling. Section 3 provides a detailed description of the simulation environment and defines the problem at hand. Section 4 elaborates on the model creation and implementation aspect of using the Simio Neural Network tool in the flow shop scheduling model. Section 5 compares the results of the experiments conducted for both the baseline and trained model. Finally, in Sect. 6, the paper concludes by summarizing the work and discussing potential future applications of this project.

2 Literature Review

This section examines literature from two different angles: (i) the current trend of integrating simulation software with AI-powered tools, and (ii) the utilization of neural networks and simulation for predictive analysis.

2.1 Integrated AI with Simulation Environments

An effective simulation software should integrate the simulation environment and AI model, allowing the programmer to train and test the model, extract insights from the collected or generated data, and suggest optimal strategies [12]. In recent years, commercial software has bridged the connection between simulation environments and AI methods such as neural network models for prediction and reinforcement learning (RL) to add autonomous capabilities. The aim is to empower simulation users to improve their decision-making models using AI-based ML algorithms, without the need for prior experience in ML coding or external software.

As part of the trend toward integrating AI with simulation environments, Simio is a pioneer in discrete event simulation software that provides built-in support for neural networks. This tool provides a training data collection interface and a platform for training the neural network model. The neural networks implemented in Simio is feedforward, meaning that data flows from the input neurons to the output neurons without looping back to the input neurons. Flexsim HCE simulation software works with 3D pick-and-drop logic for healthcare use cases and can be used to simulate an ED unit in a Hospital. This software has been used in a study to estimate the number of patients treated and patient waiting time through the analysis of healthcare resources cost and use of machine learning models Random Forest, Gradient Boosting, and AdaBoost [13]. In an attempt to merge RL with simulation environments, Pathmind introduced one of the initial instances of incorporating RL into the simulation environment.

In a study conducted by [14], Pathmind is used to apply RL in a simulated coffee shop model in AnyLogic software. Interested readers can refer to [15], which provides a review of the current literature on combining reinforcement learning with discrete event simulation, including its premise and potential applications.

Researchers have recently aimed to incorporate intelligence into simulation environments by connecting them with advanced programming languages like Python and MATLAB. This integration provides users with additional flexibility to advance simulation modeling with powerful optimization and prediction tools. For example, Dehghanimohammadabadi and colleagues used Python to couple Simio in a series of studies, including digital twin optimization [7], COVID-19 preparedness planning [16], multi-objective supply chain modeling [17], and supplier selection [18]. In [10] and [19], a novel simulation optimization approach called IOS (Iterative Optimization Simulation) was developed by integrating Simio and MATLAB. This approach involves triggering optimization within the simulation environment to provide an optimal decision during simulation runs. Furthermore, they utilized an API to establish a connection between Simio and Python for training an RL agent to prioritize jobs in a scheduling scenario [12]. Q-learning and Double Deep Q-Network (DDQN) models were implemented to address the problem, and the outcomes were compared. An external RL software is also used with Simio to determine the best route for robots in a factory moving from one physical location to another avoiding collisions with fixed barriers [20].

Peyman et al. conducted a study described in [21], which presents a tutorial on connecting Python with different simulation software to develop Simheuristics (metaheuristic algorithms that use simulations to solve complex optimization problems). The tutorial provides step-by-step instructions and emphasizes the advantages of using Simheuristics. Furthermore, the tutorial presented in [22] demonstrates the process of integrating Python with FlexSim for the development of Simheuristics, which refers to heuristic algorithms utilized to solve intricate optimization problems through simulation. The tutorial entails a systematic guide with practical examples.

In conclusion, incorporating AI with simulation software is a rapidly developing trend that has the potential to enhance decision-making capabilities across multiple industries. This integration brings intelligence to simulation modeling and presents a promising area of research for future development. Furthermore, digital twin models can benefit from this development by supplementing their technologies with cutting-edge decision-making tools.

2.2 Predictive Analysis Using Neural Network and Simulation

Integrating simulation models with neural networks enables the development of precise predictive models, recognizing system behaviors and patterns, and optimizing decision-making. This can result in enhanced performance and efficiency across different industries. A recent paper by Ye et al., [3] proposes a context-aware manufacturing system that utilizes AI algorithms such as Multi-Layer Perception (ANN), J48 Tree-based classifier, Naïve Bayes, and LIBSVM with POLY kernel/RBF Kernel. The system aims to help users capture the most relevant information and make timely decisions. A test manufacturing scenario was simulated using Simio to verify the feasibility of applying these algorithms in context-aware manufacturing systems. The algorithm performance was tested for the basic training set, missing value training set, noisy training set, and large-scale data to show the necessity of AI algorithm selection. The author then recommends focusing on deep learning and incremental learning to improve the proposed solution to the complex context-aware applications. Another study by Pinho et al., [23] proposes a SimPy package based framework for optimizing the forest fuels supply chain to improve its performance, resilience, and flexibility. The framework involves coordinating planning levels and event-based models, utilizing a closed-loop system based on forecasts provided by a discrete-event model to solve the challenging problem.

Different research conducted by Carbonneau et al., [24] investigates how well-advanced AI techniques can forecast distorted demand signals in the extended supply chain. Here Matlab Simulink software is used to simulate the scenario and compare various techniques. We specifically see that the recurrent neural network, support vector machine, and neural network perform much better and are more accurate than simpler techniques including moving average, naive, and trend methods. So, having advanced AI techniques is critical to solving complex supply chain problems. Serrano et al., [25] proposes a framework to improve the efficiency of the job shop scheduling process using the approach defined by Smart manufacturing scheduling. This framework provides greater flexibility, and enhanced rescheduling capabilities and uses a deep reinforcement learning agent. The study shows how complex neural networks can be used to schedule various jobs and increase efficiency. One challenge is the need to choose a DES system and Deep Reinforcement Learning (DRL) framework that are not only efficient but also compatible and interoperability in order to support the virtual job shop simulation environment. In addition, there is a difficulty in transferring information during the training process about the randomness of the job shop's productive environment to the neural network that models the Deep Q-Network (DQN) algorithm, which is necessary for the algorithm to effectively distinguish suitable action selections based on an accurate estimation of the value function, while also maintaining a stable level within an acceptable range.

In the following section, a Simio-based framework will be discussed, capable of generating training data, running the neural network independent of external software, and presenting various real-world scenarios to effectively solve the flexible flow shop problem by minimizing the complex network's makespan time.

3 Problem Definition

The aim of this paper is to showcase how a neural network model can be utilized in a discrete simulation setting to facilitate decision-making. This is achieved by developing a flexible flow shop model in Simio that incorporates a trained neural network as a "Selection Rule" to allocate jobs to machines in a way that optimizes efficiency and reduces the makespan or the total time needed to finish all jobs. This showcase is described in detail in the following sections.

3.1 Simulation Environment

The simulation model is a flexible flow shop scheduling problem. As shown in Fig. 1 this model consists of two Zones, denoted Zone 1 and 2, respectively. In this model, orders must be processed by one of the servers available in Zone 1 and Zone 2 before exiting the system. Zone 1 has four (4) servers, while Zone 2 has two (2) servers, each with a capacity of one (1), and specific reliability settings that are outlined in Table 1. These reliability features are included to make the scheduler more robust in handling uncertainties. There are three (3) different types of orders in this model, with varying processing times on each server in both zones (Table 2).

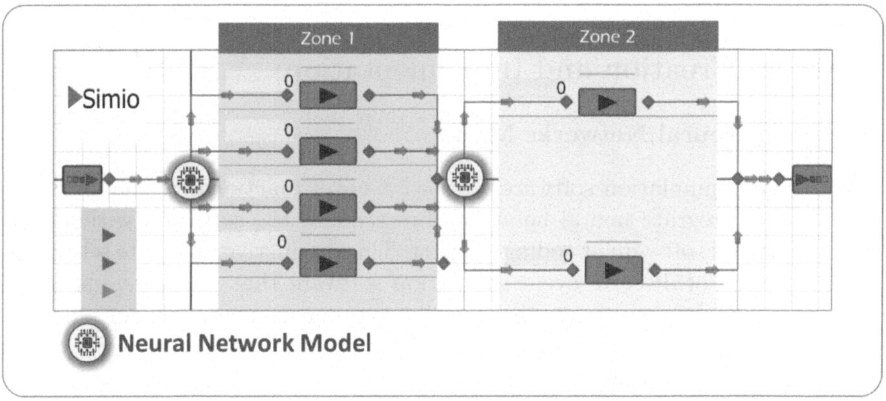

Fig. 1. Snapshot of the simulated flexible flow shop scheduling problem in Simio.

Table 1. Process server state distribution.

Process Server	Zone	Live State	Fail State (hr)	Repair State (hr)
Process1-1	Zone 1	Zone1.ProcTime	–	–
Process1-2	Zone 1	Zone1.ProcTime	Exponential (20)	Triangular (0.5, 1.0, 2.0)
Process1-3	Zone 1	Zone1.ProcTime	–	–
Process1-4	Zone 1	Zone1.ProcTime	Exponential (20)	Triangular (0.5, 1.0, 2.0)
Process2-1	Zone 2	Zone2.ProcTime	–	–
Process2-2	Zone 2	Zone2.ProcTime	–	–

To optimize the system's performance and achieve a desirable makespan, efficient routing logic is crucial for assigning orders to servers upon their arrival in each zone. While heuristic rules based on server or order characteristics are one option, they often provide sub-optimal solutions due to their greedy nature. Alternatively, optimization approaches can provide more optimal solutions but are not easily adaptable to different model settings and lack protective analytics. Neural Networks, on the other hand, can incorporate multiple factors and provide protective analytical models to support decision-making. By training a network

Table 2. Order type distribution for zone 1 and zone 2.

Order Type	Order Mix (%)	Zone 1 Process Time (min)	Zone 2 Process Time (min)
Order A	30	Triangular (05, 10, 20)	Triangular (06, 07, 08)
Order B	40	Triangular (07, 08, 15)	Triangular (05, 08, 11)
Order C	30	Triangular (10, 20, 25)	Triangular (04, 05, 06)

to predict completion times based on various variables, the model can estimate completion times for each server and select the one with the lowest predicted value to process the order. The upcoming section describes the application of the Simio Neural Network tool for routing orders to parallel servers in this system.

4 Model Creation and Implementation

4.1 Simio Neural Networks Model

In 2022, Simio simulation software released a novel functionality enabling programmers to integrate neural networks directly into the software, without the need for external software or coding support. This distinctive attribute sets Simio apart as a type of discrete event simulation software that supports embedded neural networks. It grants users the capacity to use neural networks for inference in model logic, capture training data, and establish an interface with the neural network model for training objectives.

Simio's new features allow for the creation of neural networks entirely within the simulation software, without requiring external support. The Simio Neural Network Model tools were utilized to train two networks, one for Zone 1 and another for Zone 2. The designed network is a feedforward neural network (FNN), where the information flows in only one direction, from the input nodes through the hidden nodes (if any) to the output nodes. The FNN architecture is well-suited for solving many real-world problems, including the one at hand. As illustrated in Fig. 2, the FNN comprises of three layers:

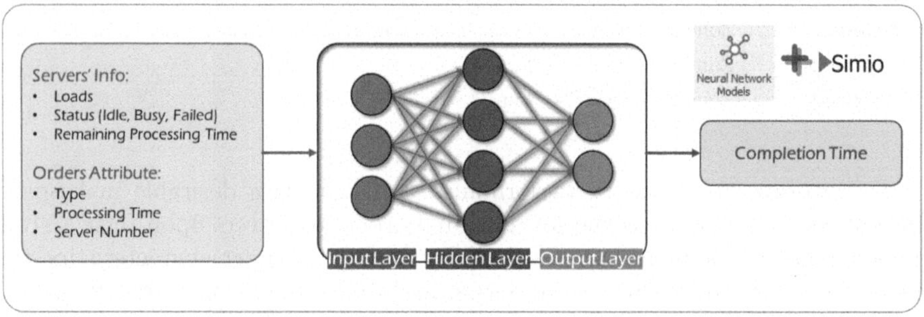

Fig. 2. The architecture of the applied neural network model with its input and output layers data.

- **Input Layer:** This layer receives the initial data for further processing by subsequent artificial neuron layers. The number of neurons in the input layer is equal to the number of input variables in the data being processed. In this case, the input data consists of two feature sets, which are Servers' Information and Order Attributes.
  - **Servers' Information:** contains the server load, server status (i.e., idle, busy, failed), and server expected completion time for existing orders.
  - **Order Attributes** includes the order type, order processing time, and the order server, which refers to the server ID responsible for processing the order.
- **Hidden Layers:** These are essential for modeling complex relationships between input and output variables, and the number of hidden layers and neurons in each layer are hyperparameters that can be adjusted to optimize performance. In Simio Neural Network Models, the user can specify the number of hidden layers and nodes by setting an array. As an example, if the array is [5, 10], it means that the Simio Neural Network Model has two hidden layers, with 5 nodes in the first layer and 10 nodes in the second layer.
- **Output Layers:** The primary objective of the network used in this case is to accurately predict the completion time of each order within the respective zone. To achieve this, the output layer is configured to track the *Completion Time* of each order after it leaves the zone. This time includes both the waiting time in the queue and the processing time on servers. When trained appropriately, the FNN model can predict the completion time of each job for each server, allowing the selection of servers with the shortest completion time to be joined.

The designed network can effectively handle order processing by utilizing Simio's latest neural network capabilities. This model can consider various factors that affect completion times. The following section provides details on generating and training data for the FNN model, as well as instructions on applying the trained model to the simulation environment.

4.2 Data Generation, Training, and Predictions

The neural network's capacity to learn lies in its ability to adjust its weights and recognize underlying patterns in the data. This process demands significant labeled training data and computing power to undergo multiple training cycles. Simulation modeling can aid in neural network training in two ways.

1. Simulation-generated data is of high quality because it is generated in a controlled environment with known inputs and outputs. This data can be used to train neural networks, improving their performance and validating their predictions.
2. Simulation modeling captures the system's behavior in detail, allowing neural networks to learn from this behavior and make accurate predictions. By training neural networks on simulation data, researchers, engineers, and analysts can make better decisions and obtain more precise predictions.

The Neural Networks tool in Simio simplifies the integration of neural networks into simulation models, enabling users to benefit from its capabilities. This process involves four steps, illustrated in Fig. 3:

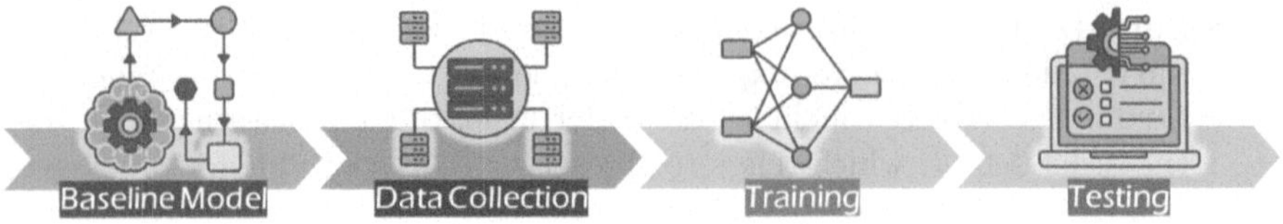

Fig. 3. Steps to apply neural network model

- **Baseline Model:** The simulation model is developed, verified, and tested to ensure that it behaves as expected. The above-mentioned flexible flow shop model is created in Simio, with appropriate attributes, variables, and add-on processes.
- **Data Collection:** In this step, adequate labeled data for neural network training is generated. Simio's "Data Generation" tab in the Neural Network Model tool simplifies the process of generating a large amount of labeled data required for neural network model training. Users can specify the desired number of observations, and Simio simulates the system until that number is reached. This step is highly efficient and can produce abundant data within a reasonable timeframe. In this case, the Baseline Model is utilized to generate 400 h of training data for the neural network, with a maximum of 500,000 records generated for each zone.
- **Training:** Simio's built-in algorithm facilitates the training step within the simulation environment. After adequate data is collected, users switch to the trading tab, where they set the network architecture, number of epochs, and the neural network optimizer. This step sets the network parameters to obtain the best-fitted model to predict the outcome. The training execution time depends on the model complexity, network architecture, and provided observed data. The applied neural networks' training set in this study is listed in Table 3.
- **Testing:** Once training is complete, the neural network can be utilized as a part of the model to support decision-making. In this example, trained models are added to the entry node of each zone (Fig. 1) to route orders to servers with the lowest expected completion time. All predictions made by the network were logged to confirm that it is able to learn and reduce error times, thereby increasing its accuracy and effectiveness.

The neural network architectures employed in the flexible flow shop scheduling problem were meticulously crafted to strike a balance between complexity and performance. By fine-tuning the number of nodes in the hidden layers and the learning rate of the optimizer, accurate and efficient models were created, enhancing the effectiveness and efficiency of flow shop scheduling.

Table 3. Training information for the neural network at zone 1 and zone 2.

Zone	Architecture	Hidden Layers	Optimizer	Learning Rate	convergence	Best
Zone 1	Feedforward NN	36, 20, 9	Adam	5e−5	270 epochs	145.4
Zone 2	Feedforward NN	20, 12, 5	Adam	5e−5	13 epochs	24.9

Both networks underwent training to minimize the mean squared error (MSE). As shown in Fig. 4, the network successfully acquired knowledge on the intricate connections among input variables and projected completion times of each server in Zone 1, as the MSE was minimized during the training process.

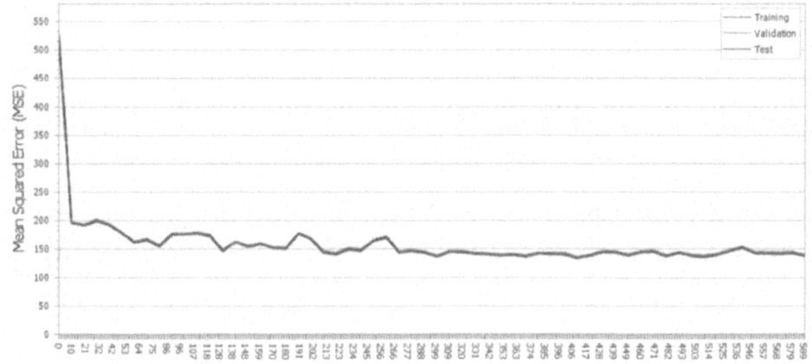

Fig. 4. Training performance for zone 1 neural network.

The difference in the number of epochs and MSE values between the two neural networks can be attributed to the differences in the complexity of the two zones. Zone 1 had more servers and input features, requiring a more complex neural network architecture to capture the relevant relationships. In contrast, Zone 2 had fewer servers and input features, which allowed for a simpler neural network architecture.

4.3 Predicting Optimal Order Path

A greedy approach is utilized to select the server with the least waiting time at each stage to find the optimal path for each order in the flexible flow shop problem. The waiting time for each server in the next zone is determined by using the neural network and the current servers' and zones' states. The input parameters for the neural network include the current order type, server numbers, and their corresponding waiting times, the number of jobs in each server's queue, and the utilization and state of each server. Subsequently, the neural network generates the predicted waiting time for each server in the next zone.

To ensure precise waiting time estimation, the predicted waiting times are converted from hours to minutes. This approach is adopted because the waiting time variance is more significant in minutes than in hours, and using minutes enables capturing more variability in the data. By selecting the server with the minimum predicted waiting time in each zone, the makespan time for each order is minimized (Fig. 5).

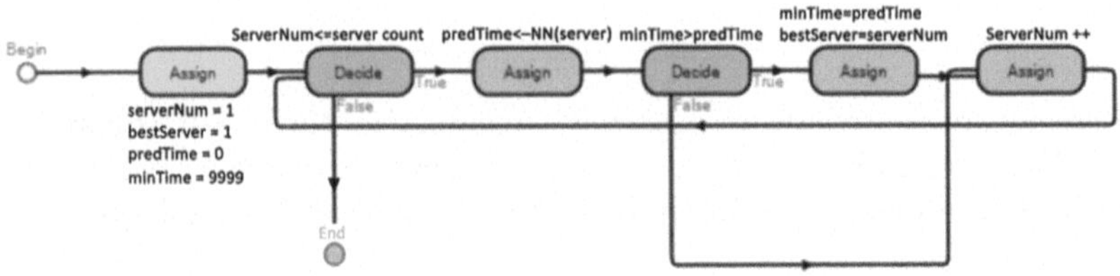

Fig. 5. Server selection from predicted waiting time.

5 Experiments and Results

The purpose of the experiment is to evaluate the efficacy of the *Trained model* in comparison to the *Baseline model*. The Baseline model employs a random uniform distribution to allocate orders among servers, whereas the Trained model dispatches orders based on the minimum predictive Cmax provided by the neural networks. To enhance accuracy and minimize random variations and errors, each model is replicated 30 times, resulting in a more precise estimation of their respective performances. The performance metrics used to evaluate the models are:

- The average time in the system
- The maximum time in the system
- The number in the system (measuring the traffic in the model)
- The utilization of resources

5.1 Average Time in System

Time in the system refers to the overall duration that a job or entity remains within the system, starting from the moment it enters the system until it completes all of its processing stages and exits the system. For orders A, B, and C, the average time (in hrs) in the system using the neural network approach is 0.408, 0.39, and 0.48, respectively. In contrast, the baseline model yields average times of 0.708, 0.693, and 0.7763, respectively. This represents a significant improvement of around 45% in the average time for the neural network approach (Fig. 6). Efficient scheduling algorithms can help minimize the time that jobs spend in the system, leading to better overall system performance.

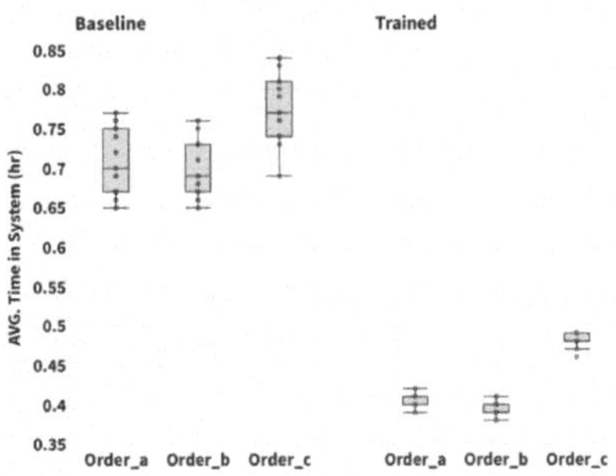

Fig. 6. Average time in system.

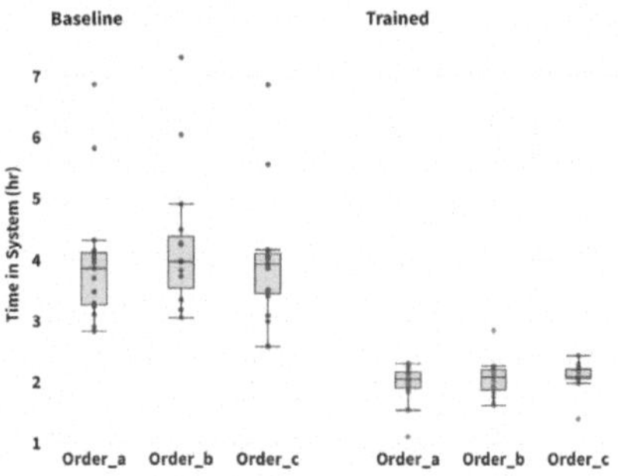

Fig. 7. Time in the system for the worst case scenario.

5.2 Maximum Time in the System

The trained neural network approach also outperformed the baseline model in the worst-case scenario for each order category. The Trained model yielded time in the system of 2.02 h, 2.11 h, and 2.234 h for the worst-performing orders of categories A, B, and C, respectively. These values represent a consistent 55% improvement over the random uniform assignment in the baseline model. These results further reinforce the effectiveness of the proposed neural network approach for flexible flow shop scheduling (Fig. 7).

5.3 Number in System

In addition to the time in system metrics, the proposed neural network approach and the baseline model were also evaluated based on the average number of orders in the system (NIS). This metric was used to assess the models' performance and their ability to handle the workload and traffic within the system. Analyzing the NIS provides insights into the system's overall efficiency and effectiveness and identifies potential areas for improvement. The results showed that the Trained model achieved a NIS of 5.1, which is significantly lower than the NIS of 8.7 obtained with the baseline model (Fig. 8). Hence for all the orders combined including order A, B, and C, there is an improvement of 41% in the average number of orders in the system.

5.4 Resource Utilization

The proposed flow shop model is designed to handle the potential failures of some servers. As the trained model should prioritize servers that are more resilient and less likely to fail. Before training, the resource utilization among servers was roughly the same. However, after training the neural network, the more resilient servers i.e. servers which are less likely to fail were assigned more tasks. It should be noted that in Zone 1, the resource utilization for server 1 and 3

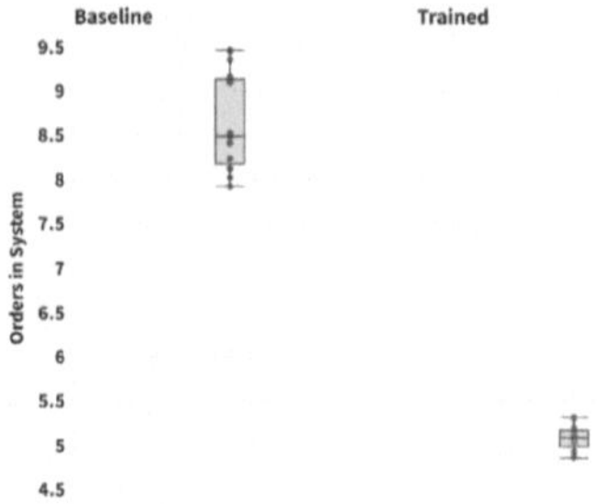

Fig. 8. Number in system.

are subsequently more than server 2 and 4 as server 2 and 4 are likely to fail (Table 1). This is demonstrated in Fig. 9, where the more resilient servers have a higher resource utilization compared to the less resilient ones (Fig. 10).

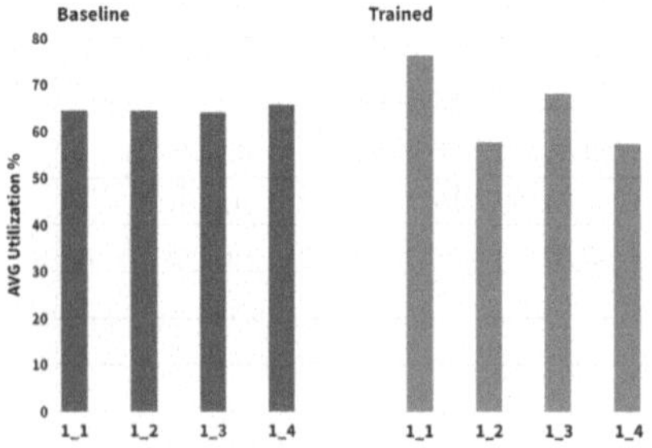

Fig. 9. Utilization by zone 1 servers.

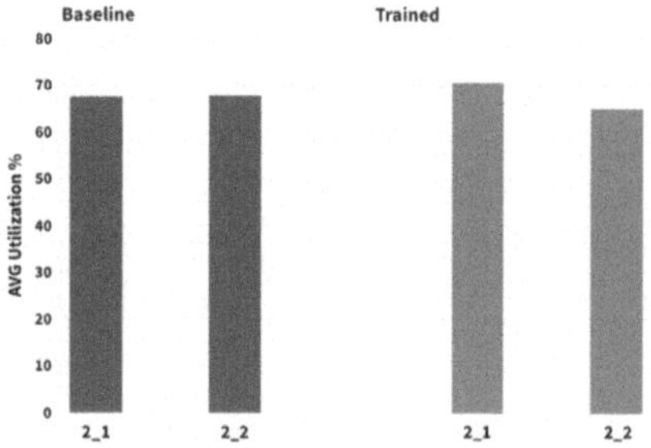

Fig. 10. Utilization by zone 2 servers.

6 Conclusions and Future Works

The simulation model presented in this study is a flexible flow shop problem that includes multiple order types, zones, and server machines with different characteristics. The simulation environment has two zones. The objective of the simulation model is to schedule various jobs to the complex network of servers and minimize the makespan orders. Using the Simio Neural Network tool, two feed-forward neural networks were trained to develop an intelligent selection rule for assigning orders to servers. The results of the experiments indicate that the proposed neural network approach outperformed the baseline model by a significant margin in terms of the evaluated metrics. This suggests that the proposed approach is more effective and efficient in managing and processing orders within the system. The average time in the system was reduced by 45%, the worst-case scenario was improved by 55%, and the number in the system was reduced by 41%. Additionally, the resource utilization was skewed towards the more resilient servers, which can increase the system's overall reliability. These results demonstrate the effectiveness of the proposed neural network approach for flexible flow shop scheduling and suggest its potential for real-world applications.

This study aims to promote the use of Neural Network models in adding intelligence to simulation environments, with the hope of drawing the attention of researchers and practitioners to the potential of AI-enabled tools in simulation modeling for achieving excellence and automation. As digital twin models and simulation environments become more complex, the potential for extending this work increases significantly. The use of AI-enabled tools like Neural Networks and Reinforcement Learning become even more evident in these advanced scenarios.

A possible direction for advancing this study is to apply neural networks to simulation models with greater operational complexity, providing valuable insights into the extent to which neural network models can support decision-making in complex environments. Furthermore, Simio supports Open Neural Network Exchange (ONNX), which enables users to export/import their trained neural network models from/to Simio to other ONNX-supported tools such as TensorFlow and PyTorch. This means that the data collected by simulation can be exported externally and trained using various neural network architectures. Consequently, users can train different networks and import them into the Simio environment for testing and comparison.

References

1. Danylenko, A., Kessler, C., Löwe, W.: Comparing machine learning approaches for context-aware composition. In: Apel, S., Jackson, E. (eds.) SC 2011. LNCS, vol. 6708, pp. 18–33. Springer, Heidelberg (2011). https://doi.org/10.1007/978-3-642-22045-6_2
2. Matei, O., Rusu, T., Bozga, A., Pop-Sitar, P., Anton, C.: Context-aware data mining: embedding external data sources in a machine learning process. In: Martínez de Pisón, F.J., Urraca, R., Quintián, H., Corchado, E. (eds.) HAIS 2017. LNCS (LNAI), vol. 10334, pp. 415–426. Springer, Cham (2017). https://doi.org/10.1007/978-3-319-59650-1_35
3. Ye, Y., Hu, T., Nassehi, A., Ji, S., Ni, H.: Context-aware manufacturing system design using machine learning. J. Manuf. Syst. **65**, 59–69 (2022)
4. Brunoe, T.D., Soerensen, D.G.H., Nielsen, K.: Modular design method for reconfigurable manufacturing systems. Procedia CIRP **104**, 1275–1279 (2021)
5. Latsou, C., Erkoyuncu, J.A., Farsi, M.: A multi-objective approach for resilience-based system design optimisation of complex manufacturing systems. Procedia CIRP **100**, 536–541 (2021)
6. Choudhary, A.K., Harding, J.A., Tiwari, M.K.: Data mining in manufacturing: a review based on the kind of knowledge. J. Intell. Manuf. **20**, 501–521 (2009)
7. Dehghanimohammadabadi, M., Belsare, S., Thiesing, R.: Simulation-optimization of digital twin. In: 2021 Winter Simulation Conference (WSC), pp. 1–10. IEEE (2021)
8. Tao, F., Cheng, J., Qi, Q., Zhang, M., Zhang, H., Sui, F.: Digital twin-driven product design, manufacturing and service with big data. Int. J. Adv. Manuf. Technol. **94**, 3563–3576 (2018)
9. Taylor, S.J.E., Johansson, B., Jeon, S., Lee, L.H., Lendermann, P., Shao, G.: Using simulation and digital twins to innovate: are we getting smarter? In: 2021 Winter Simulation Conference (WSC), pp. 1–13. IEEE (2021)

10. Dehghanimohammadabadi, M., Keyser, T.K.: Intelligent simulation: integration of SIMIO and MATLAB to deploy decision support systems to simulation environment. Simul. Modell. Pract. Theory **71**, 45–60 (2017)
11. Creighton, D.C., Nahavandi, S.: Optimising discrete event simulation models using a reinforcement learning agent. In: Proceedings of the Winter Simulation Conference, vol. 2, pp. 1945–1950. IEEE (2002)
12. Devanga, A., Badilla, E.D., Dehghanimohammadabadi, M.: Applied reinforcement learning for decision making in industrial simulation environments. In: 2022 Winter Simulation Conference (WSC), pp. 2819–2829. IEEE (2022)
13. Atalan, A., Şahin, H., Atalan, Y.A.: Integration of machine learning algorithms and discrete-event simulation for the cost of healthcare resources. In: Healthcare, vol. 10, p. 1920. MDPI (2022)
14. Farhan, M., Göhre, B., Junprung, E.: Reinforcement learning in anylogic simulation models: a guiding example using pathmind. In: 2020 Winter Simulation Conference (WSC), pp. 3212–3223. IEEE (2020)
15. Belsare, S., Badilla, E.D., Dehghanimohammadabadi, M.: Reinforcement learning with discrete event simulation: the premise, reality, and promise. In: 2022 Winter Simulation Conference (WSC), pp. 2724–2735. IEEE (2022)
16. Maghoulan, S., Ondemir, H.M., Dehghanimohammadabadi, M.: COVID-19 supply chain planning: a simulation-optimization approach. In: 2022 Winter Simulation Conference (WSC), pp. 533–544. IEEE (2022)
17. Dehghanimohammadabadi, M., Kabadayi, N.: A two-stage AHP multi-objective simulation optimization approach in healthcare. Int. J. Analytic Hierarchy Process **12**(1) (2020)
18. Kabadayi, N., Dehghanimohammadabadi, M.: Multi-objective supplier selection process: a simulation–optimization framework integrated with MCDM. Annals Oper. Res. 1–23 (2022)
19. Dehghanimohammadabadi, M., Keyser, T.K., Cheraghi, S.H., An effective tool to optimize system's performance: A novel iterative optimization-based simulation (IOS) framework. Comput. Industr. Eng. **111**, 1–17 (2017)
20. Greasley, A.: A simulation of autonomous robot movement directed by reinforcement learning. In: 32nd European Modeling and Simulation Symposium, EMSS 2020, pp. 10–15 (2020)
21. Peyman, M., Copado, P., Panadero, J., Juan, A.A., Dehghanimohammadabadi, M.: A tutorial on how to connect python with different simulation software to develop rich simheuristics. In: 2021 Winter Simulation Conference (WSC), pp. 1–12. IEEE (2021)
22. Leon, J.F., et al.: A tutorial on combining flexsim with python for developing discrete-event simheuristics. In: 2022 Winter Simulation Conference (WSC), pp. 1386–1400. IEEE (2022)
23. Pinho, T.M., et al.: Routing and schedule simulation of a biomass energy supply chain through SimPy simulation package. Appl. Comput. Inform. **17**(1), 36–52 (2021)
24. Carbonneau, R., Laframboise, K., Vahidov, R.: Application of machine learning techniques for supply chain demand forecasting. Eur. J. Oper. Res. **184**(3), 1140–1154 (2008)
25. Serrano-Ruiz, J.C., Mula, J., Poler, R.: Development of a multidimensional conceptual model for job shop smart manufacturing scheduling from the industry 4.0 perspective. J. Manuf. Syst. **63**, 185–202 (2022)

Modeling Dynamics of the Biogas Process Under Uncertainty: A Fuzzy Set Theory Approach

Zahir Barahmand(✉) and Gamunu Samarakoon

University of South-Eastern Norway, 3918 Porsgrunn, Norway
{Zahir.Barahmand,gamunu.arachchige}@usn.no

Abstract. Global concern about climate change has led to a growing interest in exploring renewable energy sources, such as the biological conversion of biomass to methane in anaerobic environments. Anaerobic digestion model no. 1 (ADM1) is the most widely used platform for biogas process modelling. To describe the biochemical processes involved in anaerobic digestion, ADM1 employs a set of 35 differential equations. These equations can be solved simultaneously by knowing the appropriate initial values for the concentration variables. Biogas production has significant uncertainties in nature, as with many real-life applications, resulting in challenges in modelling and simulation. For instance, ADM1 is constructed on an ensemble of anaerobic biochemical reactions. Due to the possible variations and fluctuations in independent variables and parameters, the reaction rates cannot be calculated with absolute certainty. There are different approaches to classifying uncertainties in a system. The present paper proposes a mathematical framework to model biogas production's uncertainties. Among different approaches and tools describing the system's ambiguity, a possibilistic approach using fuzzy set theory has been employed. Based on the results, it is evident that the fuzzy model is able to capture and reflect the defined system uncertainties quite effectively. Using this mathematical tool, designers and decision-makers can gain a deeper understanding of the system.

Keywords: uncertainty modelling · anaerobic digestion · ADM1 · fuzzy set theory

Nomenclatures

S_z	Soluble substrate concentrations in $kg.COD.m^{-3}$
q_{in}	Influent volumetric flow rate in m^3d^{-1}
V_{liq}	Volume of the liquid in m^3
$S_{z,in}$	Concentration of the influent soluble substrate in $kg.COD.m^{-3}$
$\mathcal{F}$	Represents different functions and variables defined in [16]
ρ_a	Represents 19 biochemical process rates, 6 acid-based rates, and 3 gas transfer rates in d^{-1}
$f_{product,substrate}$	Yield (catabolism only) of the product on a substrate in $kg.COD.(kg.COD)^{-1}$
N_c	Nitrogen content of component c in $k, mole.N.(kg.COD)^{-1}$

M. Mujica Mota and P. Scala (Eds.): EUROSIM 2023, CCIS 2032, pp. 307–322, 2024.
https://doi.org/10.1007/978-3-031-68435-7_22

Y_d	Yield of biomass on a substrate in $kg.COD.(kg.COD)^{-1}$
C_e	Carbon content of component e in $k.mole.C.(kg.COD)^{-1}$
X_z	Concentration of particulate component (biomass) z in $kg.COD.m^{-3}$
$X_{z,in}$	Concentration of influent biomass z in $kg.COD.m^{-3}$
S_{cat^+} and S_{an^-}	Concentration of cation and anion respectively
$S_{cat^+,in}$, $S_{an^-,in}$	Input cation and anion all in $k.mole.m^{-3}$
ρ_A	Acid-base rates
$S_{gas.u}$	Concentration of gases ($h_2.ch_4.co_2$) in $kg.COD.m^{-3}$
$\rho_{T,l}$	Gas transfer rates in d^{-1}
q_{gas}	Gas flow in Nm^3d^{-1}
V_{gas}	Gas volume in m^3

1 Introduction

The rise in global population, industrialization, and quality of life has driven up energy demand in recent years [1]. Anaerobic digestion is an efficient and popular method for treating waste and producing biogas. This alternative energy strategy has the potential to fulfill the demand for climate mitigation and energy security. In contrast, developed by the international water association (IWA), the anaerobic digestion model no.1 (ADM1) [2] is the most common platform for modelling and simulating anaerobic digestion processes. ADM1 is a structured mathematical model and covers complex processes involved in converting organic substrates into (biogas) methane, CO_2, and inert byproducts [3]. Although the ADM1 does not cover all the processes that take place during the AD process (such as precipitation of solids and sulfate reduction), the goal of ADM1 is to be a tool to provide as accurate predictions as possible that can be used in all phases of anaerobic digestion plants such as process development, operation, and optimization [4]. ADM1 includes several steps: disintegrating complex solids into soluble and particulate inerts, carbohydrates, proteins, lipids, and inert substances (a non-biological step), is the first step in ADM1 [5]. Next, the products of disintegration are hydrolyzed in the enzymatic step. The products are sugars, amino acids, and long-chain fatty acids (LCFA). Following the fermentation of sugars and amino acids, volatile fatty acids (VFA), hydrogen gas, and carbon dioxide are produced because of acidogenesis. The final step in methanogenesis is the production of methane, whether it results from acetoclastic methanogenesis (cleavage of acetate into methane) or hydrogenotrophic methanogenesis (reduction of carbon dioxide by molecular hydrogen into methane) [6].

There are different uncertainty sources and approaches to coping with them in anaerobic digestion process modelling. In order to identify and assess the sources and methods of coping with uncertainty in anaerobic digestion processes, Barahmand and Samarakoon [7] conducted a literature review. Depending on the preferences, timing, and framing of the alternative scenarios, among other factors, designers might cope with the model uncertainty in a variety of ways [8]. It is essential to distinguish between the sources of uncertainty such as data, choices, and relations, and types of uncertainty such as data variability inconsistency across alternatives [9]. Generally, uncertainty results from a lack of information: either there is no data available, or the data that is available is inaccurate or vague. However, variability can also be viewed as a necessary component of heterogeneity in data [10] or the multiple values of a quantity at different locations, periods, or instances [11]. Barahmand et al. [12] described the most used uncertainty modelling techniques, including probabilistic, possibilistic, robust optimization, interval-based analysis, and information gap decision theory (IGDT). These methods are mostly used to evaluate the impact of unknown input parameters on system output parameters. The key difference between these approaches is that they explain ambiguity in input parameters differently. There is no doubt that in many fields of study, probabilistic approaches and, in particular, Monte Carlo simulations are the most commonly used methods for quantifying and analyzing uncertainty [13, 14]. However, they are subject to limitations, such as a lack of knowledge about defining the proper distribution for uncertain parameters or the need for costly simulations.

In order to overcome the limitations of probabilistic approaches, the present study proposes a mathematical framework to model uncertainty in ADM1 using a possibilistic approach, namely fuzzy set theory. Using fuzzy set theory, this study implements anaerobic digestion model no. 1 in Python (based on [15]).

This section provides the basics of ADM1 mathematical structure, the concept of fuzzy set theory, and numerical solutions to differential algebraic equations (DAEs). Afterward, through a numerical example, the mathematical framework of fuzzy AMD1 will be introduced. The remainder of the article is organized as follows: Sect. 2 discusses the basic mathematical model of ADM1, fuzzy set theory, and Runge-Kutta order 8 as a solution to differential equations. Section 3 develops the fuzzy mathematical model, establishes uncertain parameters, and sets the simulation up. In Sect. 4, the results of the sensitivity analysis and the simulation are discussed, and Sect. 5 concludes the article.

2 Background

2.1 ADM1 Mathematical Framework

In its original form, ADM1 consists of 35 differential algebraic equations (DAE), grouped into five groups to model the concentration rates of various species in the liquid and gas phases [16]. Based on the mathematical equations in [16], Barahmand [17] grouped these 35 equations into the five general forms. For example, the first 12 differential equations for soluble matter have the following general form:

$$\frac{dS_z}{dt} = \frac{q_{in}}{V_{liq}}\left(S_{z,in} - S_z\right) + \mathcal{F}(\rho_a, f_{product,substrate}, N_c, Y_d, C_e) \tag{1}$$

Particulate matter differential Eqs. 13–24 generally follow this format:

$$\frac{dX_z}{dt} = \frac{q_{in}}{V_{liq}}\left(X_{z,in} - X_z\right) + \mathcal{F}(\rho_a . f_{x,y} . Y_d) \tag{2}$$

In the general form of differential Eqs. 25–26, the term $\mathcal{F}$ does not appear in the equations for cations and anions.

$$\frac{dS_{cat^+/an^-}}{dt} = \frac{q_{in}}{V_{liq}}\left(S_{cat^+/an^-,in} - S_{cat^+/an^-}\right) \tag{3}$$

Furthermore, six differential equations are available for the ion states of acid dissociation (27–32). In general, the form is as follows:

$$\frac{dS_w}{dt} = -\rho_{A.j} \tag{4}$$

where, w and j are defined in Table 1.

Table 1. Defining w and j in Eq. 4 (based on [17])

w	j	w	j	w	j
va^-	4	pro^-	6	ac^-	10
bu^-	5	pro^-	7	nh_3	11

Three of the differential Eqs. 33–35 are related to the gas phase and have the general form of:

$$\frac{dS_{gas.u}}{dt} = -\frac{S_{gas.u}q_{gas}}{V_{gas}} + \rho_{T,l}\frac{V_{liq}}{V_{gas}} \tag{5}$$

The ADM1 model consists of over 100 model parameters and 26 input state variables, listed in [17], that directly affect the process performance [16]. Many of the parameters have precise and deterministic values, although others are uncertain, for example, the composition of the influent substrate (uncertain quality of the feedstock). On the other hand, input state variables also suffer considerable ambiguity in real-life applications. Therefore, effectively operating a biogas production unit necessitates a thorough understanding of the system's uncertainties. It should be noted that experts are responsible for defining the uncertainty parameters and the degree of vagueness associated with them when considering the condition of the problem.

2.2 Fuzzy Set Theory Basics

Due to the necessity for dealing with information derived from the computational perception that is uncertain, imprecise, vague, ambiguous, or lacking specific limits, Zadeh [18] first proposed a mathematical technique dubbed fuzzy set theory [19]. This section introduces fuzzy sets theory's basic concepts and definitions to facilitate uncertainty in anaerobic digestion modelling.

The notation and concepts introduced by [20] are used in this section. There are some important definitions below [12]:

Definition 1: fuzzy set, which is denoted by $\tilde{F}$ is characterized by a membership function (MF) $\mu_{\tilde{F}}(x)$, where $x \in X$.

$$\tilde{F} = \left\{(x, \mu_{\tilde{F}}(x))|x \in \tilde{F}\right\}, \mu_{\tilde{F}}(x) : X \rightarrow \{0,1\} \tag{6}$$

where, $\mu_{\tilde{F}}(x)$ is the degree of membership of X in $\tilde{F}$. The closer the value of $\mu_{\tilde{F}}(x)$ is to 1, the more x belongs to $\tilde{F}$ and conversely for 0.

Definition 2: Let $\tilde{F}$ be a fuzzy set in X. Then the support of $\tilde{F}$ ($Supp(\tilde{F})$), is the crisp set given by,

$$Supp(\tilde{F}) = \{x \in X : \mu_{\tilde{F}}(x) > 0\}, \tag{7}$$

Definition 3: Let $\tilde{F}$ be a fuzzy set in X. The height of $\tilde{F}$ $h(\tilde{F})$ is defined as,

$$h(\tilde{F}) = \sup_{x \in X} \mu_{\tilde{F}}(x) \tag{8}$$

Definition 4: If the height of $\tilde{F}$ is equal to one, $h(\tilde{F}) = 1$, then the fuzzy set $\tilde{F}$ is called a normal fuzzy set.

Definition 5: A fuzzy number $\tilde{F}$ is a fuzzy set on the real-line $\mathcal{R}$, which possesses the following properties [20].

(1) A is a normal, convex fuzzy set on $\mathcal{R}$,
(2) The $\tilde{F}(\omega)$ is a closed interval for every $\omega \in (0,1]$,
(3) The membership function is an upper semi-continuous, and
(4) Support of $\tilde{F}$ is bounded.

Definition 6: As shown in Fig. 1, a triangular fuzzy number $\tilde{F} = \left[f_1, f_2, f_3, h(\tilde{F})\right]_{(m,n)}$ is a generalized triangular fuzzy number having orders of m and n if the membership function is given by,

$$\mu_{\tilde{F}}(x) = \begin{cases} 0 & x \le f_1 \\ h(F)\left(\frac{x-f_1}{f_2-f_1}\right)^m & f_1 \le x \le f_2 \\ h(F)\left(\frac{x-f_3}{f_2-f_3}\right)^n & f_2 \le x \le f_3 \\ 0 & x \ge f_3 \end{cases} \tag{9}$$

Definition 7. An $\omega - cut$ (interval of confidence) which is denoted by $\tilde{F}(\omega)$, is the crisp set of elements x$\in\mathcal{R}$ whose degree of belonging to the fuzzy set $\tilde{F}$ is at least ω [21].

$$\tilde{F}(\omega) = \left\{x \in \tilde{F} | \mu_{\tilde{F}}(x) \ge \omega \in (0,1]\right\} = \left[f^1_{(\omega)}, f^2_{(\omega)}\right] \tag{10}$$

Putting membership functions in Definition 6, equal to α, and by finding x, the Eq. (9) will be reached.

$$\tilde{F}(\omega) = \left[f^l_{(\omega)}, f^r_{(\omega)}\right] = \left[f_1 + (f_2 - f_1)\left(\frac{\omega}{h(\tilde{F})}\right)^{\frac{1}{m}}, f_3 + (f_2 - f_3)\left(\frac{\omega}{h(\tilde{F})}\right)^{\frac{1}{n}}\right] \tag{11}$$

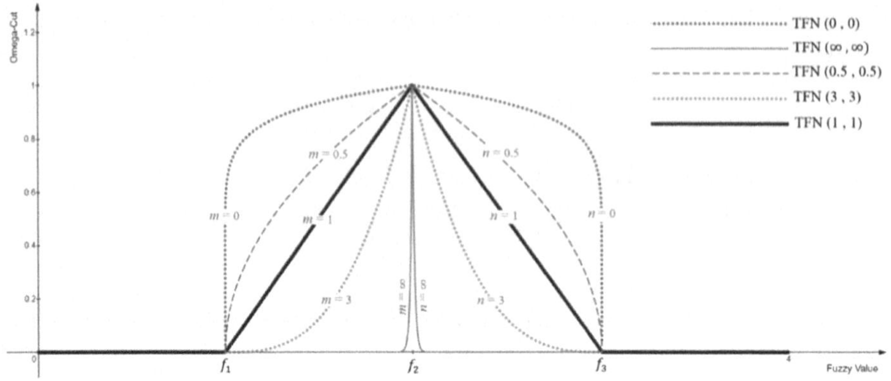

Fig. 1. The generalized triangular fuzzy number applying for different orders

2.3 8th-Order Runge-Kutta

The general form of the first-order ODE can be written as below:

$$y' = f(x, y(x)) \tag{12}$$

Having the initial condition of this ODE is $y(x_0) = y_0$ for $x_0 \leq x \leq x_n$, the unknown grid functions $y_1, y_2, \ldots, y_n$ can be calculated using the 8th-order Runge–Kutta method [20–24].

Thus, the 8th-order method is obtained by solving the 200 equations in 11 steps. The general numerical solution is given by the formula,

$$y_{i+1} = y_i + h\left(\sum_{s=1}^{11} b_s k_s\right) \tag{13}$$

where k_s can be calculated by:

$$k_s = f\left(x_i + c_s h, y_i + h\sum_{j=1}^{s-1} a_{s,j} k_s\right), x_{i+1} = x_i + h \tag{14}$$

where, h is the time step, and c_s, b_s, and $a_{s,j}$ are the coefficients which can be calculated by the Butcher tableau of the RK8 method [25].

3 Fuzzy ADM1 Model

3.1 Mathematical Model

Sadrimajd et al. [15] implemented ADM1 in Python, named PyADM1. PyADM1 was evaluated against benchmark simulation model no. 2 (BSM2) [16] as a robust Python implementation of the ADM1 model for anaerobic digestion modeling and simulation. The present study applies the system uncertainties employing fuzzy set theory methodology on PyADM1. In Python, there are different numerical methods to cope with DAEs. PyADM1 used an explicit Runge-Kutta method of order 8[1] to have a more accurate

[1] scipy.integrate.DOP853.

result. Fuzzy differential equations are used to create uncertain models for a wide variety of scientific and engineering challenges, and there are many studies on this topic, specifically Runge-Kutta methods [26–29].

To simplify and illustrate fuzzy calculations, the 4th-order Runge-Kutta equation is used as a case study. For example, the first differential equation (state no. 1) is [16]:

$$\frac{dS_{su}}{dt} = \frac{q_{in}}{V_{liq}}\left(S_{su,in} - S_{su}\right) + \rho_2 + \left(1 - f_{fa.li}\right)\rho_4 - \rho_5 \tag{15}$$

where,

$$\rho_2 = k_{hyd,ch}X_{ch}, \quad \rho_4 = k_{hyd,li}X_{li}, \quad \rho_5 = k_{m.su}X_{su}\frac{S_{su}}{S_{su} + K_{S.su}}I_5 \tag{16}$$

In this example, S_{su} is the dependent variable which is going to be calculated numerically. Among other factors, variables, and functions, $V_{liq}, f_{fa.li}, K_{S.su}$, and I_5 are considered deterministic with a certain value. In contrast, other parameters and variables are fuzzy and have uncertain values. It is important to note that X_{ch}, X_{su}, and X_{li}, same as S_{su}, are dependent variables for other differential equations (state no. 14 and 16), and each time step should be calculated and inserted into (16). By applying (16) into (15) and following the notation in Definition 1, the fuzzy form of (15) is:

$$\tilde{S}'_{su} = \frac{\tilde{q}_{in}}{V_{liq}}\left(\tilde{S}_{su,in} - \tilde{S}_{su}\right) + \tilde{k}_{hyd,ch}\tilde{X}_{ch} + \left(1 - f_{fa.li}\right)\tilde{k}_{hyd,li}\tilde{X}_{li} - \tilde{k}_{m.su}\tilde{X}_{su}\frac{\tilde{S}_{su}}{\tilde{S}_{su} + K_{S.su}}I_5 \tag{17}$$

where based on Definition 6, the triangular forms of fuzzy numbers are:

$$\tilde{S}'_{su} = \left(S_{su1}, S_{su2}, S_{su3}.h(\tilde{S}'_{su})\right)_{m\times n}$$
$$\tilde{q}_{in} = \left(q_{in1}, q_{in2}, q_{in3}, h(\tilde{q}_{in})\right)_{m\times n}$$
$$\tilde{S}_{su,in} = \left(S_{su,in1}, S_{su,in2}, S_{su,in3}, h(\tilde{S}_{su,in})\right)_{m\times n}$$
$$\tilde{k}_{hyd,ch} = \left(k_{hyd,ch1}, k_{hyd,ch2}, k_{hyd,ch3}, h(\tilde{k}_{hyd,ch})\right)_{m\times n}$$
$$\tilde{X}_{ch} = \left(X_{ch1}, X_{ch2}, X_{ch3}, h(\tilde{X}_{ch})\right)_{m\times n}$$
$$\tilde{k}_{hyd,li} = \left(k_{hyd,li1}, k_{hyd,li2}, k_{hyd,li3}, h(\tilde{k}_{hyd,li})\right)_{m\times n}$$
$$\tilde{X}_{li} = \left(X_{li1}, X_{li2}, X_{li3}, h(\tilde{X}_{li})\right)_{m\times n}$$
$$\tilde{k}_{m.su} = \left(k_{m.su1}, k_{m.su2}, k_{m.su3}, h(\tilde{k}_{m.su})\right)_{m\times n}$$
$$\tilde{X}_{su} = \left(X_{su1}, X_{su2}, X_{su3}, h(\tilde{X}_{su})\right)_{m\times n}$$

After identifying the system uncertainties, the fuzzy differential equation can be solved as below [28]. According to the procedure described in [30] and Def. 7, the upper and lower ω –cuts of $\tilde{S}'_{su}$ (triangular fuzzy number) can be written as (19–21):

$$\tilde{S}'_{su} = \left(\tilde{S}'_{su_1}.\tilde{S}'_{su_1}.\tilde{S}'_{su_1}.h(\tilde{S}'_{su})\right)_{m\times n} \tag{18}$$

$$\tilde{S}'_{su} = \left[\tilde{S}'^{\,l}_{su_{(\omega)}}.\tilde{S}'^{\,r}_{su_{(\omega)}}\right] \tag{19}$$

where,

$$
\begin{aligned}
\tilde{S}'^{\,l}_{su_{(\omega)}} = & \left(\frac{q_{in_1}}{V_{liq}}\left(S_{su,in_1} - S_{su_1}\right) + k_{hyd,ch_1}X_{ch_1} + \left(1 - f_{fa.li}\right)k_{hyd,li_1}X_{li_1}\right. \\
& \left. - k_{m.su_1}X_{su_1}\frac{S_{su_1}}{S_{su_1} + K_{S.su}}I_5\right) \\
& + \left(\left(\frac{q_{in_1}}{V_{liq}}\left(S_{su,in_2} - S_{su_2}\right) + k_{hyd,ch_2}X_{ch_2} + \left(1 - f_{fa.li}\right)k_{hyd,li_2}X_{li_2}\right.\right. \\
& \left. - k_{m.su_2}X_{su_2}\frac{S_{su_2}}{S_{su_2} + K_{S.su}}I_5\right) \\
& - \left(\frac{q_{in_1}}{V_{liq}}\left(S_{su,in_1} - S_{su_1}\right) + k_{hyd,ch_1}X_{ch_1}\right. \\
& + \left(1 - f_{fa.li}\right)k_{hyd,li_1}X_{li_1} \\
& \left.\left. - k_{m.su_1}X_{su_1}\frac{S_{su_1}}{S_{su_1} + K_{S.su}}I_5\right)\right)\left(\frac{\omega}{h(\tilde{S}'_{su})}\right)^{\frac{1}{m}}
\end{aligned}
\tag{20}
$$

$$
\begin{aligned}
\tilde{S}'^{\,l}_{su_{(\omega)}} = & \left(\frac{q_{in_3}}{V_{liq}}\left(S_{su,in_3} - S_{su_3}\right) + k_{hyd,ch_3}X_{ch_3} + \left(1 - f_{fa.li}\right)k_{hyd,li_3}X_{li_3}\right. \\
& \left. - k_{m.su_3}X_{su_3}\frac{S_{su_3}}{S_{su_3} + K_{S.su}}I_5\right) \\
& + \left(\left(\frac{q_{in_2}}{V_{liq}}\left(2 - S_{su_2}\right) + k_{hyd,ch_2}X_{ch_2} + \left(1 - f_{fa.li}\right)k_{hyd,li_2}X_{li_2}\right.\right. \\
& \left. - k_{m.su_2}X_{su_2}\frac{S_{su_2}}{S_{su_2} + K_{S.su}}I_5\right) \\
& - \left(\frac{q_{in_3}}{V_{liq}}\left(S_{su,in_3} - S_{su_3}\right) + k_{hyd,ch_3}X_{ch_3}\right. \\
& + \left(1 - f_{fa.li}\right)k_{hyd,li_3}X_{li_3} \\
& \left.\left. - k_{m.su_3}X_{su_3}\frac{S_{su_3}}{S_{su_3} + K_{S.su}}I_5\right)\right)\left(\frac{\omega}{h(\tilde{S}'_{su})}\right)^{\frac{1}{n}}
\end{aligned}
\tag{21}
$$

In (20) and (21), the height of the $\tilde{S}'_{su}$ in the minimum height of all other fuzzy numbers defined above. Now, for the 4th order Runge-Kutta, the following steps were done:

$$\tilde{S}_{su}\big|_{t+1} = \tilde{S}_{su}\big|_{t} + \frac{1}{6}\left(\tilde{K}_1 + 2\tilde{K}_2 + 2\tilde{K}_3 + \tilde{K}_4\right) \tag{22}$$

where,

$$\tilde{K}_1 = h\tilde{f}(t.x) = \left[K1^{l}_{(\omega)}.K1^{r}_{(\omega)}\right] \tag{23}$$

$$\tilde{K}_2 = h\tilde{f}\left(t + \frac{h}{2}.x + \frac{K_1}{2}\right) = \left[K2^{l}_{(\omega)}.K2^{r}_{(\omega)}\right] \tag{24}$$

$$F\tilde{K}_3 = h\tilde{f}\left(t + \frac{h}{2}.x + \frac{K_2}{2}\right) = \left[K3^l_{(\omega)}.K3^r_{(\omega)}\right] \quad (25)$$

$$\tilde{K}_3 = h\tilde{f}(t + h.x + K_3) = \left[K4^l_{(\omega)}.K4^r_{(\omega)}\right] \quad (26)$$

where, h is a time step.

3.2 Uncertain Parameters and Input Variable

As mentioned above, the standard ADM1 [2] has 104 model parameters and 26 input variables. Among these variables, all input variables and 26 parameters out of 104 were considered fuzzy numbers by experts. Although the highest uncertainty belongs to the five parameters related to the substrate influent composition, these parameters suffer from different types of uncertainty that are better handled by probabilistic approaches such as Monte Carlo simulation. There are two possible conditions for input variables. Although dynamic input variables describe time variability, the uncertainty can be applied to both dynamic and static input variables. Jeppsson and Rosen [16] defined the value of all the parameters and steady-state input variables for ADM1. Table 2 provides the fuzzy values (normal triangular fuzzy numbers with linear membership functions) of 21 uncertain parameters defined by experts in the field. All the values were assumed to testify to the mathematical framework. For simplicity, all the fuzzy values were normal fuzzy numbers with a height equal to unity. The values of steady-state input variables were listed in [16]. To generate dynamic inputs, random numbers with normal distribution were generated. The normal distribution had a mean equal to the reported value for steady-state and a standard deviation of 10 percent of steady-state values. Furthermore, to have a fuzzy dynamic input variable, the lower and upper limits of fuzzy numbers are considered as $\pm 10\%$ of each random dynamic input variable.

Table 2. The uncertain parameters and fuzzy values

	Value	Fuzzy Value	Unit		Value	Fuzzy Value	Unit
k_{dis}	0.5	(0.1, 0.5, 2)	d^{-1}	$k_{m.h2}$	35	(32, 35, 40)	d^{-1}
$k_{hyd.ch}$	10	(5, 10, 20)	d^{-1}	$k_{dec,Xsu}$	0.02	(0.01, 0.02, 0.08)	d^{-1}
$k_{hyd.pr}$	10	(5, 10, 20)	d^{-1}	$k_{dec,Xaa}$	0.02	(0.01, 0.02, 0.08)	d^{-1}
$k_{hyd.li}$	10	(5, 10, 20)	d^{-1}	$k_{dec,Xfa}$	0.02	(0.01, 0.02, 0.08)	d^{-1}
$k_{m.su}$	30	(27, 30, 36)	d^{-1}	$k_{dec,Xc4}$	0.02	(0.01, 0.02, 0.08)	d^{-1}
$k_{m.aa}$	50	(25, 50, 60)	d^{-1}	$k_{dec,Xpro}$	0.02	(0.01, 0.02, 0.08)	d^{-1}
$k_{m.fa}$	6	(5, 6, 8)	d^{-1}	$k_{dec,Xac}$	0.02	(0.01, 0.02, 0.08)	d^{-1}
$k_{m.c4}$	20	(18, 20, 24)	d^{-1}	$k_{dec,Xh2}$	0.02	(0.01, 0.02, 0.08)	d^{-1}
$k_{m.pro}$	13	(111, 13, 16)	d^{-1}	T_{op}	308	(305, 308.15, 311)	K
$k_{m.ac}$	8	(7, 8, 10)	d^{-1}	$k_l a$	200	(190, 200, 210)	d^{-1}
And all 26 input variables (as seen in Table 4)[a]							

[a] Random normally distributed numbers with a mean equal to the reported values and a standard deviation of 10% of its value.

3.3 Simulation Setup

The simulation time was set to 270 days. Therefore, the dynamic input variables should be generated for this period. To find the simulation time step, it was assumed that every 15 min, the values of input variables were recorded for 270 days. The time step can be calculated as below,

$$Time\ step = \frac{1\,day}{24\,hours \times 4} = 0{,}010417 \tag{27}$$

And this means that all 35 differential equations and over 100 equations related to ADM1 should be solved simultaneously. Each differential equation solves over 200 equations related to the 8th-order Runge-Kutta method for over 25,000 time steps.

4 Results and Discussion

4.1 Sensitivity Analysis

Sensitivity analysis is one of the most important steps in uncertainty analyses of anaerobic digestion [31]. Like many other cases, sensitivity analysis is tied to fuzzy set theory. To have a more trustable and realistic model, it is crucial to define the fuzzy values accurately for the parameters with higher sensitivity [17]. Moreover, in complex systems such as ADM1, many inputs and outputs interact with each other. Some input parameters have a positive coloration with the output parameters, and others are negative. It is important to be aware of these correlations for fuzzy calculations and arithmetics. A local sensitivity analysis has been conducted for the uncertain parameters in ADM1. For example, Table 3 shows the percent change in the output variable *pH* by changing each input parameter and its correlation.

Table 3. Sensitivity analysis (% change and sensitivity index) of input parameters on *pH*

	from	to	%	SI		from	to	%	SI
k_{dis}	0.5	5	0.005▲	0.0	$k_{m.h2}$	35	70	0.000▲	0.0
$k_{hyd.ch}$	10	50	-0.003▼	0.0	$k_{dec,Xsu}$	0.02	0.5	0.056▲	0.0
$k_{hyd,pr}$	10	50	0.019▲	0.0	$k_{dec,Xaa}$	0.02	0.5	0.106▲	0.0
$k_{hyd.li}$	10	50	0.000▲	0.0	$k_{dec,Xfa}$	0.02	0.5	-0.084▼	0.0
$k_{m.su}$	30	60	-0.001▼	0.0	$k_{dec,Xc4}$	0.02	0.5	-0.043▼	0.0
$k_{m.aa}$	50	100	0.001▲	0.0	$k_{dec,Xpro}$	0.02	0.5	-1.374▼	0.6
$k_{m.fa}$	6	12	0.001▲	0.0	$k_{dec,Xac}$	0.02	0.5	-32.78▼	**13.7**
$k_{m.c4}$	20	40	0.003▲	0.0	$k_{dec,Xh2}$	0.02	0.5	0.023▲	0.0
$k_{m.pro}$	13	26	0.004▲	0.0	T_{op}	308	313	-0.004▼	**2.7**
$k_{m.ac}$	8	16	0.108▲	**1.1**	$K_{l}a$	200	300	0.011▲	0.2

Although the percent change factor gives information about the correlations and a broad overview of sensitivity, it can not show the priorities because the changes are not homogenous. To solve this problem, the Sensitivity Index (SI) can be used [12].

$$SI = \frac{\%\,change\,in\,output\,variable}{\%\,change\,in\,parameter} \tag{28}$$

The results in Table 3 indicate that among these parameters, by far $k_{dec,Xac}$ has the highest sensitivity to pH, followed by T_{op}, $k_{m.ac}$, and $K_{l}a$. $k_{dec,Xac}$ Are the decay rate constant of acetate-degrading organisms. T_{op} is the operating temperature and $k_{m.ac}$ is the maximum uptake rate for acetate-degrading organisms.

4.2 Initial Conditions

The numerical solution of DAEs needs to have initial values. The ADM1 is very sensitive to initial values like many differential equation systems. There are two steps to finding initial values. First, a simulation was conducted with the preliminary initial values (PIV) reported [15]. Second, the steady-state outputs variables were considered the final initial values (FIV) for input variables. Table 4 gives the initial values used in the simulations.

Table 4. Preliminary and final initial values at steady-state

Parameter	PIV (BSM2)	FIV	Parameter	PIV (BSM2)	FIV
S_{su}	0.012	0.012	X_{c4}	0.43	0.437
S_{aa}	0.0053	0.006	X_{pro}	0.14	0.139
S_{fa}	0.099	0.103	X_{ac}	0.76	0.749
S_{va}	0.012	0.012	X_{H2}	0.32	0.310
S_{bu}	0.013	0.014	X_{I}	25.6	25.601
S_{pro}	0.016	0.016	S_{cat}	0.04	0.040
S_{ac}	0.2	0.213	S_{on}	0.02	0.020
S_{h2}	2.30E−7	2.44E−07	S_{h+}	3.423E−8	3.45E−08
S_{ch4}	0.055	0.055	S_{va-}	0.011	0.012
S_{IC}	0.15	0.153	S_{bu-}	0.013	0.014
S_{IN}	0.13	0.130	S_{pro-}	0.016	0.016
S_{I}	0.33	0.320	S_{ac-}	0.2	0.212
X_{xc}	0.31	0.315	S_{HCO3-}	0.14	0.143
X_{ch}	0.028	0.029	S_{CO2}	0.0	0.010
X_{pr}	0.1	0.108	S_{NH3}	0.0041	0.004
X_{li}	0.029	0.026	S_{NH4+}	0.0	0.126
X_{su}	0.42	0.421	$S_{gas.H2}$	1.02E−5	1.06E−05
X_{aa}	1.18	1.194	$S_{gas.CH4}$	1.63	1.622
X_{fa}	0.24	0.210	$S_{gas.CO2}$	0.014	0.014

4.3 Simulation Results

This section provides fuzzy modeling of ADM1 based on ADM1 implementation within the BSM2 framework [16] and Python implementation of ADM1 [15]. All input variables were considered fuzzy dynamic variables, and 21 parameters were defined as uncertain. The simulation was conducted for 270 days with a time step of 15 min. To calculate DAEs, the 8th-order Runge Kutta (DOP853) solver was called in Python [32]. There are 35 different output variables listed [16]. The local sensitivity analysis showed that the responses of output variables are different. Therefore, to analyze the uncertainty of each output variable, separate simulations should be conducted. As a numerical example, *pH* was chosen as the main output variable. Using the sensitivity analysis, the upper and lower limits of fuzzy numbers in Table 2 were used to provide extreme conditions for *pH*.

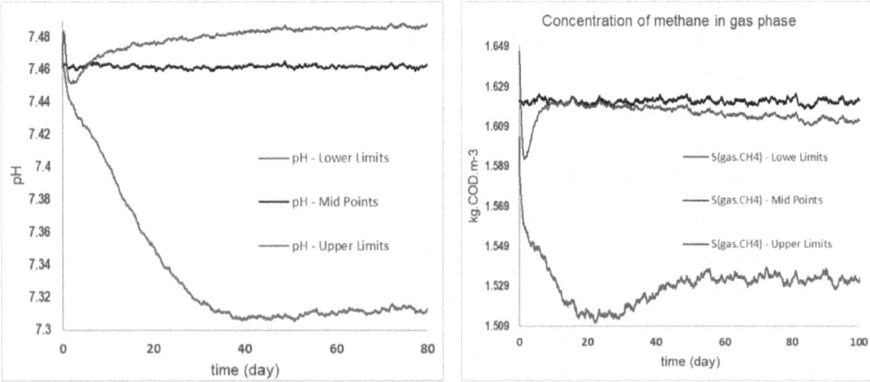

Fig. 2. Left) Fuzzy *pH* considering the extreme condition for *pH*, right) Fuzzy $S_{gas.\ CH4}$ (CH_4 concentration in the gas phase) considering the extreme condition for *pH*.

At each time step, a triangular fuzzy *pH* was calculated associated with *pH,* and Fig. 2 (left) gives a top view of these triangles. This figure indicates that under the defined uncertainty (Table 2), the *pH* will not exceed the upper (7.49) and lower limits (7.31), and the highest belonging degree belongs to mid-points (7.46).

Defining reasonable upper and lower limits is a crucial step in the fuzzy simulation. For example, Fig. 3 illustrates the same results as Fig. 2 (right) but with different limits. In the lower limits of *pH*, the production of biogas was adversely affected by a drop below 5.5. Compared to the case in Fig. 2 (left), the lower limit for biogas concentration experienced a dramatic fall of 65 percent, which is not favorable.

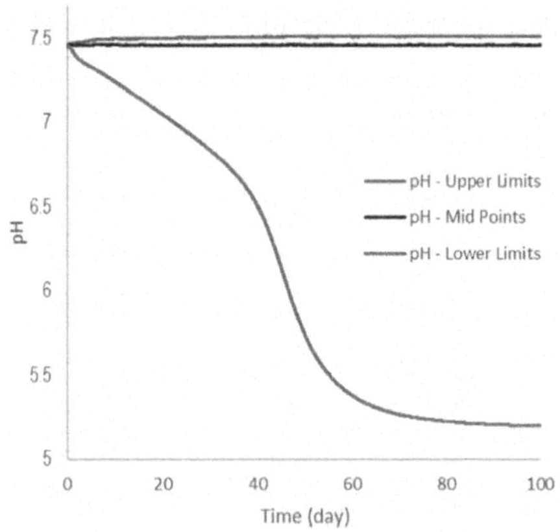

Fig. 3. Fuzzy *pH* considering the extreme condition when fuzzy numbers are not defined properly.

Table 5 provides the fuzzy values of output variables at a steady state. Some parameters have a positive correlation with each other, while others have a negative correlation. For example, there is a negative correlation between *pH* and S_{ac}.

The modeling and simulation process assists in estimating plant capacities and identifying possible deviations in processes due to changes in feed substrate conditions (which are not always consistent). As such, the simulation results can be considered "virtual sensors" for the performance of the reactor or an early warning system for possible reactor failures (frequent reactor failures are one of the major problems in the biogas industry). The accuracy of the model parameter and input variables also plays a significant role in the accuracy of model predictions (e.g., simulation results). When applied to a specific case/different cases, estimating nearly 100 parameters in ADM-1 (which is still a commonly used platform for biogas process simulation) is not easy (or possible). As a result, it is common to use historic values or values from similar cases in such situations. Therefore, there is always an element of uncertainty in those values, and consequently, the outcomes predicted by the model are inaccurate. Consequently, it may not provide a correct warning of an unstable reactor operation, and thus plant operation management may not be able to take the necessary mitigation measures (reactor failures may cost the plant a significant amount of money). A parameter such as *pH* (in the reactor) can be used to monitor the stability of the reactor. Furthermore, over- or underestimation (i.e., inaccurate) of the reactor dimension may also occur due to simulation with uncertainty.

Table 5. Averaged limits of fuzzy output variables

	Lower Limits	Middle points	Upper Limits		Lower Limits	Middle points	Upper Limits
S_{su}	0.009	0.012	0.371	X_{c4}	0.215	0.437	0.542
S_{aa}	0.005	0.006	0.050	X_{pro}	0.071	0.139	0.141
S_{fa}	0.321	0.102	4.686	X_{ac}	0.341	0.748	0.785
S_{va}	0.026	0.012	0.116	X_{H2}	0.330	0.310	0.140
S_{bu}	0.030	0.014	0.133	X_I	25.424	25.591	25.793
S_{pro}	0.044	0.016	0.740	S_{cat}	0.044	0.040	0.036
S_{ac}	1.767	0.212	0.130	S_{on}	0.022	0.020	0.018
S_{h2}	2E-7	2.4E7	3.9E7	pH	7.313	7.462	7.487
S_{ch4}	0.052	0.055	0.053	S_{va-}	0.026	0.012	0.115
S_{IC}	0.119	0.153	0.163	S_{bu-}	0.030	0.014	0.133
S_{IN}	0.117	0.130	0.150	S_{pro-}	0.044	0.016	0.738
S_I	0.236	0.320	0.421	S_{ac-}	1.762	0.211	0.129
X_{xc}	1.135	0.315	0.105	S_{HCO3-}	0.108	0.143	0.153
X_{ch}	0.016	0.029	0.055	S_{CO2}	0.011	0.010	0.010
X_{pr}	0.191	0.107	0.060	S_{NH3}	0.003	0.004	0.005
X_{li}	0.047	0.026	0.015	S_{NH4+}	0.114	0.126	0.145
X_{su}	0.517	0.422	0.205	$S_{gas.H2}$	0.000	0.000	0.000
X_{aa}	1.232	1.193	0.723	$S_{gas.CH4}$	1.533	1.621	1.608
X_{fa}	0.097	0.210	0.040	$S_{gas.CO2}$	0.015	0.014	0.015

5 Conclusion

Due to the complexity of the anaerobic digestion process, there are numerous sources of uncertainty which makes the biogas production analysis and prediction quite challenging. Different tools and methods were mentioned in the literature to cope with the anaerobic digestion model's uncertainty. As the knowledge gap and in the mathematical sense, fuzzy set theory is one of the useful tools which reduces costly and time-consuming simulations (Monte Carlo). Finding the sources of uncertainty, defining appropriate fuzzy values, and conducting sensitivity analysis are the most important steps in modeling uncertainty using fuzzy set theory to have a robust model. As a numerical example, the case study results showed that under the defined uncertainty, *pH* would be laid in the range between upper and lower limits, 7.49 and 7.31, respectively, and the highest belonging degree belongs to *pH* equal to 7.46. Knowing the variations and their possibilities gives plant operation management team, designers, and analysts a better understanding of the complex system. Furthermore, future research could extend our model by applying greater uncertainties with more advanced methods such as type-2 fuzzy numbers, and integrating different tools such as Monte Carlo simulations.

References

1. Selormey, G.K., Barnes, B., Kemausuor, F., Darkwah, L.: A review of anaerobic digestion of slaughterhouse waste: effect of selected operational and environmental parameters on anaerobic biodegradability. Rev. Environ. Sci. Biotechnol. **20**, 1073–1086 (2021). https://doi.org/10.1007/s11157-021-09596-8
2. Batstone, D.J., et al.: The IWA anaerobic digestion model no 1 (ADM1). Water Sci. Technol. **45**, 65–73 (2002). https://doi.org/10.2166/wst.2002.0292
3. Parker, W.: Application of the ADM1 model to advanced anaerobic digestion. Biores. Technol. **96**, 1832–1842 (2005). https://doi.org/10.1016/j.biortech.2005.01.022
4. Ramírez, I.: ADM1 Applications for a hybrid up-flow anaerobic sludge-filter bed reactor performance and for a batch thermophilic anaerobic digestion of thermally pretreated waste activated sludge. Rev. Fac. Ingeniería Univ. Antioquia 167–179 (2012)
5. Tassew, F.A., Bergland, W.H., Dinamarca, C., Bakke, R.: Influences of temperature and substrate particle content on granular sludge bed anaerobic digestion, vol. 15 (2019). https://doi.org/10.3390/app10010136
6. Fezzani, B., Cheikh, R.B.: Implementation of IWA anaerobic digestion model no. 1 (ADM1) for simulating the thermophilic anaerobic co-digestion of olive mill wastewater with olive mill solid waste in a semi-continuous tubular digester. Chem. Eng. J. **141**, 75–88 (2008). https://doi.org/10.1016/j.cej.2007.10.024
7. Barahmand, Z., Samarakoon, G.: Anaerobic digestion process modeling under uncertainty: a narrative review. Int. J. Energy Prod. Manag. **8**, 79–98 (2023). https://doi.org/10.2495/EQ-V8-N1-79-98
8. Heijungs, R., Lenzen, M.: Error propagation methods for LCA - a comparison. Int. J. Life Cycle Assess. **19** (2014). https://doi.org/10.1007/s11367-014-0751-0
9. Cherubini, E., Franco, D., Zanghelini, G.M., Soares, S.R.: Uncertainty in LCA case study due to allocation approaches and life cycle impact assessment methods. Int. J. Life Cycle Assess. **23**, 2055–2070 (2018). https://doi.org/10.1007/s11367-017-1432-6
10. Heijungs, R., Huijbregts, M.: A review of approaches to treat uncertainty in LCA (2004)
11. Begg, S., Bratvold, R., Welsh, M.: Uncertainty vs. variability: what's the difference and why is it important? (2014)
12. Barahmand, Z., Jayarathna, C., Ratnayake, C.: Sensitivity and uncertainty analysis in a circulating fluidized bed reactor modeling. In: Proceedings of the Linköping Electronic Conference Proceedings. Linköping University Press, Finland (2021)
13. Barahmand, Z., Eikeland, M.S.: Life cycle assessment under uncertainty: a scoping review. World **3**, 692–717 (2022). https://doi.org/10.3390/world3030039
14. Barahmand, Z., Eikeland, M.S.: Techno-economic and life cycle cost analysis through the lens of uncertainty: a scoping review. Sustainability **14**, 12191 (2022). https://doi.org/10.3390/su141912191
15. Sadrimajd, P., Mannion, P., Howley, E., Lens, P.N.L.: PyADM1: a python implementation of anaerobic digestion model no. 1 (20210
16. Jeppsson, U., Rosen, C.: Aspects on ADM1 implementation within the BSM2 framework (2006)
17. Barahmand, Z.: Comprehensive Local Sensitivity Analysis of the ADM1-Based Anaerobic Digestion Process, pp. 235–246. WIT Press (2022)

18. Zadeh, L.A.: Fuzzy sets. Inf. Control. **8**, 338–353 (1965). https://doi.org/10.1016/S0019-9958(65)90241-X
19. Balaman, Ş.Y.: Chapter 5 - uncertainty issues in biomass-based production chains. In: Balaman, Ş.Y. (ed.) Decision-Making for Biomass-Based Production Chains, pp. 113–142. Academic Press (2019). ISBN 978-0-12-814278-3
20. Carlsson, C., Fullér, R.: On possibilistic mean value and variance of fuzzy numbers. Fuzzy Sets Syst. **122**, 315–326 (2001). https://doi.org/10.1016/S0165-0114(00)00043-9
21. Thavaneswaran, A., Appadoo, S.S., Frank, J.: Binary option pricing using fuzzy numbers. Appl. Math. Lett. **26**, 65–72 (2013). https://doi.org/10.1016/j.aml.2012.03.034
22. Butcher, J.: On Runge-Kutta processes of high order. J. Aust. Math. Soc. **4**, 179–194 (1964). https://doi.org/10.1017/S1446788700023387
23. Curtis, A.R.: An eighth order Runge-Kutta process with eleven function evaluations per step. Numer. Math. **16**, 268–277 (1970). https://doi.org/10.1007/BF02219778
24. Hippolyte, S., Richard, A.K.: A new eighth order Runge-Kutta family method. J. Math. Res. (2019). https://doi.org/10.5539/JMR.V11N2P190
25. Séka, H., Assui, K.R.: Order of the Runge-Kutta method and evolution of the stability region. Ural Math. J. **5**, 64–71 (2019). https://doi.org/10.15826/umj.2019.2.006
26. Dubois, D., Prade, H.: Towards fuzzy differential calculus part 3: differentiation. Fuzzy Sets Syst. **8**, 225–233 (1982). https://doi.org/10.1016/S0165-0114(82)80001-8
27. Jayakumar, T., Kanagarajan, K., Indrakumar, S.: Numerical solution of nth-order fuzzy differential equation by Runge-Kutta method of order five. Int. J. Math. Anal. **6**, 2885–2896 (2012)
28. Pederson, S., Sambandham, M.: The Runge-Kutta method for hybrid fuzzy differential equations. Nonlinear Anal. Hybrid Syst **2**, 626–634 (2008). https://doi.org/10.1016/j.nahs.2006.10.013
29. saqib, M., Akram, M., Bashir, S., Allahviranloo, T.: A Runge–Kutta numerical method to approximate the solution of bipolar fuzzy initial value problems. Comput. Appl. Math. **40** (2021). https://doi.org/10.1007/s40314-021-01535-1
30. Appadoo, S.S.: Pricing financial derivatives with fuzzy algebraic models: a theoretical and computational approach. Doctor of Philosophy, University of Manitoba, Canada (2006)
31. Barahmand, Z., Samarakoon, G.: Sensitivity analysis and anaerobic digestion modeling: a scoping review. Fermentation **8**, 624 (2022). https://doi.org/10.3390/fermentation8110624
32. Scipy.Integrate.Solve_ivp. https://docs.scipy.org/doc/scipy/reference/generated/scipy.integrate.solve_ivp.html#scipy.integrate.solve_ivp. Accessed 31 Mar 2022

Author Index

M. Mujica Mota and P. Scala (Eds.): EUROSIM 2023, CCIS 2032, pp. 323–325, 2024.
https://doi.org/10.1007/978-3-031-68435-7

The manufacturer's authorised representative in the EU is Springer
Nature Customer Service Centre GmbH, Europaplatz 3, 69115 Heidelberg,
Germany. If you have any concerns regarding our products, please
contact ProductSafety@springernature.com

Printed and bound by CPI Group (UK) Ltd, Croydon, CR0 4YY
01/10/2026
02225834-0009